The United States in a Warming World

Addressing the widespread desire to better understand how climate change issues are handled in the United States, this book provides an unparalleled analysis of features of the US economic and political system that are essential to understanding its responses to climate change. The introductory chapter presents a firm historical context, with the remainder of the book offering balanced and factual discussions of government, business, and public responses to issues of energy policies, congressional activity on climate change, and US government involvement in international conferences. Abundant statistical evidence illustrates key concepts and supports analytic themes such as market failures, free riders, and the benefits and costs of alternative courses of action among industry sectors and geographic areas within the USA. Written for audiences both outside and within the USA, this accessible book is essential reading for anyone interested in climate change, energy, sustainable development, or related issues around the world.

THOMAS L. BREWER is a Senior Fellow at the International Centre for Trade and Sustainable Development (ICTSD) in Geneva. He taught graduate and undergraduate students at Georgetown University for twenty-five years, and he has been a consultant to the World Bank, the United Nations Conference on Trade and Development (UNCTAD), and the Organisation for Economic Co-operation and Development (OECD). Professor Brewer maintains two websites: www.usclimatechange.com and www.TradeAndClimate.net.

The United States in a Warming World

The Political Economy of Government, Business, and Public Responses to Climate Change

THOMAS L. BREWER
Georgetown University, Washington DC

CAMBRIDGE
UNIVERSITY PRESS

University Printing House, Cambridge CB2 8BS, United Kingdom

Cambridge University Press is part of the University of Cambridge.

It furthers the University's mission by disseminating knowledge in the pursuit of education, learning and research at the highest international levels of excellence.

www.cambridge.org
Information on this title: www.cambridge.org/9781107655690

First published 2015

Printed in the United Kingdom by Clays, St Ives plc

A catalog record for this publication is available from the British Library

ISBN 978-1-107-06921-3 Hardback
ISBN 978-1-107-65569-0 Paperback

To Marianne Wirenfeldt Asmussen,
extraordinary inspiration and perfect partner,
and
to my daughters, Becky, Eva, Jennifer, and Sandy,
each of whom is making a special contribution to improving the human condition

Contents

Boxes

Figures

Tables

Maps

Preface

What to do about climate change? The question has become a salient issue on the agendas of government and business as well as a concern of publics around the world. This book focuses on responses in the United States. It presents an analysis of the changing patterns and trends in perspectives about climate change, preferences about a broad array of actions that can be undertaken to address it, and the record of government and business responses to the problem. The analysis is also, in part, an account of how and why business and government in the USA have fallen behind efforts to address the problem in many other countries. The book thus addresses a wide range of questions: What does the public want government and business to do and not do? What are the specific measures that have and have not been taken by government and business? Why have government and business in the United States been laggards in their responses, compared with governments and firms in other countries? What are the economic and political constraints that need to be overcome for them to respond more effectively? What could be done to provide more leadership, domestically and internationally, on the issues?

The book offers a political economy perspective that answers these and other questions on the basis of an analytic framework with the following themes:

- *In order to understand the responses to climate change in the USA, it is necessary to understand the distinctive patterns in the interests, ideologies, institutions, and influence in the US political system.* Much of the book is about the economic geography and the political geography of the interests at stake, and how business, government, and the public have responded within the institutional context of the political system.
- *In order to succeed, efforts to address climate change must overcome two sets of market failures.* The first set is represented by costs in the form of negative externalities in the widespread use of fossil fuels and other activities that release greenhouse gases. The second set is represented by benefits in the form of positive externalities that lead to under-investment in energy efficiency and renewable energy projects with significant "public goods" payoffs or benefits in the form of positive externalities from the sequestration of greenhouse gases. The analysis thus includes an explicit

recognition of the importance of a range of market failures which are endemic features of the problem, and many of the potential solutions. A central challenge for government is to address these market failures with cost-effective and politically viable policies.

- *The responses of business, government, and the public vary across clusters of issues.* One cluster is focused on *pricing carbon dioxide and other greenhouse gases* by addressing the negative externalities associated with greenhouse gas emissions; and on measures that can be undertaken to internalize those costs, including market-based cap-and-trade systems and taxes. A second cluster of issues is focused on *facilitating energy technology innovation and diffusion* through regulations, subsidies, and other arrangements to capture the positive externalities associated with energy efficiency and low-carbon energy sources. The third cluster of issues involves *strengthening international cooperation* to cope with the limitations of a decentralized international political system and free-rider problems.

Structure of the book and content of the chapters

The Introduction offers a brief chronological overview of the evolution of climate change issues in the United States starting in the 1950s. Chapter 1 presents data about industry sources of greenhouse gas emissions and the socio-economic impacts of climate change. There is an emphasis on the regional variations in both the sources and the impacts – and thus the economic geography – of the interests at stake. The first chapter therefore establishes the broad national, sectoral, and regional economic contexts within which climate change issues are addressed. The first chapter also puts the US emissions and the economics of the issues in an international comparative context.

Chapter 2 presents a more fine-grained analysis of business interests, attitudes, actions, and inactions. It emphasizes the differences among firms within industries as well as the differences among industries and business associations – differences which reveal significant gaps between firms that are leaders and firms that are laggards on climate change issues. A focus of the chapter is the evolution of the splits among firms, industries, and associations, as well as industry patterns and nationality patterns among leaders and laggards.

Chapter 3 presents data from survey research about the patterns and trends in public perceptions of the problem and preferences for policies that could address it. The effects of partisan identities, ideologies, and regions receive special attention.

Chapter 4 focuses on the roles and policies of governmental institutions at the local, state, and regional levels. In addition, court cases that have emerged from the relationships across levels of government in a federal political system receive special attention, as does the role of "swing states" in presidential elections.

Chapters 5, 6, and 7 focus on national and international policy issues, especially three principal issue clusters: using market-based cap-and-trade systems or taxes to internalize negative externalities in the form of the costs of greenhouse gas emissions and/or subsidies for sequestration; using subsidies, regulations, and other measures to facilitate energy technology innovation and diffusion; and strengthening international cooperation to overcome the inherent limitations of a decentralized international political system.

Chapter 8 discusses the challenges of leadership in government and business in the face of the extraordinarily difficult circumstances posed by multiple economic and political system constraints. It highlights economic and political realities based on the empirical analyses of Chapters 1–7, and it identifies key leadership issues and considers pathways to the future for addressing them.

The research for the book was mostly completed in January 2014.

Audiences

I hope that students and instructors in university programs in political economy, political science, economics, public policy, law, business administration, and international relations, as well as environmental studies of course, will find the book useful to their particular needs.

Although the book is about the United States, I have consciously written it for students and other audiences outside the USA as well as for US audiences. I have learned from classroom presentations to students, as well as academic conferences, international climate change conferences, business groups, and citizens' groups in many countries that there is a widespread desire to gain a better understanding of how climate change issues are addressed in the USA and why those issues have been such difficult challenges in the USA.

The most obvious professional audience for the book – regardless of country or vocation – is specialized professionals in both the public and private sectors, not only those who are directly involved in climate change issues as part of their daily responsibilities, but also scientific or technological professionals whose interests extend to the political and economic contexts of their work. In addition, others with an interest in the politics and economics of climate change – or even more generally the politics and economics of the USA – may gain a better understanding of the US response to one of the principal challenges for government and business of the early twenty-first century.

Author disclaimer

All of the materials and comments in the book are entirely my own personal responsibility as an independent scholar; nothing in the book should be attributed in any way to any of the organizations with which I have been affiliated. This disclaimer should be especially noted in regard to the Intergovernmental Panel on Climate Change (IPCC), for which I was a Lead Author in Working Group III for the Fifth Assessment Report (AR5). Nothing in this book should be construed to be a position of the IPCC. A similar disclaimer applies to my position on the Panel of Experts of the Council on Environmental Cooperation of NAFTA. Nor should any of the views expressed here be attributed to the International Centre for Trade and Sustainable Development (ICTSD) in Geneva, where I am a Senior Fellow, nor to the Centre for European Policy Studies (CEPS), an independent think tank in Brussels, where I am an Associate Fellow.

Acknowledgments

During the summer of 2000, I was fortunate to be a participant in the Harvard Seminar on Climate Change, which was organized by Robert Stavins, Jeffrey Sachs, Theodore Panayotou, and Kelly Sims Gallagher. I am indebted to them and to the other participants for an intensive and stimulating two weeks. The experience was instrumental in encouraging me to pursue my incipient interest in climate change issues and in helping me to formulate more clearly the directions I wanted to take my research. Many of my resulting publications are cited in the book; there are also reports and other analyses of related specific topics available for downloading at my websites, www.usclimatechange.com and www.TradeAndClimate.net.

In late 2010, I participated in a climate leadership workshop at MIT organized by Climate Interactive and SEED. The experience not only gave me ideas and information for the concluding chapter, it also helped me see many climate-related issues in a different light. I am thus appreciative of the work of the organizers, Sara Schaley, Drew Jones, Travis Frank, and Stephanie McCauley – and the other participants in the workshop.

I have been affiliated with five organizations in Europe: currently as Senior Fellow at the International Centre for Trade and Sustainable Development (ICTSD) in Geneva, and as Associate Fellow at the Centre for European Policy Studies (CEPS) in Brussels; and previously as Research Director for Climate Strategies, which is a non-profit international network of researchers based in Cambridge, UK; Visiting Research Fellow at Oxford University's Smith School of Enterprise and the Environment; and Schöller Foundation Senior Research Fellow at the Friedrich-Alexander-Universität Erlangen-Nürnberg. I am indebted to them all for their professionally stimulating environments. A generous grant from the Schöller Foundation in Germany was instrumental in facilitating progress on the book and other related research; I am deeply appreciative of their support.

As for individuals, while conducting the research for the book, I became indebted to many people – some of them directly through conversations and conference panel discussions, and some of them only indirectly through their writings. As for people with whom I have had direct contact while pursuing my research interests, some must remain anonymous because of the sensitivities of their positions in government or business. They include people in the US government, as well as corporations, environmental organizations, and experts in think tanks and universities in the USA. In addition, I have discussed these issues with professionals in the capital cities and commercial centers of numerous countries. Among the people in think tanks and NGOs that I can mention by name, I would like especially to thank Ricardo Melendez-Ortiz of ICTSD, Christian Egenhofer of the Centre for European Policy Studies, Andreas Falke of Friedrich-Alexander-Universität Erlangen-Nürnberg, Michael Grubb of Climate Strategies, and Sir David King of Oxford University. At ICTSD, in addition to its Chief Executive Ricardo Melendez-Ortiz, my colleagues have included Ahmed Abdel Latif, Christophe Bellmann, Andrew Crosby, Natalia Cubilla, Caroline Imesch, Ingrid Jegou, Joachim Monkelban, Pedro Roffe, Mahesh Sugathan, and Deborah Vorhies. The ICTSD team is one of the world's finest human resources on climate change issues. At Climate Strategies, I benefited from working with its co-founders Michael Grubb and Benito Müller, Managing Directors Jon Price and Richard Folland, Board members Hans-Jurgen Stehr and Michele Colombier, Research Director Neil Hamilton, Research Manager Dora Fezekas, as well as Antonia Baker, Birgit Berry, Simone Cooper, Heleen de Coninck, Susanne Droege, Angela Köppl, Anna Korppoo, Michael Mehling, Axel Michaelowa, Karsten Neuhoff, Misato Sato, Stefan Schleicher, Grace Stobbart, Andreas Turk, and Peter Wooders.

Christian Egenhofer has been a conference co-organizer with me, a research report co-author, a tutor-mentor on EU issues, a publication draft reviewer and a friend – always with a seriousness of purpose and good sense of humor. Also at CEPS, Staffan Jerneck, Michael Wriglesworth, and Noriko Fujiwara were helpful sounding boards on various research projects on climate change issues.

Sir David King and his colleagues at Oxford University's Smith School of Enterprise and the Environment provided a friendly and informed atmosphere in which to pursue and discuss my research. During my time there I got to know other fellows, including Robert Hahn, Stephanie Richards, and Kenneth Richards – all of whom contributed to my understanding of the US responses to climate change.

Similarly, Andreas Falke, Matthias Fifka, Daniel Gossel, and Wolfgang Ramsteck at Friedrich-Alexander-Universität Erlangen-Nürnberg and its Schöller Research Institute were friendly and stimulating colleagues. Barbara Häffner helped enormously to facilitate the research in numerous ways. Maria Drabble and Sarah Beringer were excellent research assistants.

In Washington, DC, I have benefited from the ideas and information of a wide range of people: Kathleen Kelly and Thomas Legge at the German Marshall Fund of the United States; Michael Mehling at Ecologic of the United States; David Campbell of ThyssenKrupp; Sasha Golub and Annie Petsonk at Environmental Defense Fund; Rob Bradley and Jake Werksman at the World Resources Institute.

Elsewhere, the work and interest of Carlo Carraro, Frank Convery, John Drexhage, Aaron Cosby, Kirsty Hamilton, Ans Kolk, David Levy, Muthukumara Mani, Jonatan Pinkse, and Takahiro Ueno have informed and encouraged me. My participation in the annual conferences of the Swedish Network for European Studies in Economics and Business (SNEE), organized by Lars Oxelheim and Jens Forssbaeck, has provided many opportunities for presenting, revising, and clarifying my thoughts on a wide range of climate change issues. At UNCTAD in Geneva, I was involved in the preparation of the 2010 *World Investment Report*, which focused on climate change issues, and I profited from my work with James Zhan, Director, and the staff of the Division on Investment and Enterprise.

There have been scores of other professionals, including colleagues at Georgetown University and other academic institutions in many countries, who have also contributed directly or indirectly to my efforts. At Georgetown University, Gerald T. West read drafts of several chapters and suggested many ways to make them more mellifluous. Dorothy Sykes helped revise the manuscript more times than either one of us would want to count, and she was always remarkably patient. Naielia Allen and Marcia Blake were also helpful as the manuscript progressed. Among the faculty and administrators at Georgetown, Alan Andreasen, Vicky Arroyo, Tim Beach, Spiros Dimolitsas, Rob Grant, Nathan Hultman, Joanna Lewis, John Mayo, Stanley Nollen, Bob Parker, Dennis Quinn, Pietra Rivoli, Fabienne Spier, David Walker, and Charles Weiss deserve special mention. The research assistance of Amitabh Gupta on this and other projects was extraordinary; indeed, he read several portions of the manuscript in draft form and found many ways to improve it. Graduate student James Hopper also read chapter drafts and suggested useful revisions. Undergraduates Katherine Lee and Roberto Salas helped with tables and figures.

Several friends made distinctive contributions. DeWitt John of the Environmental Studies Program at Bowdoin College generously offered to use several chapter drafts in his course on climate change issues and made several useful comments on the basis of that experience. I am indebted to him and his students for sharing their thoughts. Glasgow University professor Stephen Young, with whom I have collaborated many times in research on international trade and multinational corporations, has taught me much about those topics. Economist Monty Graham and I had many opportunities to share thoughts about the topic of climate change and the political economy of responses to it, as well as the facts and follies of political life in Washington, DC. American historian Stephen Kurtz read early drafts of several chapters and pointed out ways to make the book more reader-friendly for non-specialists and at the same time meet professional standards of scholarly discourse.

Chris Harrison, Claire Wood, Vania Cunha, Ziqian Chan, and their colleagues at Cambridge University Press have been enormously helpful throughout the review, production, and marketing processes. Thank you!

Permissions

I am grateful for permission to reprint materials as follows:

Chapter 2, Figure 2.1, "Number of individual lobbyists on congressional climate change legislation"; from Center for Public Integrity (2009), "Number of Lobbyists on Climate Change by Sector, 2003 and 2008." Accessed at www.publicintegrity.org on August 24, 2010.

Chapter 3, excerpt concerning the attitudes of young people; from Climate Change Communications, for Lauren Feldman, Matthew C. Nisbet, Anthony Leiserowitz, and Edward Maibach (2010), "The Climate Generation? Survey Analysis of the Perceptions and Beliefs of Young Americans." Joint Report of American University's School of Communication, The Yale Project on Climate Change, and George Mason University's Center for Climate Change Communication. Accessed at www.climatechangecommunication.org on May 9, 2010.

Chapter 4, Appendix 4.3, "Insurer Climate Risk Disclosure Survey by state insurance commissioners"; from National Association of Insurance Commissioners (2010), "Insurer Climate Risk Disclosure Survey, Adopted Version, March 28, 2010." Accessed at www.namic.org on January 5, 2011.

Chapter 4, Box 4.1, "US Conference of Mayors' Climate Protection Agreement"; from US Conference of Mayors (2005), "Climate Protection Agreement (As endorsed by the 73rd Annual US Conference of Mayors meeting, Chicago)." Accessed at www.usmayors.org on March 10, 2010.

Chapter 4, Box 4.4, "The Environmental Protection Agency (EPA) and the state of Texas at odds"; from Greenwire (2011), "CLIMATE: Texas Faces Uphill Legal Battle against EPA," January 5. Accessed at www.eenews.net on January 6, 2011.

Chapter 4, Figure 4.1, "Urban versus rural temperature trends in the USA"; from Cambridge University Press, for Brian Stone, Jr. (2012), *The City and the Coming*

Climate: Climate Change in the Places We Live. Cambridge: Cambridge University Press.

Chapter 4, Map 4.1, "Cities in the US Conference of Mayors' Climate Protection Agreement"; from US Conference of Mayors (2010), "Cities that Have Signed On." Accessed at www.usmayors.org on March 10, 2010.

Chapter 5, Table 5.4, "Potential ultimate beneficiaries of allowances and auctions in Kerry–Lieberman bill"; from Belfer Center for Science and International Affairs, Harvard University, for Robert Stavins (2010), "An Economic View of the Environment: The Real Options for US Climate Policy," July 23. Accessed at http://belfercenter.ksg.harvard.edu on July 24.

Chapter 6, Box 6.5, "Market failures that constrain technological innovation and diffusion"; from *Ecological Economics,* for Adam B. Jaffe, Richard G. Newell, and Robert N. Stavins (2005), "A Tale of Two Market Failures: Technology and Environmental Policy," *Ecological Economics,* 54: 164–174.

Chapter 6, Figure 6.1, "McKinsey marginal cost curve for GHG abatement technologies"; from McKinsey Company, for McKinsey (2007), "Reducing US Greenhouse Gas Emissions: How Much at What Cost?" Accessed at www.mckinsey.com on February 26, 2011.

Chapter 6, Figure 6.2, "Changes in US government wind power subsidies and changes in wind industry installations"; from Bipartisan Policy Center (2010), "Reassessing Renewable Energy Subsidies." Accessed at www.bipartisanpolicy.org on May 2, 2011.

Chapter 7, Appendix 7.2, "Simulation model used to measure the gap between the target and the pledges"; from Climate Interactive (2010), "C-SPAN." Accessed at www.climateinteractive.org on December 10, 2010.

Chapter 7, Box 7.6, "Criteria for a multilateral climate change agreement"; from Cambridge University Press, for Joseph E. Aldy and Robert N. Stavins (2007), "Architectures for an International Global Climate Change Agreement: Lessons for the Policy Community," in Joseph E. Aldy and Robert N. Stavins, editors, *Architectures for Agreement: Addressing Global Climate Change in the Post-Kyoto World.* Cambridge: Cambridge University Press.

Chapter 7, excerpt of meeting during COP meeting in Copenhagen; from *New York Times* Syndicate, for Tobias Rapp, Christian Schwägerl and Gerald Traufette (2010),

"The Copenhagen Protocol: How China and India Sabotaged the UN Climate Summit," *Der Spiegel,* May 5, 2010. Accessed at www.spiegel.de on October 24, 2010.

Chapter 8, Appendix 8.3, "Policy recommendations and quotes by Republicans for Environmental Protection"; from ConservAmerica, for Republicans for Environmental Protection (2011), "Energy and Climate Change. Policy Paper. Executive Summary." Accessed at www.rep.org on May 12, 2011; and Republicans for Environmental Protection (2011), "Why Conservation *is* Conservative As Expressed by History's Preeminent Conservative Minds: A Comprehensive Collection of Conservative Quotations." Accessed at www.rep.org on May 12, 2011.

Chapter 8, Appendix 8.4, "Statement by a group of leading scholars"; from Vox, for "Thinking Through the Climate Change Challenge, An Open Letter," January 16, 2011.

Chapter 8, Appendix 8.5, "Spectrum of government-market mixes"; from *Oxford Review of Economic Policy,* for Cameron Hepburn (2010), "Environmental Policy, Government, and the Market," *Oxford Review of Economic Policy,* 26: 117–136.

Chapter 8, Appendix 8.7, "Discussion of international institutional venue issues"; from Belfer Center for Science and International Affairs, Harvard University, for Daniel Bodansky (2010), "The International Climate Change Regime: The Road from Copenhagen," Harvard Project on International Climate Agreements. Accessed at www.belfercenter.ksg.harvard.edu on April 10, 2013.

Introduction: a chronological overview

Responses in the USA to climate change have extended over more than a half century – starting with concern among a small number of scientists in the 1950s. Over time the number and diversity of people who have become attentive to the problem have increased dramatically. This Introduction highlights landmark events and other developments along the way, including those that affected the context in which climate change issues have arisen as well as those that have entailed direct action.

The chronology emphasizes US presidential and congressional decisions, but it also includes public opinion, business, court cases, state and local governments, international conferences, and particularly significant climate science developments and extreme weather events. The context and significance of the events are discussed in much more detail in various chapters.

1950s–1960s

After nearly a century of attention to climate change by mostly European scientists, climatologists in the United States began to give it more serious attention in the 1950s. As part of the Eisenhower administration's contribution to the International Geophysical Year in 1957, scientist Charles Keeling took repeated measurements of atmospheric carbon dioxide concentrations in Hawaii; he found a trend over a few years of increasing concentrations and seasonal patterns of relatively high concentrations in the fall and winter and low in the spring and summer, when trees absorb more carbon dioxide (Keeling 1960). The Keeling curve depicting these measurements has since become an icon of climatology, and it has been regularly updated (see Figure I.1). In addition, the work of Roger Revelle and Hans Suess (1957) contributed to increased interest among American scientists during this period.

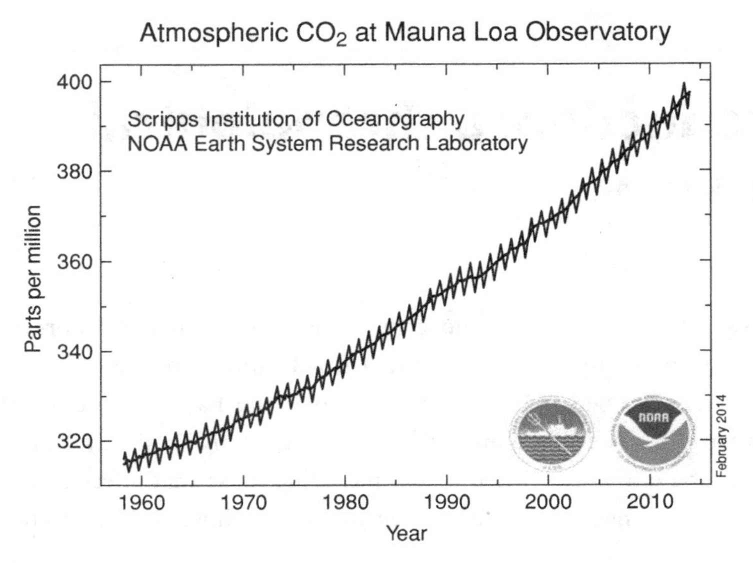

Figure I.1 The "Keeling curve" of atmospheric carbon dioxide concentrations
Source: Scripps Institution of Oceanography (2013).

1970s

There were several pertinent US government organizational developments during the 1970s, and they have directly affected the institutional context in which climate change issues have been addressed by the national government. The Environmental Protection Agency (EPA) began operations in 1970 after its creation was proposed by the Nixon administration and agreed by Congress because of concerns about air and water pollution problems. Soon thereafter, energy issues also became much more salient, starting with the quadrupling of world oil prices in 1973–1974. A series of reorganizations by the Nixon and Ford administrations led eventually to the creation of the cabinet-level Department of Energy during the Carter administration in 1977. Thus, concerns about a wide range of environmental and energy issues by the late 1970s were important elements of the circumstances in which climate change began to receive more international attention.

In 1979 the First World Climate Change Conference estimated that if there were a doubling of atmospheric carbon dioxide concentrations over pre-industrial levels, there would be an increase in the global mean surface temperature compared with pre-industrial levels of 1.4–4.5 degrees centigrade (i.e. 2.5–8.1 degrees Fahrenheit).

1980s

Nearly a decade later, in 1987, the US and other governments agreed to the Montreal Protocol on Substances that Deplete the Ozone Layer – some of which substances, such as chlorofluorocarbons (CFCs), are also powerful and long-lasting greenhouse gases that contribute to climate change. The Protocol was negotiated by the Reagan administration and ratified with bipartisan support and a unanimous Senate vote in 1988. The same year the Intergovernmental Panel on Climate Change (IPCC) was established as an international agency to review climate change research for input into international negotiations and governmental decision-making.

A national US public opinion survey in the spring of 1988 found that 63 percent worried "a great deal" or "fair amount" about "global warming."[1] A few months later, the US Senate Committee on Energy and Natural Resources held hearings on climate change; at those hearings James Hansen, Director of the Institute for Space Studies of the National Aeronautics and Space Administration (NASA), testified that "The greenhouse effect has been detected, and it is changing our climate now" (Sheppard 2008). He thus became the first US government scientist to state publicly that climate change was a problem. That same summer, there were widespread droughts and record-breaking heat waves in many areas of the USA.

A year later the federal government established the Global Change Research Program to coordinate and integrate federal research on climate change and other issues.

1990s

An amendment to the Clean Air Act in 1990, proposed by President George H.W. Bush and approved by Congress, established a cap-and-trade program to limit sulfur dioxide (SO_2) emissions from power plants and thus reduce the problem of acid rain, which had become especially severe in southern California. That system – in light of its widely recognized cost-effective success – became the precursor to subsequent proposals to create a cap-and-trade system to limit greenhouse gas emissions.

In 1992 President George H.W. Bush signed the UN Framework Convention on Climate Change (FCCC) at an international conference in Rio de Janeiro. There was bipartisan support for it, and the Senate ratified the FCCC by a unanimous vote.

1 The survey results reported in the Introduction are all based on annual Gallup polls about environmental issues taken in the spring at the time of Earth Day. The annual timing and question are thus the same in each of the polls reported.

The agreement, which remains the legal basis for many international climate change programs, established the goal of stabilizing greenhouse gas concentrations "at a level that would prevent dangerous anthropogenic (human-induced) interference with the climate system."

During the 1992 presidential election campaign, between the incumbent George H.W. Bush and challenger Bill Clinton, climate change was an explicit national campaign issue for the first time. Public opinion surveys at the time found that between 57 and 62 percent of the public worried "a great deal" or "a fair amount" about "global warming."

In 1993, the recently elected Clinton administration proposed a tax on carbon emissions, specifically on the basis of the heat content of fossil fuels; this was the first formal proposal in the USA for national legislation that would have affected the price of carbon. The proposal was passed by the House of Representatives by a narrow majority – but with a strongly partisan division and in spite of strong opposition from agricultural and industrial lobbyists. In the Senate, the American Petroleum Industry, the National Association of Manufacturers, and other organized business lobbies opposed it. It was abandoned in the Senate without a formal floor vote, as senators from fossil-fuel-intensive states such as Texas, Louisiana, and West Virginia expressed their opposition.

In 1995 the Second Assessment Report of the Intergovernmental Panel on Climate Change (IPCC) concluded that there was a "discernible" human impact on the climate.

In 1997 the Senate passed the Byrd–Hagel amendment, which was a non-binding sense of the Senate resolution that the United States should not be a signatory to an international agreement regarding the UN Framework Convention on Climate Change that would limit US greenhouse gas emissions unless it also limited the greenhouse gas emissions of developing countries. The Senate vote was unanimously in favor of the resolution. The same year, the Clinton administration signed the Kyoto Protocol, but did not submit it to the Senate for ratification. The Protocol set a goal of reducing industrialized countries' greenhouse gas emissions to 5 percent below 1990 levels during the first commitment period of 2008–2012. The Senate vote on the Byrd–Hagel amendment and the Clinton administration's decision not to submit the Protocol for ratification reflected an impasse on climate change issues, at least at the national level.

In 1998, 50 percent in a national public opinion survey said they worried "a great deal" or "a fair amount" about "global warming," which was about a 10 percent decline from previous years.

As climate science research continued to accumulate at an increasing rate, an article by Mann, Bradley, and Hughes (1999) presented a curve of northern hemisphere

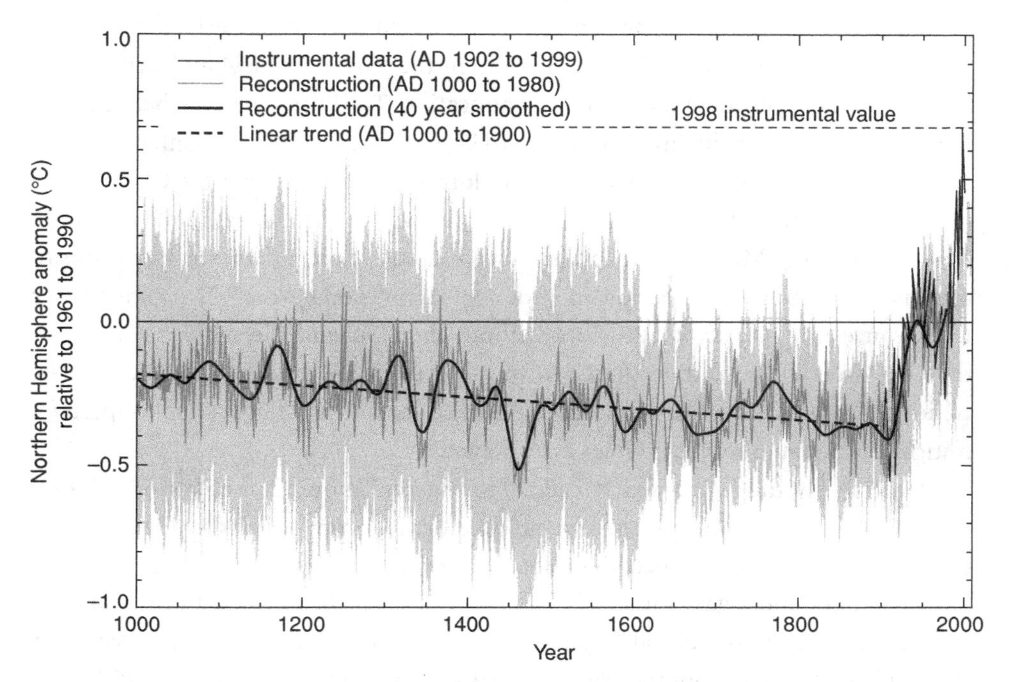

Figure I.2 The "hockey stick" curve of long-term temperature trends

Source: Mann, Bradley, and Hughes (1999).

temperatures from the eleventh century to the late twentieth century showing a decline during the fifteenth to nineteenth centuries and then a sharp increase in the nineteenth and twentieth centuries. Because of its shape, it came to be known as the "hockey stick" curve, and like the Keeling curve of atmospheric carbon dioxide concentrations, it became an iconic graph (see Figure I.2).

2000–2008

During the 2000 presidential election, climate change was an issue, with candidate Al Gore advocating more serious action on it and candidate George W. Bush expressing less enthusiasm about doing so. In that year, 72 percent in a national public opinion survey said they worried "a great deal" or "a fair amount" about "global warming" – which was 20 percentage points higher than two years previously. However, four years later, in the presidential election year of 2004, only 51 percent in a national public opinion survey worried "a great deal" or "a fair amount" – a decline of about 20 percentage points.

In the meantime, in early 2001, the recently elected George W. Bush administration announced that it opposed US participation in the Kyoto Protocol and that it would not submit the Protocol to the Senate for ratification. The Protocol nevertheless entered into force internationally in 2005, without US participation. At the Conference of the Parties to the UNFCCC (COP) meeting in Montreal, Canada – after the US delegation walked out because the meeting was also serving as the Meeting of the Parties to the Kyoto Protocol (CMP) – the COP agreed to establish two parallel tracks to consider actions in the post-2012 period in order to accommodate the US position – one track concerning the Kyoto Protocol without the USA and the other a track under the UNFCCC and including the USA. In 2007, the USA agreed to the COP-13 Bali Action Plan, which identified four main elements for future international cooperation: mitigation, adaptation, technology, and finance.

Domestically, during the first decade of the new millennium, the state of California adopted numerous energy efficiency and renewable energy policies and programs. In particular, in 2006 it adopted the Global Warming Solutions Act with the goal of reducing greenhouse gas emissions to 1990 levels by 2020; a key part of the program is a cap-and-trade system that began operations in 2012. Also during the decade, on the other side of the country, the Regional Greenhouse Gas Initiative (RGGI) was established with a cap-and-trade system for the carbon dioxide emissions of electric power plants in nine Northeastern states beginning in 2009. Also during the decade, many other states in several regions of the country began to develop their own climate change programs, as did numerous cities and other local government agencies.

During the period 2001–2008 at the national level, the initiative in addressing climate change shifted to the Congress, where Senators McCain and Lieberman proposed a mandatory national cap-and-trade system for greenhouse gases in 2003. Known as the Climate Stewardship Act, it would have established a nation-wide system covering all greenhouse gases and most sectors of the economy. It was defeated in the Senate by a vote of 43 to 55. The Senate considered cap-and-trade legislation again in 2008 – this time known as the Climate Security Act, with Senator John Warner of Virginia being a co-sponsor with Senator Lieberman. It failed to receive enough votes for a cloture of debate and thus was not further considered.

A Supreme Court decision in 2007 in the case of *Massachusetts v. EPA* significantly changed the place of climate change on the national agenda. In a 5–4 decision the court ruled that the EPA could regulate greenhouse gases through the Clean Air Act. Although there were still many administrative and legal processes to come into play, the decision marked a turning point because it created a new legal basis for the executive branch to develop regulations and because it pushed climate change issues more deeply into the federal court system.

During the presidential election year of 2008, 66 percent in a national public opinion survey said they worried "a great deal" or "a fair amount" about "global warming." Presidential candidate Senator McCain said very little about climate change during the campaign, despite his long efforts to address the problem in the Senate; President Obama also said little about the issue, despite having supported cap-and-trade legislation as a senator as a co-sponsor of the McCain–Lieberman bill in 2003.

2009–2012

In 2009 the House of Representatives took the initiative on climate change in the form of the American Clean Energy and Security Act; introduced by Representatives Waxman of California and Markey of Massachusetts, it would have established an extensive cap-and-trade system and substantially expanded the government's energy efficiency and renewable energy programs. The bill was passed by a 219-to-212 vote, with 83 percent of the Democrats voting in favor and 95 percent of the Republicans voting against. In the Senate, a year later, however, there was not sufficient support for a similar cap-and-trade bill to gain the 60-vote super-majority vote needed for cloture, though there appeared to be a simple majority of votes in favor of passage if a cloture motion could have been passed.

Business lobbies were split on the issue. There was substantial support for cap-and-trade legislation by key segments of the electric power industry, by energy efficiency and renewable energy firms, and by some other major individual firms. However, the coal industry, the oil and gas industry, the chemical industry, and perhaps most importantly umbrella business organizations such as the Chamber of Commerce and the National Association of Manufacturers, as well as industry-specific organizations such as the American Petroleum Institute, all continued to oppose all cap-and-trade proposals.

At the COP-15 meeting in Copenhagen in 2009 President Obama reached agreement with a small subset of the nearly 200 governments represented at the conference: along with Brazil, China, India, and South Africa, the USA eschewed making formal commitments to emission targets within the context of the multilateral FCCC and instead agreed to a schedule of "pledges" to be made more specific later.

In the United States, following the 2010 congressional election, the Tea Party caucus in the House of Representatives supported legislation that would have prevented the government from funding a high-level climate advisor on the White Staff and would have prevented the Department of State from having a Chief Negotiator at climate change international conferences. The proposals passed the House but were defeated in the Senate.

At the COP-18 in Doha near the end of 2012, there was an agreement to extend the Kyoto Protocol with a commitment on the part of industrial countries to reduce greenhouse gas emissions by at least 18 percent below 1990 levels between 2013 and 2020. The USA was not a party to this agreement.

2013

After having mentioned climate change briefly in his victory speech on election night in November 2012, President Obama gave climate change more extensive attention in his January 2013 inauguration speech and then yet more attention in February in his State of the Union speech. Two other events marked 2013 as a potential landmark year in the evolution of US government climate change policymaking. The first was a speech by President Obama (US White House 2013a), who noted a series of measures the administration could take without congressional actions endorsing them in the context of a "new national climate action plan." The measures included new standards for carbon dioxide; emissions standards for existing and new electric power plants, with flexibility for how individual states implement them; increased production of natural gas, with measures to prevent methane emissions; allowing additional wind and solar installations on government land managed by the Interior Department and expanding the use of renewable power on Defense Department bases; continuing to strengthen fuel efficiency standards for trucks, buses, and vans as well as cars; increased federal government use of renewable energy to 20 percent of its electricity by 2020; and increased federal government budget support for state and local adaptation projects to protect against droughts, floods, and wildfires. In their totality the measures could contribute to significant reductions in US greenhouse gas emissions. It was unlikely, however, that these measures would be sufficient to reduce emissions enough to meet the administration's stated goal of a 17 percent reduction from 2005 levels by 2020 – even in combination with other measures already undertaken by the administration such as increases in motor vehicle fuel efficiency standards. Nevertheless, the speech did signal the administration's intention to address climate change across a wide range of domestic policies. Importantly, however, such measures would be subject to industry challenges in court cases, state- and local-level opposition in some instances, and attempts by climate deniers and others in Congress to try to undermine them.

In terms of becoming more engaged internationally, in the speech the president expressed his intentions to increase government-private partnering for increased investment in natural gas in other countries; end government support for coal-fired electric power plants; undertake negotiations to reduce barriers to international trade in clean energy technologies; and cooperate more actively in bilateral and other international venues. There were several international agreements reached subsequently during 2013: two bilateral agreements with China (US Department of State 2013; US White House 2013a) and an agreement with the G-20 and other countries (US White House 2013b). In the first bilateral agreement, China and the USA agreed to undertake cooperative actions in five areas: emissions and black carbon from heavy-duty trucks; financing of energy efficiency in buildings; carbon capture and storage projects; improving greenhouse gas data collection and measurement; and the development of electricity "smart grids." Like the speech on domestic policies, the agreement with China represented the potential for substantial gains in mitigating emissions – between them China and the USA account for more than 40 percent of total global greenhouse gas emissions. The actual implementation of the agreement, however, and thus its climate change impacts, remain to be seen. The same could be said for the other 2013 international agreements: China and the USA agreed in September to phase down the production and consumption of hydrofluorocarbons (HFCs), which are powerful, long-lived greenhouse gases used in refrigerators, air conditioners, and industrial processes as substitutes for chemicals that have been prohibited by the Montreal Protocol on Substances that Deplete the Ozone Layer. At the same time as this China–US bilateral agreement, the USA signed on to a similar agreement with the G-20 and other countries.[2]

Also during 2013, however, the atmospheric carbon dioxide concentration level at the iconic Mauna Loa Observatory in Hawaii reached 400 parts per million (ppm), which was 26.2 percent higher than the 317 ppm level in the observatory's initial

2 The G-20 includes Argentina, Australia, Brazil, Canada, China, France, Germany, India, Indonesia, Italy, Japan, Korea, Mexico, Russia, Saudi Arabia, South Africa, Turkey, the United Kingdom, the United States, and the European Union. In addition, the following had signed the agreement by early September 2013: Ethiopia, Spain, Senegal, Brunei, Kazakhstan, and Singapore.

measurements in 1958 and the highest concentration level for the earth in about three million years.

2014

In the State of the Union Speech, President Obama (2014) noted that "a changing climate is already harming western communities struggling with drought, and coastal cities dealing with floods" and highlighted the administration's "work with states, utilities, and others to set new standards on the amount of carbon pollution [emitted by] power plants"; he also noted that the "shift to a cleaner energy economy won't happen overnight, and it will require tough choices along the way. But the debate is settled. Climate change is a fact."

In technical and tangible ways, the administration was moving ahead, for instance with the increased use of the "social cost of carbon" (US Interagency Working Group on Social Cost of Carbon 2013) in assessing the costs and benefits of regulatory climate change mitigation and adaptation programs and in allocating financial resources in energy budgets through the annual budget cycle. These and other administration initiatives on climate change and energy policies, however, were encountering continuing challenges among some groups in the Congress, particularly in the House of Representatives, and in court cases being pursued by some industry groups and state governments.

These developments and issues are considered in detail in the remainder of the book, as follows: Chapter 1 establishes the national, sectoral, and regional economic contexts of climate change issues within the United States. The chapter emphasizes regional variations within the United States in the sources and the impacts of climate change; it is in part, therefore, an analysis of *economic geography*. The chapter also puts the US emissions and the economics of the issues in an *international comparative context*. Chapter 2 focuses on *business* interests, attitudes, actions, and inactions. It emphasizes differences between firms that are leaders and firms that are laggards on climate change issues; the oil and gas, automotive, and insurance industries receive special attention because of their importance and because of the tendency of US firms to be laggards within those industries. In Chapter 3 concerning *public opinion* there is much data about perceptions of the problem and preferences for policies. Differences in opinions according to party identities, ideologies, and regions receive special attention. In Chapter 4 there is a focus on *governments at the local, state, and regional levels* – including the role of states in court cases in a federal political system and the role of "swing states" in presidential elections.

US national government policy issues in three principal *issue clusters* are the topics of Chapters 5, 6, and 7. Chapter 5 emphasizes issues concerning market-based cap-and-trade systems, taxes, and other *regulatory measures* to reduce greenhouse gas emissions. Chapter 6 concerns *technological approaches* to mitigating greenhouse gas emissions – government subsidies, regulations, and other measures to facilitate energy technology innovation and diffusion. Chapter 7 is about a variety of modes of *international cooperation*, and it traces the evolution of international efforts to address a wide range of climate change issues.

Chapter 8 adopts a more normative approach, which identifies key leadership issues and potential pathways that could be pursued for more effective actions – within the economic and political constraints that are analyzed in the empirical analyses of Chapters 1–7.

PART I

ISSUES

CHAPTER 1

Questions, analytic framework, and context

We live in an era of global warming. That much is certain.

Michael Oppenheimer (2009)

We can't solve problems by using the same kind of thinking we used when we created them.

Albert Einstein (no date)

Climate change is not a new issue, and it is not an issue that will go away. It has been widely recognized by scientists as a problem since at least the 1980s. It has received much greater attention in recent years as governments and businesses have begun to address it more tangibly, and as publics around the world have begun to consider more seriously the nature of the problem, its implications, and approaches to trying to solve it. This book addresses questions about what government and business in the USA have done and not done, what the public believes and wants, or does not believe or want. Several of the key questions concern change or the lack of it: Why have business and government in the United States been laggards in their responses to climate change? Other questions concern variations in attitudes and policies among industries and regions about what to do. Yet others concern how and why government policies have been affected by business lobbying. In short, the book covers a wide array of questions about government, business, and public responses to climate change issues in the United States.

Although the focus is the USA, many questions and answers are discussed in a comparative international context. There are numerous indicators and studies comparing the USA with other countries in terms of the nature and extent of any leading or lagging. For instance, the USA ranked seventh among the G-8 countries in a detailed comparative analysis by Ecofys (2009) of government policies as well as emission levels and trends. The USA followed Germany, the United Kingdom, France, Japan, Italy, and Russia; the USA ranked ahead of only Canada.

This initial chapter presents a political economy perspective on the issues. It introduces an analytic framework that includes market failures, which are central to understanding the *problem* of climate change and understanding the constraints on many of the *solutions.* The chapter also presents data about the industry sources of greenhouse gas emissions that contribute to climate change and data about the socio-economic impacts of climate change. The analysis of this data emphasizes regional variations in the economic interests at stake in both mitigation and adaptation, and thus the importance of economic geography to an understanding of the political economy of the issues. (For an earlier analysis of the economic geography of US climate change issues, see Brewer 2005b.)

1.1 Analytic framework and themes: the political economy of climate change

Market failures are fundamental, enduring conditions that must be addressed successfully in order for the problem of human-induced climate change to be solved or significantly ameliorated. The market failures are represented by (1) the negative externalities associated with the widespread use of fossil fuels and other activities that release greenhouse gases, and (2) the positive externalities associated with investment in energy efficiency and renewable energy or carbon sequestration projects with significant "public goods" payoffs. *Negative externalities are thus inherent in the cause of the problem, and positive externalities are inherent in the constraints on the solutions.*

The existence of externalities was initially recognized in the economic literature by Marshall (1890) and then Pigou (1920). The existence of both types of market failures – negative externalities and positive externalities – was explicitly recognized by Milton Friedman (1955, 1; italics added):

> In . . . a free private enterprise exchange economy, government's primary role is to preserve the rules of the game by enforcing contracts, preventing coercion, and keeping markets free. Beyond this, there are only three major grounds on which government intervention is to be justified. One is 'natural monopoly' . . . *A second is the existence of substantial 'neighborhood effects,' i.e., the action of one individual imposes significant costs on other individuals for which it is not feasible to make him compensate them or yields significant gains to them for which it is not feasible to make them compensate him – circumstances that again make voluntary exchange impossible.* The third [is] . . . paternalistic concern for children and other irresponsible individuals.

It should be added, in a more contemporary climate change context, that the "neighborhood effects" recognized by Friedman are *global* in scope.

Government responses to address these market failures include policies to internalize the costs of greenhouse gas emissions in the prices of transactions, such as those in producing and consuming electricity, in order to mitigate the emissions. Government policies also include subsidies and other efforts to incentivize businesses and consumers in order to increase investments in technologies that can reduce greenhouse gas emissions. However, for such efforts to succeed, they must overcome key endemic features of the political economy of climate change issues – not only the market failures just noted, but also two systemic political challenges. First, because of the well-known global "public goods" nature of measures to reduce greenhouse gas emissions, there is a systemic political challenge represented by the difficulties of addressing free-rider problems.

The concepts of *externalities*, *public goods*, and *free riding* – and their relationships – are summarized in Box 1.1.

There is another systemic phenomenon that constrains political responses to the climate change problem – namely the difficulties posed by the asymmetric distributions over time and among groups of the costs and benefits of climate change mitigation efforts. In particular, these distributions are often as conceptualized as in Panel A of Figure 1.1; mitigation costs occur in the short term and tend to be concentrated among specific groups within the society, such as those associated with particular industries that are greenhouse-gas-intensive; the benefits of mitigation, on the other hand, tend to be widely distributed among many groups and occur in the distant future.

These cost and benefit distribution issues have been noted by Kamieniecki and Kraft (2009, xi) as follows: "[T]he benefits of acting on climate change are uncertain over the long term and broadly distributed to the public, whereas the costs tend to be more certain, imposed in the short term, and concentrated on particular groups or economic sectors eager to avoid them." For instance, regulations such as limits on the emissions from coal-fired electric power plants concentrate the costs of emissions abatement among owners of such facilities. However, to the extent that the electricity producers can pass along the increased costs to the industrial and residential consumers of electricity from those plants, the costs become more widely distributed; yet, the costs are still somewhat concentrated among coal-based electricity consumers, who are a subset of the entire population and do not include producers or consumers of electricity from hydro and other lower-carbon sources. In contrast, in any case, the benefits of such a greenhouse-gas mitigation measure are widely dispersed among the entire population.

Box 1.1 Externalities, public goods, and free riding

The three concepts of *externalities, public goods*, and *free riding* are related. *Externalities* refer to the costs (negative externalities) or benefits (positive externalities) of a transaction that affects people who are not participants in the transaction. In the context of climate change, for instance, when one person uses electricity that has been generated in a plant using coal, oil, or natural gas as a fuel and the carbon dioxide that is emitted contributes to climate change, there are negative externalities – namely, the costs borne by different people, who pay for construction of increased barriers against sea level rises. There can also be positive externalities: when a coal-fired power plant shifts to natural gas, there are positive externalities in the form of the benefits to others who do not have to pay so much for protection against sea level rises. Analogous effects of carbon dioxide emissions and other greenhouse gas emissions can be observed in transportation and many other industries and human activities.

The externalities represent market failures in the sense that the prices of the market transactions do not fully reflect the benefits or costs that accrue to society as a whole from a private, market-based transaction. The market failures are a rationale for government intervention – a rationale based on economic efficiency. The government policies – which can address either negative or positive externalities – provide for instance less carbon-polluted air, which is a *public good*. Government regulations that limit the negative externalities of carbon dioxide emissions of electric power plants or motor vehicles are obvious examples. There are also government policies that produce public goods in the form of augmented positive externalities, such as government-supported research and development on solar or wind energy technologies as alternatives to fossil fuel plants that produce carbon dioxide emissions; the public good is thus the additional benefit to the public that comes from the reduction of carbon dioxide emissions.

Because government policies can yield public goods that are potentially available to anyone, whether they have paid for them through taxes or fees, there are opportunities for *free riding*. Again in the context of climate change issues, an example at the international level is that people in countries whose governments have not adopted substantial climate change mitigation policies nevertheless enjoy the benefits of the mitigation measures of countries that have done so. For instance, people living in the USA have been free riding on the emission reduction policies of the European Union countries, and people living in China have been free riding on the emission reduction policies of the state of California in the USA. (At the same time, people in California have been experiencing the negative externalities of the emissions of Chinese coal-fired electric power plants.)

Obviously, these are important conceptual issues, which must be clearly understood in order to address intelligently and effectively many of the underlying issues about the problem of climate change and the solutions to it. Conceptual clarity, in short, is thus necessary to understand the issues; but it is not sufficient. For many of the issues are economic issues about the sizes of the costs and benefits of market transactions and government policies: how much are the costs and benefits? There are also diverse political issues about the distribution of the costs and benefits: who pays, and who gains? There are, furthermore, wide-ranging ethical issues about fairness: who is or should be responsible for the costs of the problem, and who benefits from the solutions?

Panel A. Conventionally-Assumed Pattern for Cap-and-Trade Systems

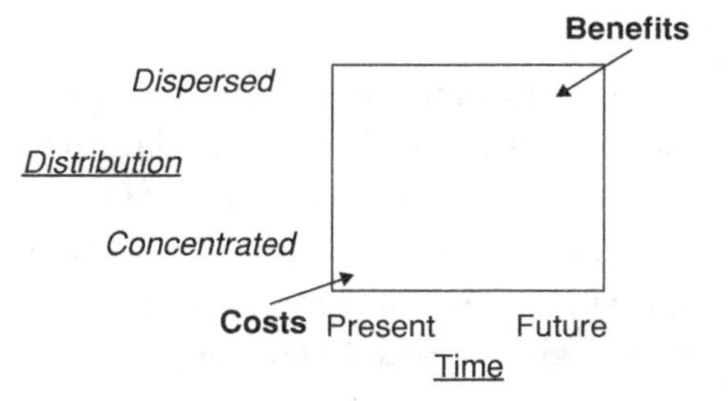

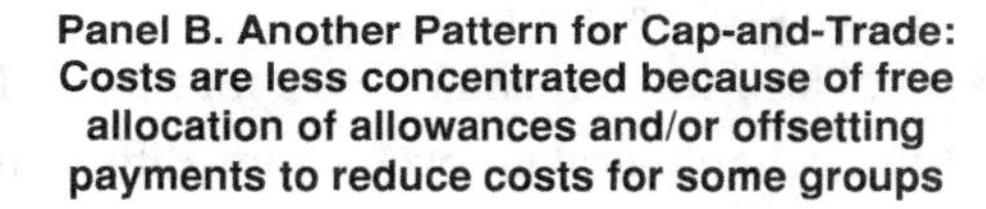

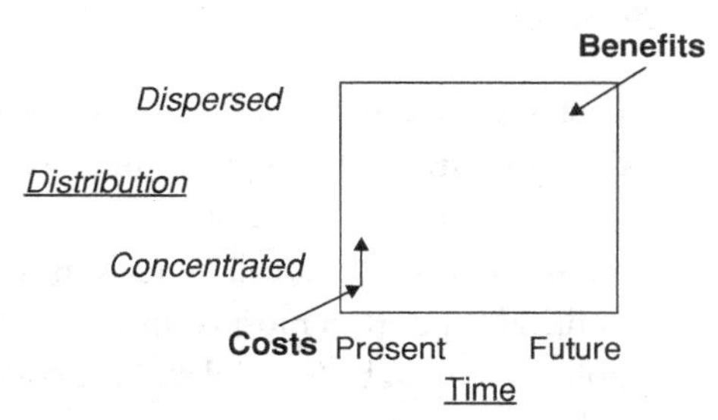

Panel C. Pattern for Technology Subsidies: There are current direct benefits for selected producers and/or consumers; costs are deferred to the future through government deficit financing

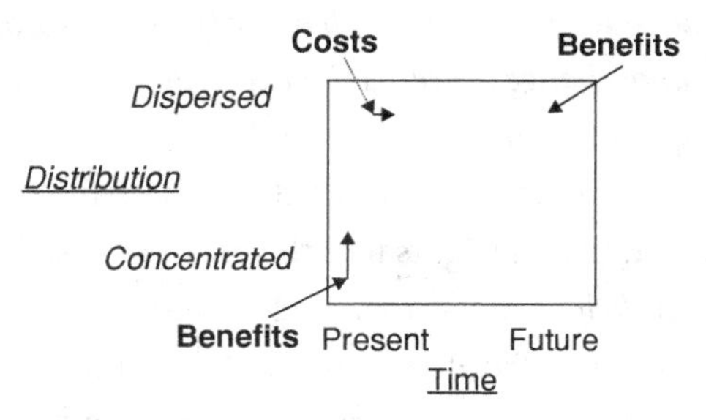

Figure 1.1 The distributions of benefits and costs of mitigation measures

These insights and illustrations highlight important features of the (widely presumed) distributions of economic costs and gains associated with climate change and its mitigation; however, there are significant additional analytic complications. One complication is that the benefits of mitigation are not evenly distributed. Some groups such as coastal residents and farmers in drought-prone areas, for instance, benefit disproportionately from mitigation, as in Panel B of Figure 1.1. Furthermore, as we shall see in Chapter 6 in detail, some government policies, such as energy technology subsidies, can increase and concentrate the short-term economic benefits from mitigation measures among groups such as renewable energy industries or purchasers of certain types of products such as wind generators. These types of *distributive* effects of course make them more popular among the beneficiaries of the subsidies and accordingly increase political support for them. The costs of the subsidies, meanwhile, can be widely dispersed among a large number of taxpayers, and the costs can be spread over the long-term future through government deficit spending and borrowing, as in Panel C of Figure 1.1.

These fundamental facts about the political attractiveness of government subsidies for technological innovation and diffusion have been noted by Jaffe, Newell, and Stavins (2005, 169):

> One might reasonably ask... why governments who are unwilling to impose costs on the economy to reduce greenhouse gas emissions directly should be more willing to invest resources in improving energy technologies to reduce greenhouse gases? One reason is purely political: policies subsidizing technology do in fact receive considerable political support in most countries... An explanation is that the benefits of such policies tend to be focused and the costs dispersed, giving rise to favourable political-economic conditions.

The distributions of costs and benefits are thus at the core of the political economy issues associated with climate change. Indeed, much of the story of the domestic political economy of US responses to climate change concerns these distributions, efforts to change them, and efforts to change perceptions of them.

While *necessary* to an understanding of the underlying challenges of addressing climate change, however, these insights into the importance of the distribution of costs and benefits are *not sufficient* to gain an understanding of the political economy of the problem or solutions to it. For there are additional political economy perspectives that are also necessary to developing a more nuanced and specific understanding of the variations in the responses of individual governments and businesses. In particular, US responses to climate change issues have been determined by a combination of the distinctive patterns in the economic interests and ideologies, as well as the institutions and patterns of influence, in the US political economy.

The patterns of *economic interests* include, for example, the relative importance of the oil and gas, coal, automotive, and electric power industries in the national economy generally and in particular regions. The key *ideological feature* is variations in support for (or opposition to) government intervention in the economy – ideological variations that coincide to some degree with regional differences, but that also vary over time and across specific issues about the forms of government intervention. The dominant *institutional features* of the US political economy are the decentralization of influence in a federal political system and in a national government marked by the overlapping authority of the Congress, the executive, and the courts. There is thus a strong pluralistic tendency in the *distribution of influence* in the US economic and political systems, a tendency that imposes numerous obstacles to effective action on climate change (Skolnikoff 1990; Lee 2001; Brewer 2009a). At the same time, though, the institutional pluralism facilitates action by one set of institutions when others are inactive. *Understanding the full range of US responses to climate change thus requires a disaggregation of the national economic and political systems. Much of the book is therefore devoted to an analysis of firms, industries, and regions in the economy, as well as individuals, groups, and institutions in the political system.*

Finally, the analytic framework emphasizes differences in the patterns of political economy across *three issue clusters*: (1) issues concerning how to address the problem of the market failures in the form of negative externalities associated with greenhouse gas emissions, especially measures such as market-based cap-and-trade systems or taxes that put a price on carbon dioxide and other greenhouse gases, (2) issues concerning how to address the problem of the "spillovers" and other positive externalities that inhibit private-sector energy technology innovation and diffusion, and (3) issues that emerge from an international political system that is highly decentralized and encourages free riding and other behaviours that constrain cost-effective responses to climate change issues.[1] These three principal issue clusters overlap with one another, thus creating a total of seven issue clusters, as depicted in the Venn diagram of

1 Although the emphasis of the book is on mitigation issues, there are also a variety of adaptation issues, including in the international issue cluster, as discussed in Chapter 7.

Table 1.1 Issue clusters with examples

Issue cluster[a]	Example
1. Pricing carbon dioxide and other greenhouse gases	Establishment of cap-and-trade system
2. Facilitating energy technology innovation and diffusion	Funding of climate-friendly R&D programs
3. Strengthening international cooperation	International negotiations that establish the functions, programs, and procedures of the UNFCCC, Major Economies Forum, and other institutional arrangements
4. Overlap of 1 and 2	Use of cap-and-trade auction revenues for energy R&D funding
5. Overlap of 1 and 3	International leakage/competition effects of differences in countries' cap-and-trade systems
6. Overlap of 2 and 3	Policy barriers to international transfers of clean energy/climate-friendly technologies
7. Overlap of 1, 2, and 3	International carbon offset programs that entail international technology transfers

[a] Adaptation issues are embedded in issue clusters. For instance, the use of cap-and-trade revenues and/or the use of revenues from an international aviation levy for international adaptation programs are examples of the overlaps of (1) and (2). Policies concerning the financing of international technology transfers for adaptation are examples of the overlaps of (2) and (3).

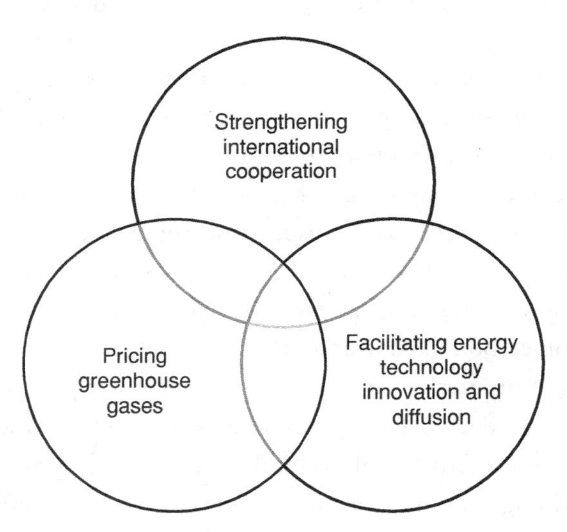

Figure 1.2 Venn diagram of issue clusters

Figure 1.2. The variety of the specific issues that are encapsulated in that diagram is illustrated in the examples of Table 1.1.

The analytic framework, which is summarized diagrammatically in Figure 1.3, focuses initial attention on the nature of the climate change problem – as represented

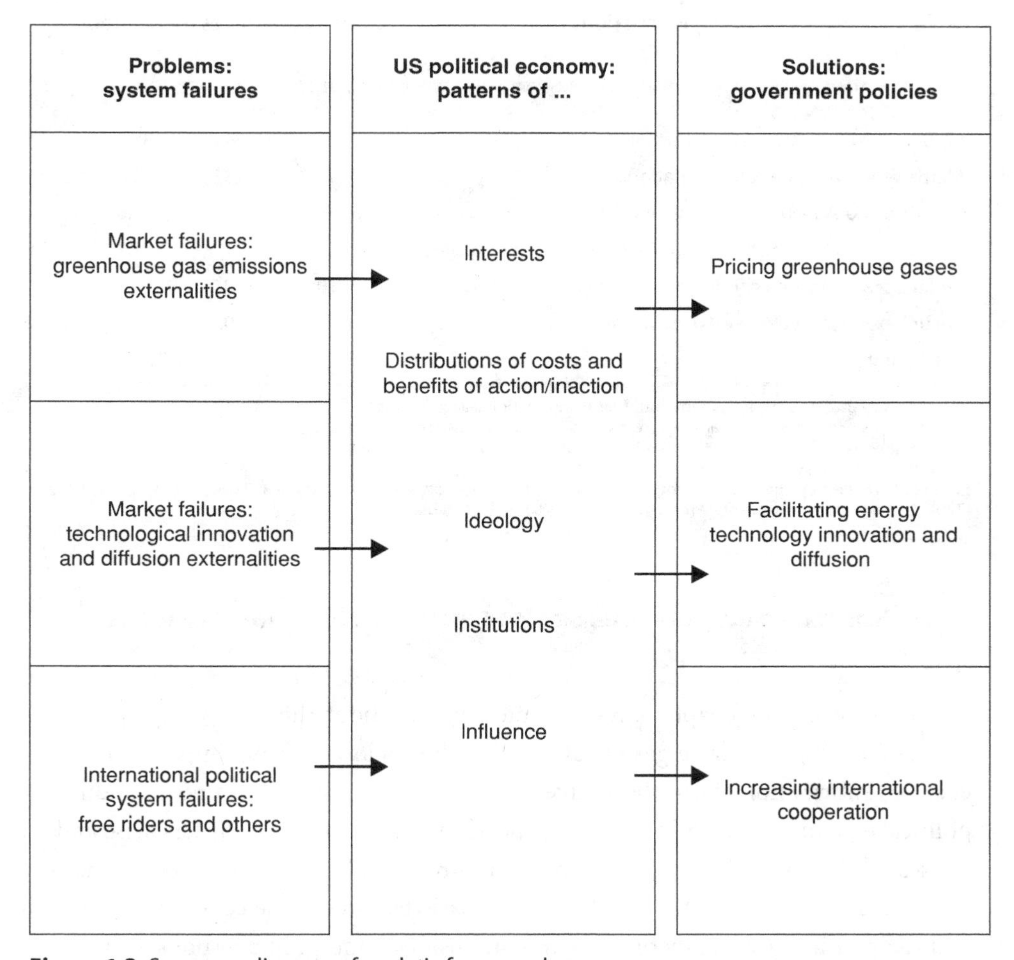

Figure 1.3 Summary diagram of analytic framework

at the left side of the figure and in this first chapter. The analysis of the problem includes both the emissions rates and concentration levels of greenhouse gases, and it includes the socio-economic impacts of diverse types of climate change. In the figure, economic and political system failures are represented in the left column as problems, and the corresponding issue clusters and government policies are represented in the right column. The intervening features of the US political economy in the middle column are the key variables determining the nature of US responses to the systemic deficiencies.

Table 1.2 US greenhouse gas emissions by types of gas and principal sources (percentage of total)

Type of greenhouse gas / Principal source	1990[a]	2000[b]	2005[c]	2010[d]
Carbon dioxide/Fossil fuel combustion	83.3	84.8	85.6	83.6
Methane/Animal enteric fermentation; landfills	10.0	8.3	7.8	9.8
Nitrous oxide/Agricultural soil management	5.3	4.9	4.6	4.5
Hydrofluorocarbons/Substitution of ozone-depleting substances	0.6	1.5	1.7	1.8
Perfluorocarbons/Semiconductor and aluminum production	0.3	0.2	0.1	0.1
Sulfur hexafluoride/Electrical transmission	0.5	0.3	0.3	0.2
Total (mmtCO_2e)	6127	7045	7133	6822

[a] Commonly used base year by many countries in international agreements.
[b] Commonly used base year in analyses of historical trends in emissions.
[c] Commonly used base year in US legislative proposals and government documents.
[d] Most recent year available at time of publication.
Source: Compiled by the author from US Environmental Protection Agency (2010), 2-3, Table 2-1; and (2012a), 2-3, Table 2-1; and US Environmental Protection Agency (2013a), 2-3, Table 2-1.

1.2 Sources of emissions: sectors and the regional locations of industries

The balance of this chapter presents information about the economic sectors that are sources of greenhouse gas emissions and the socio-economic impacts of climate change. Information about the sources and impacts is essential to understanding the political economy context in which the specific issues in the issue clusters arise and are addressed. For there are important patterns among industries and among regions of the country, and these patterns establish the basic features of the economic geography and the political geography of government, business, and public responses to climate change issues. In terms of the summary of the analytic framework in Figure 1.3, this is fundamental information about the economic interests and the distribution of costs and benefits of action or inaction. (Readers who are already familiar with the sources and impacts of greenhouse gas emissions in the USA may wish to skim the remainder of the section.)

Table 1.2 documents the predominance of carbon dioxide among the many types of greenhouse gases in terms of their contributions to total US emissions, and it indicates the principal source of the emissions for each type of gas. In recent years, carbon dioxide's share of total greenhouse gas emissions, expressed in carbon dioxide equivalents, has been more than 80 percent, overwhelmingly from fossil fuels. By comparison, methane has constituted less than 10 percent, and nitrous oxide about 5 percent or less.

Table 1.3 International comparisons of the amounts of greenhouse gas emissions (carbon dioxide equivalents)

	Cumulative amounts of emissions: 1950–2005; Percent of world total *(world rank)*	Annual amount of emissions: 2006; Percent of world total *(world rank)*	Per capita annual emissions: Metric tonnes of carbon equivalent per person, 2006 *(world rank)*	GHG intensity of economy: Metric tonnes of carbon equivalent per USD million at purchasing power parity, 2006 *(world rank)*
USA	27 *(1)*	20 *(2)*	5.3 *(1)*	124
Brazil	1	1	0.5	59
China	10 *(3)*	22 *(1)*	1.3	286 *(1)*
EU	22 *(2)*	14 *(3)*	2.3 *(3)*	83
India	3	5	0.3	136
Indonesia	1	1	0.5	132
Japan	5	4	2.7	86
Russia	9	6	3.1 *(2)*	242 *(2)*
South Africa	1	1	2.0	226 *(3)*

Source: Compiled by the author from World Resources Institute (2010c).

The combination of HFCs, PFCs, and SF_6 have together only constituted 2 percent. Clearly then, at the aggregate national level, carbon dioxide is by far the most important greenhouse gas in terms of global warming impacts; however, each of the other gases is significant in individual industries – for instance, methane in agriculture.

In an international perspective, the rates of total emissions and the rates of increase can be compared in terms of several indicators. Data for four of the commonly used ones are presented in Table 1.3. The USA ranks first, followed by the European Union, in terms of cumulative emissions over the past half century, while China ranks first in current annual emissions. China and other developing countries, of course, rank low in per capita emissions, while the USA ranks the highest, followed by Russia and the European Union. As for the greenhouse gas intensity of their economies, China, Russia, and South Africa rank first, second, and third.

In Figure 1.4 the widespread incidence of greenhouse gas emissions among key sectors of the economy is evident in Panel A; viewed from a different perspective, in which electric power production is treated as a separate sector, the relative importance of emissions in the electric power and transportation sectors is apparent in Panel B. In fact, the combination of the two sectors – electricity production and transportation – has accounted for well over half of US total carbon dioxide emissions for many years.

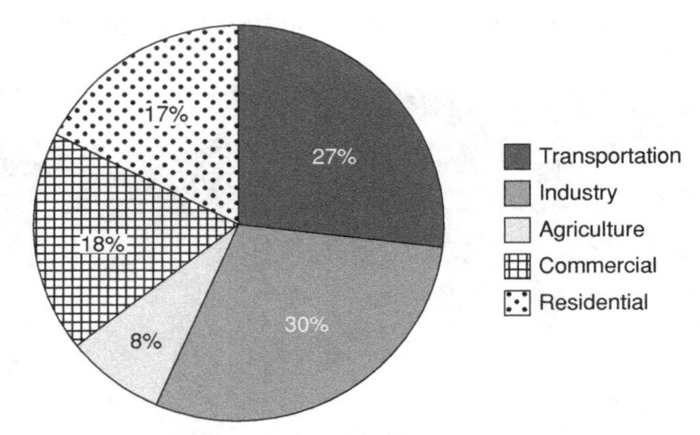

A. With electricity distributed among end-use sectors

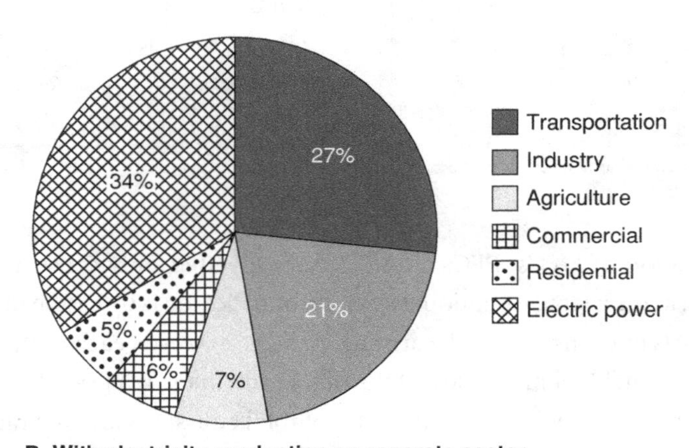

B. With electricity production as separate sector

Figure 1.4 Sectoral sources of US greenhouse gas emissions (CO_2e tonnes, percent of total)

Source: Computed by the author from US Environmental Protection Agency (2012a), ES-16, Table ES-8 for Panel A, and ES-15, Table ES-7 for Panel B; data are for 2010.

Table 1.4 highlights the predominant fuel source of emissions within each of these two sectors. The contributions to US total carbon dioxide emissions of coal in the electric power sector and cars and trucks in the transportation sector have been, respectively, about one-third and one-fourth. The two together contribute nearly 60 percent of US total carbon dioxide emissions. To a significant extent, therefore, the problem of greenhouse gas emissions in the United States is about producing electricity in coal-fired power plants and powering motor vehicles with petroleum-based fuels. *Coal-based electric power and petroleum-based road transportation are inevitably central to the political economy of climate change issues.*

Table 1.4 Carbon dioxide emissions in the US electric power and transportation sectors

	1990[a]	2000[b]	2005[c]	2010[d]	Percent change 1990–2010
Mmt CO_2e					
Electric power	1831	2311	2418	2275	+19.5
Of which: coal	1548	1927	1984	1827	+18.0
Transportation	1497	1823	1906	1744	+16.5
Of which: cars and trucks	1180	1457	1564	1544	+26.9
Percent of total CO_2e emissions					
Electric power: coal	30.3	32.2	33.1	33.8	+3.5
Transportation: cars and trucks	23.1	24.4	25.6	28.6	+5.5
Coal + cars and trucks	53.4	56.6	58.7	62.4	+9.0

[a] Commonly used base year by many countries in international agreements.
[b] Commonly used base year in analyses of historical trends in emissions.
[c] Commonly used base year in US government documents.
[d] Most recent year available at time of publication.
Source: Computed by the author from US Environmental Protection Agency (2010), 2–18, Table 2–13 and 2–21, Table 2–15; and (2012a), 2–9, Table 2–5; 2–19, Table 2–13; 2–23, Table 1–15; and US Environmental Protection Agency (2013a), 2–18, Table 2–13.

Although other sectors make smaller contributions to total greenhouse gas emissions, they are nevertheless individually and collectively important. In the manufacturing sector (designated "industry" in Table 1.2 above) chemicals, steel, oil refining, and cement are significant contributors. Services on the other hand are not generally significant direct sources of greenhouse gases, with some important exceptions such as airlines and maritime shipping, which are contributing increasing shares of the total. Also, services such as architecture and engineering are obviously important because they affect building design, and thus potentially reduce the contributions of greenhouse gases in the "residential" and "commercial" categories.

Some key greenhouse-gas-emitting industries are concentrated in a relatively small number of states, for instance the auto industry in the Midwest and Southeast. The regional patterns of these industries are depicted in Maps 1.1, 1.2, and 1.3, respectively, for the coal, oil and gas, and motor vehicle industries. These patterns are basic facts about the *economic geography* of climate change issues in the USA, patterns which in turn directly affect the *political geography* of those issues. The political consequences are examined in greater depth in later chapters.

Agriculture and forestry are important sources of greenhouse gases, but they are also key sectors in greenhouse gas sequestration patterns, potentials, and programs.

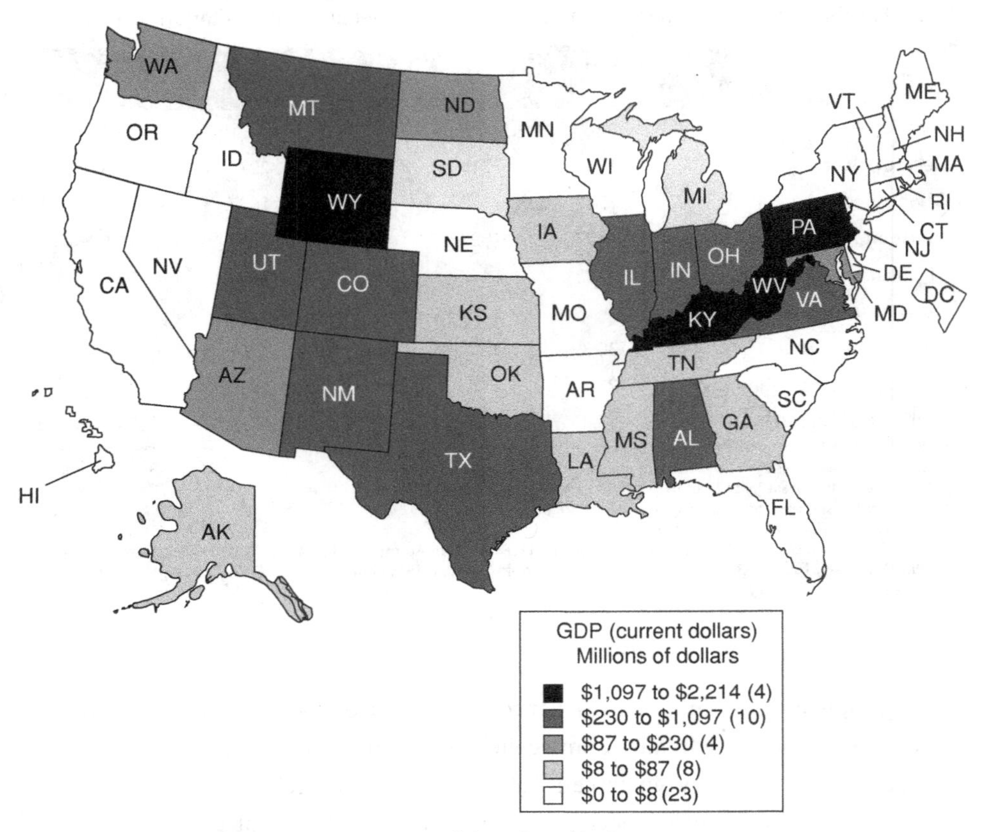

Map 1.1 Regional concentration patterns of the US coal industry

Source: US Bureau of Economic Analysis (2010a).

As the data of Table 1.5 indicate, the relative magnitudes of sequestration levels in the agricultural and forestry sectors have been about one-sixth the magnitudes of total national greenhouse gas emissions. Because the sequestration potential of these sectors in the future is much greater, government subsidies, industry practices, and other issues concerning sequestration are increasingly salient.

Appendix 1.1 presents statistics on emission levels for each individual state, and the states are arranged according to standardized regional groupings to reveal the regional patterns in the relative greenhouse gas intensity, as measured by per capita emissions. The relatively low levels of the West Coast states (except Alaska) and the Northeast states are evident.

Table 1.5 Greenhouse gas sequestration magnitudes (sinks in the form of land use, land use change, and forestry)

	1990[a]	[2000][b]	2005[c]	2010[d]
Land use change and forests (mmt CO_2e)	810	[664]	1069	1043
Of which, forest land	701	[468]	941	922
Total emissions	6187	[5593]	7238	6866
Ratio of sinks to total emissions	.13	[.12]	.15	.15

[a] Commonly used base year by many countries in international agreements.
[b] Commonly used base year in analyses of historical trends in emissions; these are unrevised estimates reported in EPA (2010), as indicated in sources below.
[c] Commonly used base year in US government documents.
[d] Most recent year available at time of publication.
Source: Compiled by the author from US Environmental Protection Agency (2010), ES-13, Table ES-5 and 2–13, Table 2-9; and (2012a), ES-4, Table ES-2; 7–1, Table 7–1; and US Environmental Protection Agency (2013a), ES-13, Table ES-5.

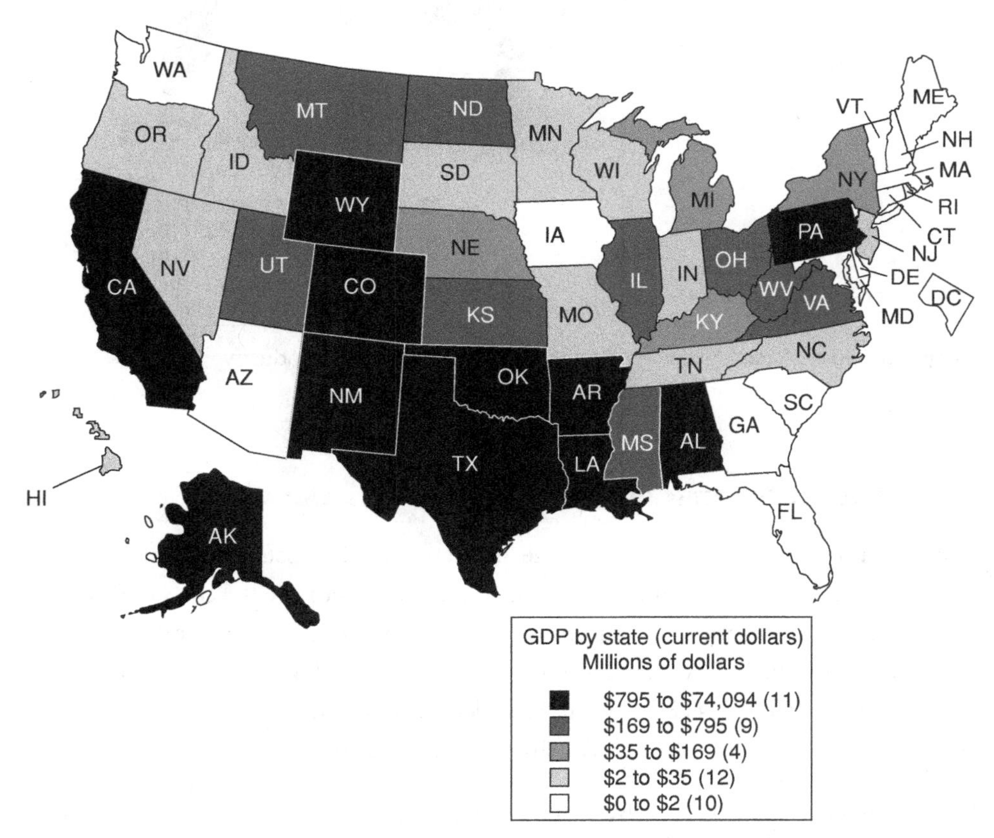

Map 1.2 Regional concentration patterns of the US oil and gas industry

Source: US Bureau of Economic Analysis (2012b).

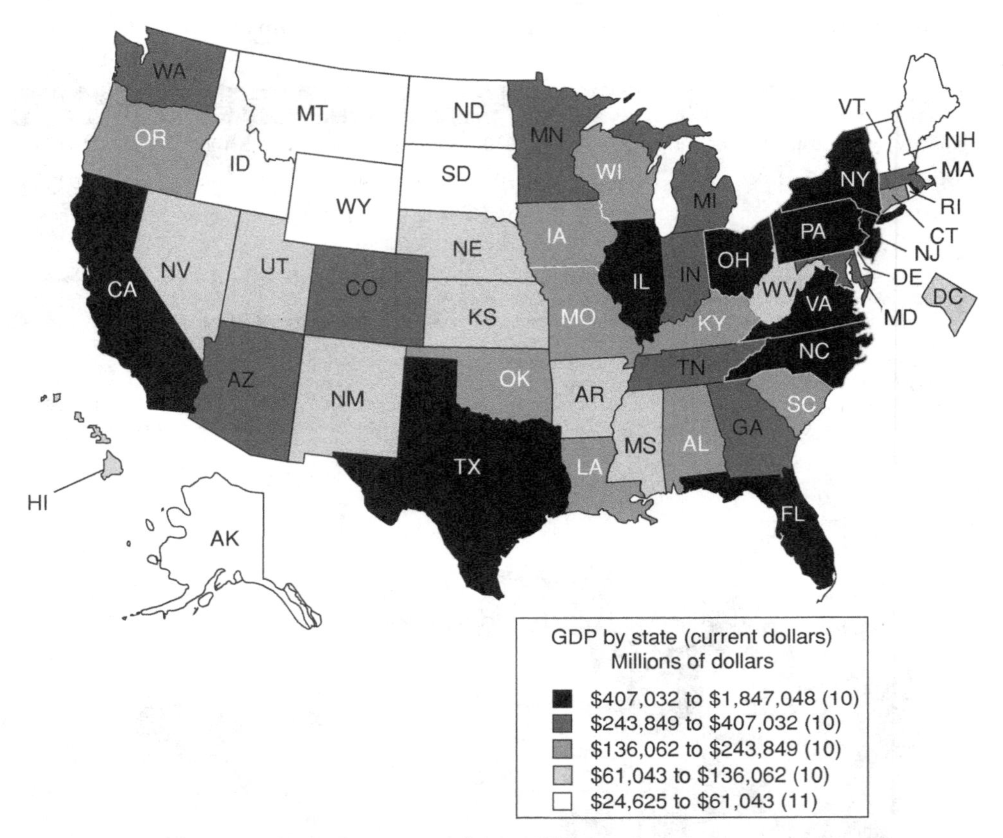

Map 1.3 Regional concentration patterns of the US motor vehicle industry

Source: US Bureau of Economic Analysis (2012c).

1.3 Socio-economic impacts: regional variations

The economic costs of climate change impacts and the costs of adapting to them are not distributed evenly and will not be in the future; nor are the costs of mitigating greenhouse gas emissions, through emissions limits or sequestration or other measures, distributed evenly. Indeed, there are major variations in all their costs among regions within the United States (Ebi, Meehl, Bachelet, Twilley, Boesch et al. 2007). A study by the US Congressional Budget Office (2009, 13) concluded:

> A changing climate will involve changes in typical patterns of regional and seasonal temperature, rainfall, and snowfall, as well as changes in the frequency and severity of extreme events, such as heat waves, cold snaps, droughts, storms, and floods...

Table 1.6 Regional and industry exposures to the economic impacts of climate change

Regions	Types of industries and other assets at risk	Examples
Northeast and Mid-Atlantic	Transportation infrastructure	Increased losses at airports, bridges, tunnels, subways in New York and Boston because of sea level rise and hurricanes
	Forestry	Major employment losses in some states because of pests, drought, and fires.
	Winter tourism	Declining winter skiing because of decreased snow falls[a]
Southeast	Forestry	Major employment losses in some states because of pests, drought, and fires.
	Fishing	Losses from increased storm activity
	Agriculture	Losses from droughts
Midwest	Great Lakes shipping	Dredging costs and/or increased shipping costs because of declining water depths
	Agriculture	Declining dairy and grain production
Great Plains	Agriculture	Declining aquifers and increasing crop pests
West and Pacific Northwest	Forestry	Increased losses from pests
	Agriculture	Declining water supplies for irrigation
	Coastal infrastructure	Increased annual maintenance costs because of sea level rise, erosion, and severe storms

[a] 138 skiing facilities (10–20 percent loss of skiing days = $410–810 million lost revenue per year).
Source: Compiled by the author from Center for Integrative Environmental Research (2007), 16–39; also see US Congressional Budget Office (2009); and US Global Change Research Program (2009).

> Regional climates in the United States are expected to become more variable, with more intense and more frequent extremes of high temperature and rainfall. In general, extreme events tend to have disproportionately greater effects: A small percentage increase in hurricane wind speeds, for example, can greatly increase the potential damage.

Other studies of the impact of climate change in the USA have also focused on regional variations in the impacts on industries and other costs; Table 1.6 summarizes the results of one such study, which incorporates the results of a wide range of preceding studies by government agencies, the insurance industry, and other organizations.

The effects in particular industries in two parts of the country – the Northeast states and California – have been analyzed by the Center for Integrative Environmental Research (2007, 6). It is evident that the agricultural and tourism sectors of the economies in those regions are particularly vulnerable:

> [I]n the Northeast, the maple sugar industry – a $31 million industry – is expected to suffer losses of between 15 and 40% ($5–12 million) in annual revenue due to decreased sap flow. The region can expect a decrease of 10–20% in skiing days, resulting in a loss of $405–810 million per year. The dairy industry is also highly sensitive to temperature changes, since the dairy cows' productivity starts decreasing above 77 °F (25 °C). In California, an annual loss of $287–902 million is expected for this $4.1 billion industry. Losses are expected to the $3.2 billion California wine industry as well, since grape quality diminishes with higher temperatures. In each case, these may be considered small niche sectors in their respective economies – accounting for less than one-tenth of gross state product – yet they are an essential element of local employment, history, culture and landscape.

These and other impact studies should be interpreted with caution. Because it is not scientifically sound to attribute any *one specific* extreme weather event and its impacts *entirely* to the long-term trend in temperatures, measurement of the exact economic impacts of climate change is problematic. However, it is appropriate to note that particular extreme events are the *types of events* that tend to become more frequent and more severe as global warming occurs. To the extent that individual events that transpire are examples of climate change patterns, trends, and their impacts, one can consider such events to be *illustrative* of the types of events that become more frequent and/or more severe as part of the progression of climate change. They are thus indicative of the kinds of economic impacts that climate change is already having and will have, to a greater degree, in the future. Furthermore, probabilistic interpretations of the relative significance of long-term climate change as a contributing factor to the occurrence of specific events such as droughts can be developed (see especially Hansen 2012).

Consider hurricanes, for instance. Hurricanes Katrina and Rita in 2005 not only increased interest in scientific issues about the linkages between global warming and hurricanes in particular, they also more generally increased interest in the socio-economic costs of such extreme weather events. Studies have found correlations between sea surface temperatures and the incidence of the most extreme hurricanes (Webster, Holland, Curry, and Chang 2005; Hoyos, Agudelo, Webster, and Curry 2006; cf. Muller 2009); however, estimating the economic costs of global warming in the form of increased damage from hurricanes, of course, requires a consideration of a broad array of factors. These factors include not only the effects of sea water warming on hurricane activity, but also real estate market prices, infrastructure damage, the costs of services such as medical care, the value of lives lost, and a variety of other direct and indirect costs. Despite such complexities, it is often possible to compute order-of-magnitude estimates; in the case of Hurricanes Katrina and Rita, estimates of the total costs exceeded $200 billion, or slightly more than 1 percent of US GDP at the time (Center for Integrative Environmental Research 2007, 5).

Other examples of socio-economic costs are the impacts on Alaskan fishing villages, which have already experienced the costs of sea level rise and anticipate substantial increases in those costs in the future. As a result, plans are being developed to relocate the villages to less vulnerable areas at greater distances from the coast. A study of the economic cost of such measures that was undertaken by the US Government Accountability Office (2004) estimated that hundreds of millions of dollars would be needed to relocate the villages a few miles inland to save them from destruction.

The costs of wildfires have also been subjected to cost estimates, as the increasing incidence and severity of wildfires in California, Arizona, New Mexico, and other states in the western region of the country can be attributed, in part, to prolonged droughts and the long-term process of desertification caused by global warming. An analysis by the US Global Change Research Program (2009, 82) concluded:

> In the western United States, both the frequency of large wildfires and the length of the fire season have increased substantially in recent decades, due primarily to earlier spring snowmelt and higher spring and summer temperatures. These changes in climate have reduced the availability of moisture, drying out the vegetation that provides the fuel for fires. Alaska also has experienced large increases in fire, with the area burned more than doubling in recent decades.

Wildfires in Southern California in 2007 resulted in more than $1 billion estimated property losses; more generally, estimates of the costs to the federal government for its fire-fighting services exceeded $1 billion per year from 2003 (Center for Integrative Environmental Research 2007, 15). The total economic consequences of such fires depend of course on real estate market prices, the estimated value of the destroyed timber, and other economic calculations; further, the extrapolation and attribution of costs to climate change depends on the frequency and severity of droughts, their causal connection to global warming, and their contribution to the marginal increase in the frequency and severity of the fires. So, again, such estimates and projections clearly need to be treated as only approximate order-of-magnitude data.

These cost estimate issues became particularly salient during the summer of 2012 when the USA experienced its hottest month on record, and the associated grain crop losses and subsequent food price increases became matters of widespread concern and controversy. Although the economic costs were thus ultimately borne nationally and internationally, they were concentrated in the Southwestern, Midwestern, and Southeastern regions of the USA.

Since regional economic interests can be determined not only in terms of the *impacts* of climate change but also in terms of the greenhouse gas intensity of local production or consumption patterns in the *sources* of greenhouse gas emissions, the interests of regions at stake in climate change issues are often mixed and conflicting. That is, some regions are significant sources of emissions and are thus economically threatened by mitigation measures, but at the same time they also have important economic interests that are threatened by the impacts of climate change. For example, Alaska and Texas rank high on both the sources and impacts dimensions because they have greenhouse-gas-intensive industries as direct or indirect sources of emissions (oil in both states), but they are also exposed to sea level rise. Other states are high on both dimensions for other reasons: the Midwestern states of Indiana and Illinois, for instance, are dependent on the carbon-intensive automotive industry and coal-fired power plants, but at the same time they have agricultural interests that are exposed to the threat of droughts. Yet other states such as Idaho and Vermont are vulnerable to the impacts of global warming as a result of declining winter tourism from shorter skiing seasons in both states, timber loss from increased pest infestation in the former, and the loss of autumn tourism as forest colour patterns migrate north to Canada in the latter. But neither has an emission-intensive economy; furthermore, both offer significant forest carbon sinks. The mixes of US domestic economic interests and their political implications are discussed in more detail in Chapter 4.

The impacts of climate change, of course, extend to the entire world (UN IPCC, Working Group II 2014). Climate change is inherently a global problem as the discussion above of externalities, public goods, and free riding makes clear. The free-riding challenges are especially problematic internationally because there is no centralized authority that can punish free riders or otherwise force them to participate in international mitigation efforts. There are, however, other ways to address free-riding issues, as the discussion of Chapter 7 on international cooperation makes clear. In any case, the basic types of impacts are similar in many countries, but with significant variations among countries – and regions within countries – in the mix and severity of the consequences. Small island countries and territories are the most vulnerable in that their existence is at stake; indeed some have already made arrangements with other countries to relocate their populations before they are inundated by sea level rise. In addition to the USA – where Florida, Texas, and the states in between have flat coastal zones – other countries with large low-lying coastal zones are also exposed; among them, as many as 100 million residents of Bangladesh face the prospect of having to be relocated to avoid inundation. Many of the world's most populous cities – e.g. Shanghai – also face coastal sea rise and increased storm damage issues. For other areas, the most threatening impacts are in the form of droughts affecting agricultural production in rural areas or water shortages affecting water supplies in urban areas. Floods in Thailand in 2011 that destroyed or damaged factories making components for computers and other electronic goods were a reminder of how quickly and extensively supply disruptions in one part of the world can spread to entire global industries with highly integrated and internationalized supply chains.

How to value the impacts on society of climate change – and thus how to value greenhouse gas emissions – are sensitive issues politically and ethically. When the US administration increased the estimate of the cost of carbon dioxide in 2013 from $21 to $36 per ton as part of a cost-benefit evaluation of new regulations, the change prompted much comment and even opposition among some members of Congress (Lieberman 2013). Such a change, of course, can alter the estimated net economic effects of regulations and therefore the economic rationale for them. In any case, more generally there are ethical issues about the focus of cost-benefit analysis of climate change policy options in view of the wide range of non-economic social impacts.[2]

1.4 Implications: economic interests and economic geography

The economic geography of industry sources of greenhouse gas emissions and the economic geography of the impacts of climate change both domestically and internationally are integral to the political economy of US responses to the problem of climate change. Variations in the regional patterns of emissions within the USA are obviously dependent on regional variations in key greenhouse-gas-emitting industries, such as

2 Attributing economic value to premature deaths – in some studies, with variable values depending on the income levels of countries – is a particularly nettlesome issue, not only in climate change studies, but also in other studies of the health consequences of social choices. It is not necessary to enter into that discussion in the present context, except to note that there are economic consequences of premature deaths and that they can be taken into consideration in estimates of the costs of global warming. However, there are obviously also ethical, social, and political considerations that are centrally important – and indeed for many observers, of transcendent importance, compared with the economic costs. Since this is a book about political economy, it focuses on economic costs. For further discussion of these issues, see for instance Kelman (1981) and Schelling (1987).

coal, oil and gas, and motor vehicles, and regional variations in their use. Furthermore, variations in the regional patterns of the impacts of climate change in the USA are also a central element of the economic geography of climate change issues. Economic geography is thus essential to an understanding of regional variations in the perceptions of the problem as well as preferences for solutions to it.

Subsequent chapters consider further the implications of economic geography in two respects. Chapter 2 on business focuses on industries and firms – and thus complements the macro-level data of the present chapter with meso-level and micro-level economic analyses. Chapters 3–5 focus on the political geography issues that emerge from the economic geography by considering, respectively, regional patterns in public opinion, state government policies, and congressional policymaking, as they are affected by the regional economic patterns.

The appendices of the present chapter provide additional details on three topics: Appendix 1.1, as noted above, presents state-level and regional-level data on per-capita emissions in the USA. The other two appendices provide additional information about climate science. Thus, readers who are interested in the history of climate science, recent developments and current questions, are invited to read Appendix 1.2, and readers who would like to know more about the "global warming potentials" of the different types of greenhouse gases are invited to read Appendix 1.3.

Suggestions for further reading and research

The periodic reports of the UN Intergovernmental Panel on Climate Change provide extensive and thoroughly vetted summaries of published studies concerning climate science, the impacts, and the policy issues. The results of the *Fifth Assessment Report* are available as the reports of Working Group I (2013) on physical science, Working Group II (2014) on impacts, and Working Group III (2014) on policies.

A series of reports published by the US National Academy of Sciences in 2010–2011 also offer balanced, comprehensive analyses of the science, impacts and policy issues: *Advancing the Science of Climate Change* (2010a), *Limiting the Magnitude of Future Climate Change (2010b)*, *Adapting to the Impacts of Climate Change (2010c)*, *Informing an Effective Response to Climate Change* (2010d), and *Climate Stabilization Targets: Emissions, Concentrations, and Impacts Over Decades to Millennia* (2010e), and especially *America's Climate Choices* (2011). A short introduction to climate science is available in Emanuel (2012). There is a readable collection of short essays on many climate change topics – with illustrative photographs – in Schmidt and Wolfe (2009). An accessible introduction to many climate science issues is provided by DiMento and Doughman (2007). Mathez (2009) offers a solid introduction to climate change and related energy issues, with good graphics and color photos and a brief "Student Companion" that includes key concepts and questions. The US National Oceanic and Atmospheric Administration (2010) includes a status report on the science.

Detailed and updated sources of data on US emissions are in the annual reports of the US Environmental Protection Agency (e.g. 2010; 2011; 2012) and US Energy Information Administration (e.g. 2012). For international comparisons, see the annual reports of the International Energy Agency (e.g. 2012) and the periodically updated interactive database of the World Resources Institute (2012). For introductions to climate change science, see Archer and Rahmstorf (2010), Abatzoglou, DiMento, Doughman, and Nespor (2007), Eggleton (2013), Houghton (2009), Oreskes (2007), Stern (2007), Wigley (1999), and World Bank (2009b). Muller (2009) provides detailed analyses of key issues in the climate science literature. An insightful analysis of the interaction of climate science and politics, especially in the United States, is available in Dessler and Parson (2010).

For keeping up with advances in scientific research on climate change, the journals *Nature*, *Science*, and *Scientific American* are accessible to non-specialists. Several websites are useful – especially www.realclimate.org, which includes extensive discussions of a wide range of scientific issues, such as linkages between ocean surface temperatures and hurricanes; also see www.climatesciencewatch.org, www.theprojectonclimatescience.org, and www.insideclimatenews.org. The websites of the UN Framework Convention on Climate Change at www.unfccc.org and the UN Intergovernmental Panel on Climate Change (IPCC) at www.ipcc.ch are both useful starting points for searches, as is the website "America's Climate Choices" sponsored by the US National Research Council of the National Academy of Sciences at www.americasclimatechoices.org.

A collection of short readings edited by McKibben (2012) covers science, impacts, and politics; also see Walker and King (2008) and Zedillo (2008). On the politics and economics of climate change, not only in the USA but in other countries and in international venues, see the numerous studies in Helm and Hepburn (2009). Recent collections of studies on diverse aspects of climate change issues, including their political and economic aspects in the United States, are provided by Driesen (2010a), Harrison and McIntosh (2010), Rabe (2010c), and Selin and VanDeveer (2009). Individual contributions to these collections are cited in subsequent chapters in reference to particular topics.

There are numerous and disparate studies of climate change issues and government policymaking in the USA. Among them, Pomerance (1989) provides an analysis at the time climate change was emerging on the public agenda in the USA. Also see the collections edited by Harris (2000; 2009), including the individual contributions to the former by Betsill (2000), Fischer-Vanden (2000), Harrison (2000), and Park (2000); and see Schneider, Rosencranz, and Niles (2000), including the contribution by Anderson (2000). Skolnikoff (1990; 1999) emphasizes the decentralization of influence in the policymaking process; also see Lee (2001) and Brewer (2009a). A wide-ranging official US government report on climate change patterns and trends in the USA and responses to them is available in US Department of State (2010). Pooley (2010) provides a detailed factual account of US national government policymaking on cap-and-trade proposals and other issues over the period from the late 1980s through 2009. Antholis and Talbott (2010) discuss the policies of the two Bush administrations and the Clinton administration, and the role of the Obama administration in the Copenhagen agreement of 2009. *Climate Change News Digest* at www.climatechangenews.org is a useful way to keep track of newsworthy events. For introductions to environmental policies and politics, especially focused on the USA, see Kraft (2007) and Vig and Kraft (2010).

There have been three major, large-scale studies of the macro-economics of climate change – by Cline (1992), Nordhaus (1994), and Stern (2007). Each has made an important contribution to understanding the issues; in recent years, the latter has been particularly influential in focusing attention on the economic costs of inaction, in comparison with the costs of action. Also see Nordhaus (2013) and Weynant (2000) for reviews of the economic issues. Jepma and Munasinghe (1998) provide an extensive integrative framework for analysis, with an emphasis on economic cost-benefit analyses of policy options. For economic and other impacts in the USA, see especially Center for Integrative Environmental Research (2007), DiMento and Doughman (2007), Ebi, Meehl, Bachelet, Twilley, Boesch, et al. (2007), McGarity (2010), and US Global Change Research Program (2000; 2009).

An excellent analysis of the roles of market failures – and government failures – in climate change issues in particular and in other environmental issues as well is available in Hepburn (2010). The nature and consequences of market failures, and the associated political economy issues, are also discussed in Brown (2001), Dijkstra (1999), Jaffe, Newell, and Stavins (2005), Lofgren (1995), Misbach (2000), Proost (1995), and Shogren and Toman (2000). Technical issues about the calculation of the social cost of carbon dioxide emissions to adjust for the market failures that do not take into account all of the costs of greenhouse gases are addressed in US Interagency Working

Group on Social Cost of Carbon (2013). The role of ideas about economics is stressed in the contributions in Driesen (2010a); see especially the chapter by Schroeder and Glicksman (2010). On the issues of free riders, common goods, and collective action, see especially Olson (1971) and Ostrom (1990).

There is of course a massive literature on political economy – some of it written by economists, some by political scientists, and some by both. The collection edited by Weingast and Wittman (2008a) and the section on political economy in Goodin and Klingemann (1998a) are particularly useful and diverse. The former includes an overview by Weingast and Wittman (2008b). The latter includes contributions by Alt and Alesina (1998), Atkinson (1998), Grofman (1998), and Offe (1998). The relationships between governments and markets, or more generally the roles of government in the economy, are examined in Lindblom (1980), Tufte (1978), and Wren (2008).

For readers who are interested in the literature on the principal elements of the analytic framework, as well as the focus of the overall framework, the following should be of interest. The emphasis of the integrative framework on government policymaking as a *problem-solving* process is evident in public policy studies (Kraft and Furlong 2010), and in some of the political science classics such as Deutsch (1963). In the political science literature Kaplan and Lasswell (1950) present a well-known statement of politics as a "value allocation" process, in which "wealth" is one of the eight human values that are at stake. That and another classic work by Lasswell (1936) focus on the *distributive effects of politics* – i.e. "Who Gets What, When, and How." The work of Lowi (1964) has been prominent in its emphasis on distributional issues. Further, many studies of the domestic politics of international trade policies focus on the asymmetries of the distributions of the costs and benefits of trade policy liberalization – or trade protectionism – for consumers and producers (see for instance Spar 2009). There are, in fact, parallels between distributional issues in trade and climate policies, though there are also significant differences, particularly in the much longer time frames involved in the effects of inaction on the climate problem. The role of *ideology*, especially in US politics, has had periods of intense scrutiny and relative neglect, just as generalizations about its importance have varied. For a review of the literature about the USA in general, see Minar (1961); for the role of values and ideologies in environmental policymaking concerning many issues, see Layzer (2011). Scholarly interest in the details of governmental *institutions* has waxed and waned – except of course among public administration specialists – with Rhodes (2009) and March and Olsen (1998) being useful reviews of the literature. The role of *influence* in government in the political science literature hardly needs mentioning, as the classical works are well known – among them Dahl (1961; also see Dahl and Stinebrickner 2002; and Willer, Lovaglia, and Markovsky 1997). Much of the literature on lobbying (Truman 1971 being an early classic study) has been concerned with business organizations, especially their influence (see especially Kamieniecki 2006; Kraft and Kamieniecki 2007a, 2007b). Finally, variations in the patterns and trends in policymaking among types of issues has been a periodic concern in public policy studies. Lowi's work (1964) has been especially conspicuous in this regard; however, the recurrent interest is also evident in Zimmerman (1973).

For readers whose principal interest is the problem of climate change but who may not have a background in political science, two excellent collections of scholarly overviews of the political science literature are available in Goodin (2009), and in Goodin and Klingeman (1998a). Similarly, for policy analysis see especially Kraft and Furlong (2010). On the importance of institutions in political systems and the literature concerning them, see especially Peters (1998), Rhodes (2009), and Shepsle (2008). For the "role of place" in politics (i.e. political geography), see Therborn (2009). For an introduction to economic geography, see Venables (2008). For a wide-ranging collection of political science studies of the international relations of climate change, see the studies in Luterbacher and Sprinz (2001).

APPENDIX 1.1

Per-capita greenhouse gas emissions of states and regions

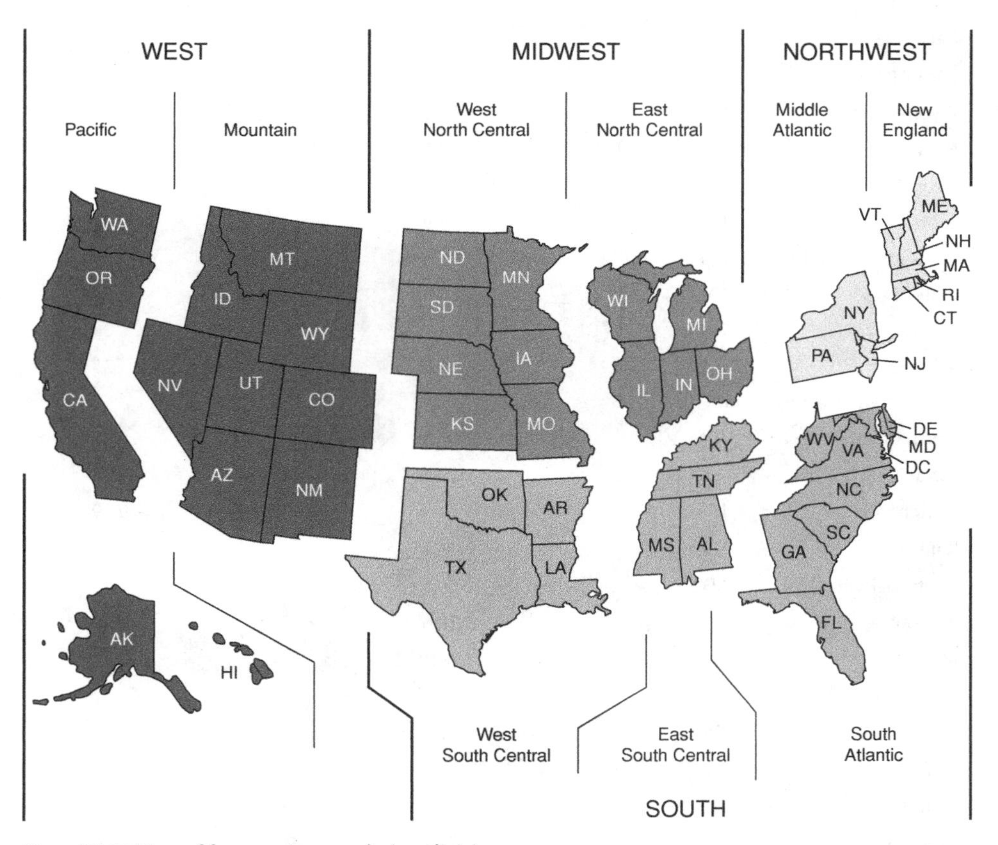

Map A1.1 Map of four regions and nine divisions

Source: US Energy Information Administration, Department of Energy (2010b).

Table A1.1 Metric tonnes of GHG emissions, CO_2 equivalent per capita: amounts and rankings

Region/Division	State abbreviation	Metric tons CO_2e per person[a]	State rank
State			
West/Pacific			
Alaska	AK	77.4	4
Hawaii	HI	21.0	32
Washington	WA	15.0	43
Oregon	OR	14.6	44
California	CA	13.1	46
West/Mountain			
Wyoming	WY	172.1	1
Montana	MT	52.6	5
New Mexico	NM	40.0	12
Utah	UT	29.9	17
Colorado	CO	25.6	22
Idaho	ID	18.8	35
Nevada	NV	18.2	36
Arizona	AZ	17.6	37
Midwest/West North Central			
North Dakota	ND	100.5	2
Iowa	IA	42.6	8
Nebraska	NE	42.5	9
Kansas	KS	41.1	11
South Dakota	SD	39.6	13
Missouri	MO	28.6	19
Minnesota	MN	24.1	24
Midwest/East North Central			
Indiana	IN	43.3	7
Ohio	OH	26.9	21
Illinois	IL	22.5	27
Wisconsin	WI	22.2	28
Michigan	MI	21.1	31
South/West South Central			
Louisiana	LA	49.4	6
Oklahoma	OK	39.0	14
Texas	TX	32.9	16
Arkansas	AR	29.6	18

Table A1.1 (*cont.*)

Region/Division	State abbreviation	Metric tons CO_2e per person	State rank
South/East South Central			
Kentucky	KY	42.4	10
Alabama	AL	37.2	15
Mississippi	MS	28.3	20
Tennessee	TN	23.9	25
South/South Atlantic			
West Virginia	WV	77.5	3
South Carolina	SC	22.6	26
Delaware	DE	22	29
Georgia	GA	21.3	30
North Carolina	NC	19.5	33
Virginia	VA	19.2	34
Florida	FL	15.9	40
Maryland	MD	15.7	41
District of Columbia	DC	6.4	51
Northeast/Middle Atlantic			
Pennsylvania	PA	24.6	23
New Jersey	NJ	16.7	39
New York	NY	12.0	50
Northeast/New England			
Maine	ME	17.2	38
New Hampshire	NH	15.1	42
Massachusetts	MA	13.4	45
Connecticut	CT	12.9	47
Vermont	VT	12.4	48
Rhode Island	RI	12.0	49

[a] Includes six GHGs; excludes land use and land use changes. Data are for 2007, which was the latest year available.
Sources: Compiled by the author from World Resources Institute (2010c) and US Census Bureau (2010).

A. *Summary of patterns*

The five *Pacific Coast* states are generally low in terms of their GHG emissions per capita – except for Alaska, which is among the most GHG-intensive states in the country. The eight *Mountain* states of the west, on the other hand, are generally more GHG-intensive – with Wyoming at more than 170 metric tonnes per capita being the highest in the country (from coal and cattle) – but the area is overall rather heterogeneous. The two "divisions" within the *Midwest* are economically different and thus have different GHG profiles in terms of their industry composition – i.e. the largely agricultural states of the West North Central division, and the heavy industry (autos and steel) and mostly coal-dependent states of the East North Central division; but both groups are GHG-intensive, as all of the states in both divisions are above the national per-capita level of emissions. In the *South*, the four states of the West South Central division (Louisiana, Oklahoma, Texas, and Arkansas) are all well above the national average, as are all four of the states of the East South Central division (Kentucky, Alabama, Mississippi, and Tennessee). The *South Atlantic* division is mostly below the national average – with the important exception of West Virginia, which is the third most GHG-intensive state in the country in terms of per capita emissions. The three populous states of the *Middle Atlantic* division range from above-average Pennsylvania, with its coal and steel, to below-average New York and New Jersey. Finally, the six states of *New England* are all below the national average, with mostly small-scale manufacturing and agricultural sectors.

Note that these are emissions data – and do not include sinks, land use, or land use change data. If sinks were included, the heavily wooded states of the Northwest (Oregon and Washington) and the Northeast (Maine, New Hampshire, and Vermont) would be even less GHG-intensive.

APPENDIX 1.2

A short history of climate science[3]

Recorded scientific work on greenhouse gases can be traced back nearly two centuries. In the 1820s the French scientist Joseph Fourier recognized that there is a natural greenhouse gas effect that keeps the earth warmer than it would otherwise be. As he wondered how the earth stays warm enough to support life in spite of the intolerably cold space around it, he speculated that some portion of the energy from the sun that penetrates the earth's atmosphere and reflects from its land mass and oceans is reflected back to the earth's surface by water vapour and other gases in the atmosphere. Papers advancing his theory were published in *Annales de chimie et de physique* in 1824 and in *Memoires de l'Academie Royale des Sciences* in 1827. At the time, because Fourier's analysis lacked supporting empirical evidence, it was not considered compelling; yet, the identification of the phenomenon now known as the greenhouse gas effect was thus first broached in the scientific literature.[4]

In the 1860s the Irish mathematician John Tyndall measured the heat-absorbing capacity of carbon dioxide, ozone, and water vapour, and thus demonstrated that these gases in the atmosphere could affect the earth's surface temperature. His reasoning, though, was concerned with the corollary of the global warming effect of such gases; he reasoned that a decline in such gases in the atmosphere could lead to an ice age, without concerning himself with the mirror-image possibility of global warming.

The American astronomer Samuel Langley, working in the 1880s, focused on the relationship between solar energy and the earth's atmosphere. He calculated that the

3 This appendix draws extensively from Weart (2008) and Christianson (1999). Also see Archer and Rahmsdorf (2010, Chapter 1).

4 Although it is not normally included in a history of climate science, an even earlier event is related to the subsequent development of climate science. While in Paris in 1783, the American scientist and diplomat Benjamin Franklin speculated that an Icelandic volcano could have accounted for the unusually cool summer in continental Europe that year. The suggestion was the beginning of the recognition among scientists that localized volcanic activity could have international climate consequences.

earth's surface temperature would be –200 degrees Celsius if the atmosphere did not absorb and reflect back to the surface of the earth some of the solar energy that would otherwise escape into space.

The Swedish physicist Svante August Arrhenius in the 1890s developed more explicitly and precisely the relationship between carbon dioxide in particular as a greenhouse gas and the earth's surface temperature. He demonstrated that changes in the concentrations of CO_2 in the atmosphere from the burning of fossil fuels could lead to significant climate changes. He undertook a large number of calculations yielding estimates of temperature changes at different latitudes that would result from different assumed levels of increased CO_2 concentrations. For instance, he estimated that a doubling of CO_2 levels would produce a 5–6 degree Celsius (9–11 degree Fahrenheit) increase in the average surface temperature. Such a prospect, however, was not particularly alarming to Arrhenius because he assumed that the warming effect would be beneficial in the far northern hemisphere where he lived, and that it would be several centuries before the emissions of fossil fuels would be great enough to yield a doubled level of concentration of CO_2 in the atmosphere.

As a result of the work of Fourier, Tyndall, Langley, and especially Arrhenius, it was possible in 1937 for an American professor of geography, Glen Thomas Trewartha at the University of Wisconsin, to present a discussion of the "greenhouse effect" in his textbook, *An Introduction to Weather and Climate.* At about the same time in the UK, Guy Callendar was collecting 1880–1934 data on temperature changes from 200 weather stations. He found that there had been a 1 degree Fahrenheit increase during a half century. He also estimated that 150 billion tonnes of CO_2 had been emitted into the atmosphere during that period and that three-fourths of it was still there. He subsequently analyzed historical data on CO_2 concentrations from several regions of the world and concluded that the concentration was 290 parts per million by volume (ppmv) in 1900, and estimated 325 ppmv for 1956.

A similar but more intensive and sustained effort of data collection was begun in the 1950s by Charles Keeling in the United States. Keeling was a chemist who took repeated measurements of CO_2 at ground level at many points in the western USA and found the measurements were consistently around 315 ppmv (and thus close to the 325 ppmv computed by Callendar). These measurements were subsequently reinforced and refined when Keeling was able to establish a monitoring station at Mauna Loa in Hawaii, where readings were similar to the 315 ppmv he had previously found in the continental US observations, though they varied from 314 to 318 ppmv depending on the time of year. For there was (and still is) a strong seasonal pattern in CO_2 concentrations: the readings tend to be low in spring and summer when trees and other vegetation absorb more CO_2, and high in the fall and winter when the leaves are dormant as sinks for CO_2 and when decaying vegetation emits carbon dioxide.

Another landmark development in the late 1950s was the publication of an article by Roger Revelle and Hans Suess (1957), who had collected data on CO_2 and temperatures and also incorporated the earlier work of Arrhenius and the work of Callendar. They concluded that there were important processes underway involving human-induced changes in the earth's carbon cycle with potentially significant consequences for climate change. This work of Revelle and Suess (1957), along with the findings of Keeling and Callendar, combined with the first International Geophysical Year in 1957 to generate much more interest in global warming among atmospheric scientists.

By the end of the 1950s, therefore, more than a century after Fourier had first pondered greenhouse gases and their link to global temperatures and more than a half century after Arrhenius had made preliminary calculations about those linkages, climate change and the role of CO_2 and other gases in it had become an officially recognized topic of serious scientific inquiry in the USA and other countries. Since the late 1950s, the resources devoted to worldwide data collection and analysis have increased. As a result of the activities of the US National Academy of Sciences (NAS) and of the UN Intergovernmental Panel on Climate Change (IPCC) as well as the work of individual scientists and research organizations, it is possible to trace trends in surface temperatures and atmospheric concentrations of CO_2 and other greenhouse gases with more confidence and precision. A long list of landmark studies and reviews by the US National Academy of Sciences (1975; 1991; 1999; 2001; 2010a; 2010b; 2010c; 2010d; 2010e; 2010f) and the UN Intergovernmental Panel on Climate Change (1996; 2001; 2007; also see Pinkse and Kolk 2009, 24–28) have reaffirmed and refined a series of core conclusions about climate change – conclusions which are supported by numerous data sets with millions of data points based on observations from diverse instruments.

There has thus been a strong consensus among environmental scientists for many years that global warming has been occurring since the Industrial Revolution; it has been occurring at an increasing rate during the past three decades; it is substantially caused by carbon dioxide and other greenhouse gases that are emitted as a result of human activity; and its consequences include more frequent and severe extreme weather events, droughts, sea level rise, diminishing biodiversity, shifting disease patterns, and associated socio-economic costs. A study by Anderegg, Prell, Harold, and Schneider

(2010) of 1372 climate scientists found that "(i) 97–98% of the climate researchers most actively publishing in the field surveyed . . . support the tenets of ACC [anthropogenic climate change] outlined by the Intergovernmental Panel on Climate Change, and (ii) the relative climate expertise and scientific prominence of the [2–3% of the] researchers unconvinced of ACC are substantially below that of the convinced researchers." A different kind of analysis was conducted by Cook et al. (2013) of 11,944 articles published during 1991–2011 in the "peer-reviewed scientific literature" that concerned "global climate change" or "global warming" according to their abstracts. They found that among the abstracts where an opinion on whether humans are causing global warming was expressed, 97.1% agreed. In a subsequent analysis in which the authors were surveyed directly, 97.2% of those expressing an opinion agreed.

In any case, the work of climatologists was challenged just before the Copenhagen international climate conference in December 2009, on the basis of emails among some climate scientists that became public. There were concerns about the data used in some studies and the transparency of the IPCC review process. There were two factual errors in the 2007 IPCC review of research on climate change and its impacts. One misstated the melting rate of glaciers in the Himalayas, indicating they would be gone by 2035 instead of 2350; the other misstated the area below sea level in the Netherlands, indicating 55 percent instead of 20 percent (Radio Netherlands [RNL] 2010). However, after much media attention for a few weeks, six separate independent reviews by professionals in the following months concluded that the factual errors were rare and did not change the principal conclusions of the individual studies, nor the patterns of findings and implications of the accumulated climate science studies over several decades (see especially InterAcademy Council 2010; also *Financial Times* 2010b; ClimateWire 2010b, 2010c; Pooley 2010). There was, though, some concern about a lack of transparency in IPCC procedures, concerns that led to a reform of the IPCC procedures (InterAcademy Council 2010).[5]

There is an extensive literature on the politicization of climate science, with a special focus on the attitudes and actions of climate "deniers." It includes Bowen (2008), Bradley (2011), Hoggan (2009), Mann (2012), Norgaard (2011), Oreskes and Conway (2010), Powell (2011), Schneider (2009), and Washington and Cook (2011).

5 The author of this book was a Lead Author in Working Group III for the IPCC AR5. However, all of the materials and comments in the book are entirely my own personal responsibility, including in particular this analysis of the controversy and conclusions about the IPCC procedures.

Finally, a topic of current special interest among climatologists: There is increasing research into possible "tipping points" that would reflect threshold events in global warming processes – such as tundra melt and the associated methane release – that probably could not be reversed (see for instance Kriegler, Hall, Held, Dawson, and Schellnhuber 2009). Such self-reinforcing cycles clearly pose the possibility of especially catastrophic scenarios.

APPENDIX 1.3

Global warming potentials, emission rates, and concentration levels

The table below indicates the global warming potential (GWP), atmospheric lifetimes, and concentration levels for six widely recognized types of greenhouse gases. Because greenhouse gases remain in the atmosphere for many years – some for centuries – there are significant lags in changes in emissions rates from the earth and changes in concentration levels in the atmosphere. Thus reductions in emission *rates* are not followed immediately by reductions in *concentration* levels; rather, there are lags in the order of decades or centuries. In addition, it should be noted that the commonly used GWP numbers are based on a reference of a 100-year average atmospheric life for carbon dioxide, and therefore underestimate the shorter-term impact of methane, with an average atmospheric life of only twelve years.

Table A1.3 Global warming potentials, emission rates, and concentration levels).

Greenhouse gases	Global warming potential (GWP)[a]	Atmospheric lifetime	Pre-industrial era atmospheric concentration level	2009 atmospheric concentration level
Carbon dioxide (CO_2)	1	50–200 years	278 ppm	385 ppm
Methane (CH_4)	34	12 +/−3 years	0.715 ppm	1.741–1.865 ppm
Nitrous oxide (N_2O)	310	120 years	0.270 ppm	0.321–0.322 ppm
Hydrofluorocarbons (HFCs)	140–11,700	2–264 years	[b]	[c]
Perfluorocarbons (PFCs)	6500–9200	3200–50,000 years	40 ppt	74 ppt
Sulfur hexafluoride (SF_6)	23,900	3200 years	0 ppt	5.6 ppt

[a] Based on 100-year lifespan in the carbon dioxide reference case. The 100-year GWP of methane has been revised upward to more than 30 in IPCC (2013), and its twenty-year GWP is approximately 80.

[b] Not available in the source documents.

[c] In addition, there are some references to nitrogen trifluoride (NF_3) as a seventh GHG. For instance, the Western Climate Initiative has included NF_3 in their Design Recommendations for the WCI Regional Cap-and-Trade Program (Western Climate Initiative 2010).

Sources: US Environmental Protection Agency (2010), 1–2, Table 1–2, and 1–7; and (2012a), ES-3, Table ES-1. Also see UN Intergovernmental Panel on Climate Change, Working Group I (2013).

PART II

DOMESTIC ECONOMICS AND POLITICS

CHAPTER 2

Business: part of the problem and part of the solution

We're the third-largest emitter of CO_2 among corporations in America because we generate 70 percent of our electricity at 20 coal-fired power plants. Of all the companies in the world, we're the 12th-largest emitter of CO_2. I share these numbers with you not to brag, but to give you a sense of my special responsibility, the daunting job in front of me.

James E. Rogers, CEO of Duke Energy (2010)

There needs to be a price for carbon, then the market will decide which technology wins.

Jeffrey Immelt, CEO of GE (2005)

This chapter addresses questions about business responses to climate change at several levels: firms, industries, and national and international business associations. It focuses on the economic interests of business, including how interests vary among industries and among firms within industries, and how firms' and associations' positions have changed over time.

There have been important splits within "the business community" in what to do about climate change. These splits have sometimes been publicly observable as long-established business associations and *ad hoc* alliances of firms have splintered over positions on specific pieces of legislation. In addition, while some firms have participated in partnerships with environmental organizations, others have eschewed such arrangements. The themes of variations in firms' interests vis-à-vis climate change and how those variations have affected firms' business practices and political activities are developed by addressing a series of questions: In Section 2.1, which emphasizes diversity in firms' interests, the key question is: What firms and industries are the leaders and laggards in responding to climate change issues – and why? Who are the winners and losers from climate change policies? In Section 2.2 the insurance industry receives special attention because of its importance as an industry that is directly affected by climate-related loses and as a major investor in many sectors of the economy. What is the evidence of the US insurance industry being a laggard compared with its international rivals? How have insurance firms responded to the concerns of state insurance regulators about the industry's attitudes toward climate change and its implications for their business? Section 2.3 examines firms' relations with their stakeholders: What are firms' "carbon disclosure" practices that inform current and potential investors? What kinds of relationships do firms have with environmental NGOs and labor unions on climate change issues? In Section 2.4: What are the positions taken by business associations on government policies? These have been considered in detail because of their importance in the policymaking process; their impact on specific policies is considered further in Chapters 5–7. In the present chapter, we present much data in Section 2.5 about business lobbying activities and campaign contributions: Who gives how much to whom? Again, the consequences of these activities in relation to specific issues are considered in subsequent chapters. The conclusion returns to the chapter themes, namely the diversity of interests on climate change issues in the "business community" and the evolution of their positions over time.

The diversity among firms in their economic interests, their corporate cultures, and their leaders' perspectives on business-government relations in general as well as their attitudes toward climate change issues in particular – all these account for the wide range in how firms approach government policy questions. The range of responses to government policies is often underscored by how differently firms respond to the following rationales for supporting certain types of government policies. Some firms readily accept them, while others reject them.

There are three key reasons for many firms to support certain government climate change policies, according to the Pew Center (2010a; italics added); these are reasons that resonate with many business leaders as well as researchers who follow business developments related to climate change issues:

- *The need for regulatory certainty.* "A clear, long-term, legislative framework for reducing GHG emissions would alleviate much of [the] uncertainty [associated with the prospects for state, regional and EPA regulations], allowing for more intelligent business planning."
- *The economic opportunities arising from climate solutions.* "Clean energy is projected to be one of the great global growth industries of the 21st century. Policy support can accelerate growth in these industries, and help US companies compete against foreign firms that are quickly establishing dominant positions in these important markets."
- *The reputational benefits of supporting public policies that combat climate change.* "Customers, shareholders, employees, and other stakeholders are increasingly pushing companies to demonstrate social responsibility and environmental stewardship."

All of these considerations are evident in firms' relations with their shareholders and other stakeholders.

The theme of diversity of interests is further developed below by noting in more detail that *there are business opportunities as well as threats to business interests involved in climate change issues.* In fact, for any one industry or even a single firm within an industry, there is a mixture of potential costs and benefits at issue; and the mixes are of course different for mitigation and adaptation measures. There is evidence of such variations at the level of individual firms as well as the level of industries, and they are noted in detail in the balance of the chapter. Of course, it is also true that circumstances change – for instance as extreme weather events become more frequent and as public opinion shifts – and thus firms' interests and responses to climate change also change from time to time. The diversity and evolution of firms' interests and responses are considered in detail in the remainder of the chapter.

2.1 Diversity of interests and responses

Many firms confront strategic threats in the issues posed by climate change; many encounter opportunities; most encounter a mixture of both.

For firms across a broad range of industries, the diversity of issues and interests at stake is evident in the following summary of patterns and trends compiled by the Carbon Disclosure Project (2006):

> Climate-driven risks [for firms] will continue to grow: Looking ahead, a series of . . . 'mega-trends' will continue to amplify the financial impacts of climate change [including] . . . strengthening evidence about the reality, gravity, and causes of climate change, . . . increase in extreme weather events, . . . further regulatory action by government at local, national, regional, and global levels, . . . continuing growth of renewable energy and clean technology markets, . . . improved understanding of the variability of company-specific impacts, . . . improved quantification of the potential financial impacts of inaction, . . . increasing exposure of investors to overseas regulatory regimes, . . . growing institutional shareholder activism on corporate carbon risks, . . . global momentum for improved disclosure on corporate risks.

Of course, the importance of such developments varies significantly among firms and industries. For any one industry or firm there is a mixture of effects; this is especially true for large multinational firms with numerous business lines and diverse business interests in many countries. Finally, there are often significant differences between short-term and long-term effects; for example, firms facing relatively strong regulatory regimes may initially experience higher compliance costs, but in the longer term the incentives to innovate in their production processes and product offerings tend to put them in stronger competitive positions.[1]

Among firms that benefit from no mitigation action are, for instance, aluminum and cement firms, the former because they consume much electricity in their production processes and the latter because their production processes are relatively intensive emitters of greenhouse gases. At the same time, aluminum producers can of course reduce their production costs by increasing their energy efficiency and simultaneously reducing their greenhouse gas emissions; and they can experience new business opportunities as lightweight aluminum is used increasingly to replace steel in motor vehicles and other products as firms reduce their greenhouse gas emissions. Commercial banks, energy engineering service companies, and consulting firms have exploited new business opportunities in services to advise corporate clients on managing new regulatory requirements, physical risks, and/or new business opportunities for themselves. Business in the agriculture and fishing sector, meanwhile, is threatened by global warming – by droughts in agriculture and by changes in water temperatures in fishing, though agricultural production of some crops and the stocks of some fish increase in some regions. Finally, the coal industry, of course, is threatened by efforts to reduce dependence on coal-fired electric utility plants because they are a principal source of greenhouse gas emissions, as we noted in detail in Chapter 1.

The central strategic challenge for industries and individual firms within them is to decide whether and how to pursue the opportunities and/or reduce the risks of climate change. These are challenges to which they have been responding with varying degrees of attention and resolve. There are, therefore, not only winners and losers among firms; there are also leaders and laggards (see especially studies by Kolk and Levy 2001; Kolk 2000; Kolk and Pinkse 2005; Lehman Brothers 2007; 2009; Levy 2005; Newell and Patterson 2010; and the Carbon Disclosure Project 2012).

There are also substantial variations among firms *within* industries. For instance, five large US-based firms have been among the industry leaders in their respective industries: Alcoa, DuPont, United Technologies (UTC), General Electric (GE), and Citigroup. The aluminum firm Alcoa changed its production processes in the 1990s in order to reduce its emissions of greenhouse gases, and it called upon US firms to

1 The enhanced competitive position of firms from countries with relatively strong environmental regulations is sometimes referred to as the Porter hypothesis, after Professor Michael Porter (1991; also Porter and van der Linde 1995). It has been the subject of many studies and much discussion (see especially Ambec, Cohen, Elgie, and Lanoie 2010; and Wagner 2003).

address the problem more aggressively.[2] Among chemical firms, DuPont decided to take a leadership role on climate change issues on the basis of its earlier experience as a manufacturer of a refrigerant that was causing ozone depletion over Antarctica, and the associated controversies surrounding the chemical industry's knowledge and responsibilities in addressing that problem. Among diversified heavy electrical and transportation equipment manufacturers, UTC was particularly responsive to climate change concerns; it reduced greenhouse gas emissions from its own operations by 23 percent per year from 2006 to 2009 (Coalition for Environmentally Responsible Economics (CERES) 2009).

At GE, which is one of the largest US-based corporations in terms of market capitalization, after several years of ignoring climate change issues while being headed by Jack Welch, the firm became a leader on the issue when his successor, Jeffrey Immelt, called for more national government action on climate change in a landmark speech in Washington, DC, in April 2005 (Immelt 2005). GE also began to increase significantly its R&D in the development of new "ecomagination" products in a wide range of industries where energy efficiency and greenhouse gases are important considerations. By 2012 it reported that its ecomagination products were yielding revenues of $100 billion and "growing at more than twice the rate of the rest of the company" (*Financial Times* 2012a).

Subsequently, many other firms and industry associations announced support for more action by the federal government. For instance, Citigroup (2007) announced that it supported "urgent action" at all levels – state, regional, national, and international.

Citigroup and GE joined with other firms, as well as environmental organizations, to form the US Climate Action Partnership (USCAP). In its "Call for Action," USCAP (2007)

> Urge[d] policy makers to enact a policy framework for mandatory reductions of GHG emissions from major emitting sectors, including large stationary sources and transportation, and energy use in commercial and residential buildings. The cornerstone of this approach would be a cap-and-trade program.... The group recommend[ed that] Congress provide leadership and establish short- and mid-term emission reduction targets; a national program to accelerate technology research, development and deployment; and approaches to encourage action by other countries, including those in the developing world, as ultimately the solution must be global.

2 Its CEO at the time, Paul O'Neill, later became the first Secretary of the Treasury in the George W. Bush administration in 2001. During an early meeting with the President on a variety of issues, O'Neill suggested that action on climate change should be a priority of the administration – but to no avail (Suskind 2004).

Table 2.1 Distributions of industry[a] leaders and laggards on climate change issues among US-based firms and non-US firms

	US-Based firms	Non-US firms
Industry[a] leaders (top 20% of firms in industry)	3 (6.1%)	12 (54.5%)
Industry[a] laggards (bottom 20% of firms in industry)	11 (22.4%)	4 (18.2%)
Total numbers of firms	49	22

[a] The industries are: airlines, auto, chemicals, food, industrial equipment, mining, oil and gas.
Source: Compiled by the author from data in Coalition for Environmentally Responsible Economics (CERES 2006), 4.

During the consideration of cap-and-trade legislation by Congress in 2010, however, three members of USCAP – BP, Caterpillar, and ConocoPhillips – dropped their membership because they did not support certain provisions in the legislation being developed in the Senate.

Of course, these particular firms and events are only a few among a wide range of firms and numerous events in the evolution of business responses to climate change. Indeed, many major industry associations have consistently taken positions and undertaken lobbying activities in efforts to prevent serious US government action on climate change issues over many years. Furthermore, despite efforts by some corporations and executives, in many industries US firms have tended to lag behind their foreign rivals. This pattern has been especially noteworthy in the auto and oil industries (Kolk and Levy 2001), but it has also been true in other industries as well. For instance, in a study of major industries in the USA which have significant foreign-owned US subsidiaries as well as US-based firms, the US-based firms were rarely industry leaders and frequently industry laggards, as compared with their European and Japanese competitors (Coalition for Environmentally Responsible Economics 2006). *In seven highly internationalized industries, only three US-based firms were among the leaders in their industries; meanwhile eleven US-based firms were laggards within their industries. In contrast, among non-US firms, twelve were leaders and only four were laggards (see Table 2.1).*

Laggards in the auto and oil industries

In the auto industry, Japanese-based firms Toyota and Honda have been industry leaders. A study by the American Council for an Energy Efficient Economy (2007) found that all ten of the most energy-efficient automobiles for sale in the United States at the time were manufactured by Japanese (or Korean) firms, and none by US-based firms.[3] By 2013 US-based firms had a few models in the top ranks, but only after many years of lagging behind their Asian rivals.[4]

In the oil industry, the European-based firms BP, Shell, Statoil, and Total were leaders in addressing climate change for many years.[5] Based in Europe but with extensive business presences in the USA, both BP and Shell began to take action to mitigate greenhouse gas emissions many years before their US-based rivals. As early as 1997, Lord John Browne (1997), the CEO of BP, which operated in the USA as BP America, gave a speech at Stanford University in which he broke ranks with the members of the Global Climate Coalition – a group that was fostering doubts about climate science and opposing government measures to address the climate change problem. Browne said the problem of climate change was serious and that oil firms had a special responsibility to respond to the challenges it posed. His speech was a landmark event in the evolution of business responses to climate change, not only in the USA but for the industry worldwide. BP and its European rival Shell subsequently developed internationalized *intra-firm* greenhouse gas emissions trading schemes. Over the next decade after 1997, other high-profile firms periodically announced new perspectives on the problem, began to advocate a wide range of government policies, and undertook business initiatives, including the modification of existing products and production processes, the development of new products, and the adoption of new internal decision-making procedures.

Oil industry leader–laggard patterns are evident in the results of surveys by the Carbon Disclosure Project (2006) and the Coalition for Environmentally Responsible Economics of the four super majors of the oil industry, as displayed in Table 2.2. Although the rankings are the same for BP (1st), Royal Dutch Shell (2nd), and ExxonMobil (4th), Chevron's position is quite different in the two rankings, as it is a distant third in one survey but tied for second in the other. These differences are partly a reflection of differences in the two scales used by the two rating organizations, but also of changes in Chevron's attitudes and actions on climate change issues that were occurring

3 As late as 2008, Robert Lutz, the General Motors Vice Chairman of the Board, who oversaw new product development, suggested that global warming is a "total crock of [****]" (Reuters 2008). Yet, ironically, he was also a key supporter within GM of the development of its new hybrid electric model, the Volt.

4 The location of production is another matter. By 2010, many of the Japanese-branded models were being assembled in the USA in the factories of Japanese firms' subsidiaries.

5 In light of the Gulf of Mexico oil spill in 2010 and previous incidents regarding BP's safety record, it should be noted that the focus in the present context is specifically on its responses to climate change issues.

Table 2.2 Four major oil corporations' ratings by CERES and Carbon Disclosure Project

Firm (home country)	CERES rating on 100-point scale (20 firms, mean = 35)	Carbon Disclosure Project rating on 100-point scale (15 firms, mean = 75)
BP (UK)	90	95
Royal Dutch Shell (UK and Netherlands)	79	85
Chevron (US)	57	85
ExxonMobil (US)	35	80

Source: Compiled by the author from data in Coalition for Environmentally Responsible Economics (CERES 2006), 4; and Carbon Disclosure Project (2006), 114.

during this period.[6] (For illustrative annual reports on the details of compiling indicators and summaries of firms' responses to climate change and for useful lists of the many dimensions along which firms' strategies and operations can be assessed, see especially Carbon Disclosure Project (2012) and Institutional Investors Group on Climate Change, Investor Network on Climate Change, and Investor Group on Climate Change (2012).)

ExxonMobil has been one of the most prominent and controversial US firms for its approach to climate change issues. Although its position has evolved, especially after a change in the ranks of its executive leadership in early 2006 and changes in the domestic politics of climate change in the USA in 2006–2009, it has had a long history of attacking climate change science and opposing many government actions (Pooley 2010, e.g. 36n, 357n; Union of Concerned Scientists 2007). Yet, by 2007 it was at least nominally supporting a national US carbon tax as a measure for addressing climate change. Of course, it was highly unlikely that such a measure would be adopted at the time, but the fact that ExxonMobil would be publicly advocating a carbon tax was symbolic of a shift in the political environment and in the way many firms approached the issue. By 2012, ExxonMobil's CEO Rex Tillerson explicitly acknowledged that increasing concentrations of carbon dioxide would have socio-economic impacts, but suggested that those impacts would be "manageable" (E&E News 2012; *Financial Times* 2012c). This was a shift from the public positions taken by his predecessor Lee Raymond, who focused on doubts about the existence of the problem. Yet, despite this shift in emphasis there was still a focus on opposing government actions to address the problem and – though at lower levels – still funding organizational activities that opposed greenhouse gas mitigation efforts (Financial Times 2012b).

6 Chevron, which has its headquarters in San Francisco, California, has since increased its Carbon Disclosure Project score.

These many pieces of evidence about differences in the responses to climate change among firms within the auto and oil industries reflect the combination of the diversity and complexity of their interests – and the multi-dimensional nature of the climate change issues that the firms confront. The patterns also reflect differences in the home countries of the parent firms of these large multinational firms. In particular, firms in the USA are accustomed to highly adversarial relations with governmental institutions – relations in which they expect to engage in extensive lobbying and frequent law suits in opposition to government policies. In European countries, Japan and South Korea, where many of the other major automotive firms have their home headquarters, there are stronger traditions of business–government cooperation across a broad range of strategic and operational business concerns – one might suggest even collusion in some instances. Of course, there are still business–government conflicts, which sometimes spill over into the public domain; however, once governments and business reach agreement on basic realities about the existence of a problem and perhaps on how to approach solutions to it, firms tend to work with their governments, at least in comparison with US-based firms, which have more adversarial traditions in their business–government relations. In the case of climate change, therefore, when their governments accepted the reality of climate change and resolved to take actions to mitigate their countries' greenhouse gas emissions, many firms in European countries, Japan, and South Korea were inclined to work with their governments. This has also been true in the oil industry, where European-based BP and Shell were much earlier than their US counterparts to adopt more cooperative attitudes and actions vis-à-vis their governments.

There has also of course been more consensus within governments outside the USA – not only about the existence of climate change and its seriousness – as compared with the presence of climate change skeptics and deniers in the US Congress. US firms opposed to government measures to mitigate greenhouse gas emissions have thus been able to find like-minded members of Congress, some of whom have been influential in legislative processes as party leaders and committee chairpersons (as discussed further in Chapter 5). Furthermore, US public opinion has been more divided about the problem and the solutions, as compared with the publics in other countries (as discussed further in Chapter 3). In short, US firms that have been inclined to be laggards have been able to do so because of the sympathetic attitudes of some political leaders and some groups in the public.

Such cross-national differences in firms' responses to climate change and in the national political contexts within which they operate have also applied in the insurance industry; US firms have been notable for their lagging responses to climate change issues, as compared with their rivals in other countries. At the same time, the regulatory context for the insurance industry is highly fragmented in the USA because of the importance of state-level insurance regulatory commissions, as we see in the next section.

2.2 Insurance industry

Insurance firms are major institutional investors in many industries, as they must manage assets in support of the pension funds and insurance policies for which they have fiduciary responsibilities. Some firms within the insurance industry have therefore not only become concerned and vocal about their own increasing losses from extreme weather events and the consequent increased uncertainty about pricing policies for their policy premiums, but also become concerned about the vulnerability to climate change of the firms and industries in which they have large investments. Such concerns, however, developed in Europe long before the USA; for instance, the world's largest reinsurance firms, Swiss Re and Munich Re, were issuing public warnings about the significance and implications of climate change – and publishing much data about the economic costs – well before US-based insurers became concerned. It was not until the aftermath of Hurricane Katrina in 2005 that some US insurance firms became seriously concerned about the issue, and even then most US insurers did not take public positions on climate change issues. The overall non-responsiveness of the US insurance industry has been described as follows: "Despite three decades of concern about the impact of climate change on the insurance industry, . . . it has been estimated that less than one [US] company in a hundred has examined climate change issues to any significant degree" (Haufler 2009, 244). Yet, of the ten largest US insurance industry losses as of 2009, eight were from hurricanes, with Katrina in 2005 leading the list at more than $40 billion (Haufler 2009, 246).

Some firms within the insurance industry have become concerned and vocal about increasing losses from extreme weather events and the consequent increased uncertainty about pricing policies for their policy premiums; yet, at the same time, they also have opportunities for new weather-related products with the prospect of more frequent and severe extreme weather events of various types. Most US insurance companies have been slow to respond to climate change issues (Schiller 2012); in one study (CERES 2011a), only eleven of eighty-eight US-based insurance companies had developed a corporate policy about how to respond to climate change as a business issue.

Additional evidence about firms' responses collected by state insurance regulators may have prompted and reflected new industry perspectives (see Chapter 4). There is some evidence (Mills 2012) that US-based insurers may have begun to catch up with non-US firms since about 2008 in their concern about climate change; in particular after many years of lagging far behind non-US firms in responding to Climate Disclosure Project surveys, by 2008 US insurance firms' response rate had climbed nearly to equal that of the non-US firms, and exceeded it in 2009–2011. Yet, at the same time, there is evidence that US firms have been relatively inactive in other respects, for instance in their non-participation in the UNEP Finance Initiative, which has attracted widespread participation by insurance firms from many regions of the world.

Finally, a study (CERES 2012a) suggests that most US firms are still not taking climate change seriously and that they are still lagging behind their foreign competitors in that respect. The study was based on the responses of 184 firms to a survey by the insurance regulatory authorities in three US states (California, New York, and Washington). It found that only 23 of the 184 firms had developed comprehensive strategies to address climate change, and that of those 23 firms, 13 were owned by non-US firms. Further, two of the largest US firms were still exhibiting "strong ambivalence about the . . . existence of climate change and what is causing it." Nevertheless, some of the firms were beginning to reduce the portions of their investment portfolios in carbon-intensive industries, some were reducing their investments in municipal bonds in cities facing the prospect of severe water shortages, and some were reducing their casualty insurance exposure in Florida.

State regulatory agencies' activities and surveys in the insurance industry are discussed in Chapter 4, Section 4.2 and Appendix 4.2.

2.3 Relations with stakeholders

Firms not only need to address a variety of internal managerial issues, they also have to address issues about their relations with a wide range of stakeholders. These issues include what they disclose about their exposure to losses – or gains – from climate change, how they respond to shareholder resolutions, and how they relate to environmental NGOs and labor unions.

Disclosure of business practices and carbon exposure

Shareholders and other stakeholders have an interest in knowing how firms respond to climate change along many managerial dimensions: Business strategy – What goods and/or services should the firm produce? Where should it produce them? And sell them? How should it produce them? Business operations – How are the costs of transportation affected by new regulations? Customer and public relations – What kind of actions have firms taken or not taken about climate change in order to have beneficial relations with customers and the public? Government relations – What positions have firms taken on climate change issues? Concern about these and other questions about firms' responses to climate change has increased among investment banks. This increasing concern has been evident in their reports on industries' and firms' exposures to climate change risks, as well as in the business opportunities they face in a new era of climate change regulations and increasing investment in energy efficiency and renewable energy segments of the broader energy sector (see, for instance, Goldman Sachs 2009; Bloomberg New Energy Finance 2011). Despite the turmoil and the changing ownership arrangements during the financial crisis of 2008–2010, the interest of the investment banking industry has continued to increase.

As noted above, the Carbon Disclosure Project (2012; 2013), which is supported by "more than 655 institutional investors representing US$78 trillion in assets," is collecting and disseminating information on firms' exposure to climate change issues; and on behalf of US-based institutional investors, the Coalition for Environmentally Responsible Economics (CERES 2013a), which "works with more than 130 member organizations [including institutional investors such as pension funds, NGOs, and unions] . . . to engage with corporations," is similarly conducting periodic surveys, the conduct and publication of which also create pressures on firms to be more attentive to their carbon exposures and other climate change issues. In addition, resolutions concerning firms' responses are now regularly presented at the annual meetings of many corporations on behalf of institutional investors in the Interfaith Center for Corporate Responsibility, with hundreds of billions of dollars in assets, which represents the pension funds of several religious organizations. Yet another group of institutional investors, the Investor Network on Climate Risk, includes the treasurers and other financial officials of states such as California, Connecticut, and New York.

By 2011 there were forty-one firms that faced sixty-six shareholder resolutions calling on the firms' boards and executives to explain what they were and were not doing in response to climate change (ClimateWire 2011; Greenwire 2010b; Investor Network on Climate Risk 2010). Of the resolutions that received votes, none passed; the average level of support was 22 percent, the highest in the several years since the resolution campaign had been launched. In the face of such resolutions, some firms have agreed to undertake measures in order to have the resolutions withdrawn. The major oil firm

ConocoPhillips, for instance, made a commitment to spend $300 million on research for low-carbon fuels. Meanwhile, the large West Coast bank, Wells Fargo, agreed to undertake evaluations of the greenhouse gas emissions by firms represented in its loan portfolios in the energy production, electric power generation, and agriculture sectors. Again in 2012 and 2013 similar types of resolutions were also filed (CERES 2013b), with many of the same major oil firms on the list but with the additional issue of hydraulic fracturing being raised (see Chapter 6 for further information about climate change issues associated with hydraulic fracturing).

Partnerships with environmental organizations

Many firms have become partners with environmental NGOs in their programs to advance particular types of global warming mitigation measures. These programs are international in scope and include multinational firms from most economic sectors and many countries. In the United States among the most active NGOs in partnerships with business are C2ES (Center for Climate and Energy Solutions – formerly Pew Center on Global Climate Change), Environmental Defense Fund (EDF), Natural Resources Defense Council (NRDC), Coalition for Environmentally Responsible Economics (CERES), and World Resources Institute (WRI). The diversity in their business partnership programs' emphases is evident in the synopses in Appendix 2.2. A study by Ecofys for WWF reported in Energy Efficiency News (2012) found that the firms participating in its Climate Savers Program focused on increasing energy efficiency had cut their carbon dioxide emissions by more than 100 million tonnes from 1999 to 2011. It further projected that continuing the same rate of improvement would yield 350 million tonnes of reductions by 2020 and that if all firms in the sixteen industry sectors involved in the program would participate, they could collectively reduce up to another billion tonnes annually.

Corporations' involvement in these partnerships gives them access to the expertise of the NGOs as well as credibility and an aura of responsibility, which in turn give them access to some government policymaking circles. The corporate memberships for two of the programs are listed in Table 2.3.

Labor unions

The AFL-CIO's President observed that "The AFL-CIO and all the unions in North America are strongly on board the global campaign to reduce carbon emissions and stabilize climate change. Working together with environmental organizations we hope to reverse practices that put our very survival at risk" (Labor Strategies 2010). Though

Table 2.3 Firms' partnerships with environmental NGOs' climate change programs

Climate Change and Energy Solutions (C2ES), Business Environmental Leadership Council	World Resources Institute (WRI), Corporate Consultative Group
Air Products	Abbott Laboratories
Alcoa	Akzo Nobel
Alstom	Alcoa
American Water	Autodesk
Areva	Best Buy
Bank of America	Bloomberg
Bayer	Caterpillar
BP	Citigroup
CBRE	Dow
Cummins	DuPont
Daimler AG	Energy Development
Delta	Exelon
Dominion	Johnson & Johnson
Dow	Kimberly-Clark
DTE Energy	Morgan Stanley
Duke Energy	NewPage
DuPont	Pfizer
Entergy	PricewaterhouseCoopers
Exelon	Rio Tinto
GE	Samsung Global Environment Research Center
GM	SC Johnson
Hewlett-Packard	Sempra Energy
Holcim	Shell International
IBM	Siemens
Intel	Sindicatum Sustainable Resources
Johnson Controls	SAS Institute
National Grid	SC Johnson
Next Era Energy	Shell International
NRG	Siemens
Pacific Gas & Electric	Staples
PNM Resources	Target
Rio Tinto	Tetra Pak International
Royal Dutch Shell	United Technologies
Toyota	UPS
Weyerhaeuser	Wal-mart
	Walt Disney
	Weyerhaeuser

Sources: Compiled by the author from Center for Climate and Energy Solutions (C2ES) (2013b); and World Resources Institute (2013).

perhaps a bit of an overstatement in its reference to the opinions in "all the unions in North America," the comment does reflect a significant shift in the attitudes of organized labor. Many unions have become engaged in climate change issues – sometimes working with firms and sometimes working with environmental organizations, sometimes opposing measures to mitigate greenhouse gas emissions and sometimes supporting them.

As stakeholders in business, labor unions and their individual members have common interests as well as conflicting interests with firms. As a result, in some instances labor unions and the firms with which they are associated have joined together to adopt common positions on issues and even undertaken joint political or legal actions. Three prominent examples concern the automotive industry, the electric power industry, and the steel industry.

In the automotive industry, the United Auto Workers union joined with the Alliance of Automobile Manufacturers in a law suit against the state of California to try to prevent it from implementing a law (State Assembly Bill 32) that imposes greenhouse gas emissions regulations on motor vehicle tailpipe emissions. In a petition to the Federal District Court of Northern California, the manufacturers and unions contended that the bill violated the interstate commerce clause of the national constitution. The petition was eventually withdrawn during the financial crisis in 2009 when the federal government bailed out General Motors.

In the electric power industry and in the steel industry, unions and firms joined forces over the issue of the international competitive implications of cap-and-trade legislation that was proposed in the 111th Congress in 2009–2010. American Electric Power (AEP), one of the largest electricity producers in the country, and the International Brotherhood of Electrical Workers (IBEW) jointly proposed including "border adjustment measures" to reflect international differences in electricity prices and hence differences in manufacturing costs as a result of differences in governments' climate change policies. Once the issue became a central concern in congressional deliberations, other industries became interested as well. Among them, the steel industry was particularly active as its industry association, the Iron and Steel Institute (ISI), together with the United Steel Workers union (USW), lobbied strongly on the issue.

Some unions have also worked with environmental organizations, including in the Blue Green Alliance. The Alliance membership, consisting of unions and environmental organizations, "unites more than eight and a half million people in pursuit of good jobs, a clean environment and a green economy" (Blue Green Alliance 2010a). In addition to its founding members, i.e. the United Steel Workers and the Sierra Club, its member organizations include the Communications Workers of America, Service Employees International Union, International Union of North America Laborers, Utility Workers Union of America, American Federation of Teachers, Amalgamated Transit Union, Sheet Metal Workers' International Association, and the environmental organization, Natural Resources Defense Council. Its members "share a common conviction that successful climate change legislation must be guided by two overriding principles: the best scientific advice on reduction targets and implementation mechanisms that rapidly put Americans back to work in an effort to reach those targets" (Blue Green Alliance 2010a). Its "principles for climate change legislation" include the following detailed elements (Blue Green Alliance 2010b):

- Their goal is to reduce US emissions by at least 80 percent from 1990 levels by 2050, and they support a renewed US effort to forge a global treaty to reduce worldwide emissions by 50 percent by the same date. To meet this goal, climate change legislation should reduce US emissions significantly below 2005 levels by 2020 (individual Alliance partners advocate targets ranging from 14–25 percent below 2005 levels by 2020);
- Legislation should create an economy-wide cap-and-trade system that accounts for international competitiveness and regional disparities – providing a variety of mechanisms that offset rising energy costs to low- and moderate-income Americans and adversely impacted regions of the country and mechanisms to account for global competition in energy-intensive industries;
- Legislation should be focused on the creation and retention of millions of new and existing, family-sustaining green jobs, particularly in manufacturing and construction;
- BGA supports complementary regulation, including standards for low-carbon fuels, renewable energy, energy efficiency resources and fuel and appliance efficiency;
- Legislation should include investments in a wide range of technologies, including carbon capture and sequestration technology, and finance the transition to a clean energy economy.

2.4 Industry associations' positions on government policy issues

Industry associations in the United States have addressed a wide range of climate-related questions, such as:

- How multinational firms should adapt their operations to a fragmented international regulatory environment in which some countries have mandatory greenhouse gas emission limits, while others do not;
- What positions local subsidiaries of foreign-based multinational firms should adopt on international issues such as the development and implementation of international agreements on climate change, when the local host government and the parent firm in the home country have taken opposite positions;
- What the competitive consequences are for a parent firm that is headquartered in a country whose regulatory systems are lagging behind those of its rivals' home countries;
- What the implications would be for firms based in the USA if other governments, under pressure from rival firms in their countries, imposed "border measures" in international trade to offset the short-term competitive advantages of US-based firms operating in a low-cost fossil-fuel economy because the US government has taken less effective measures to reduce carbon dioxide emissions;
- How to respond to the proliferation of state-level regulatory regimes, some of them involving mandatory measures for firms and others involving voluntary measures;
- What to expect from the US national judicial system from the increasing numbers of law suits involving city and state governments as well as the national government, firms, industry associations and labor unions.

Umbrella organizations

The large, general-purpose, multi-industry "umbrella" associations in the USA have been highly active as lobbyists on climate change issues – including, in particular, the Business Roundtable, the National Association of Manufacturers, the US Chamber of Commerce, and the US Council for International Business.[7] They are all on record as opposing the establishment of a cap-and-trade system and opposing US participation in the Kyoto Protocol, while favoring instead technological approaches as well as voluntary government-sponsored programs for business. Yet, each organization represents different clusters of interests, and each has given a different emphasis to its position on climate change issues, with significant changes over time in some of their positions. Their influence in climate policymaking processes warrants a focused consideration of

7 Another well-known national business organization, the Conference Board (2008), is different; it is a "non-advocacy" organization "with one agenda: to help [its] member companies understand and deal with the most critical issues of our time," including climate change in particular but without lobbying directly for any specific legislation or regulation.

their nature, goals, and policy preferences (on business influence generally on environmental policy, see especially Kamieniecki 2006; Kraft and Kamieniecki 2007a; 2007b). See Appendix 2.1 for more information about the positions taken by each of the major umbrella organizations.

Issue-specific and industry-specific organizations

Among the issue-specific organizations that have been opposed to mandatory mitigation efforts, a particularly prominent one was the Global Climate Coalition (GCC). For several years, the GCC was a major lobbying organization against the Kyoto Protocol, and it disputed the consensus view of scientists about the causes and consequences of climate change. This public denial of the scientific consensus was promulgated despite the fact that scientists within the oil industry were themselves in agreement with the consensus (Union of Concerned Scientists 2007).

Within key industries, some firms joined the GCC earlier and left later than their rivals, and some did not join at all. Within the auto industry, Ford left the GCC during 2000; DaimlerChrysler a month later; and General Motors yet later. On the other hand, Japanese auto firms and other non-US auto firms never joined. Within the oil industry, European-based Shell and BP left before Texaco, which was the first US-based firm to leave (DiPaola and Arris 2001). In an unusual and awkward development, the US subsidiary of Shell remained a member for a matter of months after the parent firm Royal Dutch Shell of the UK and Netherlands withdrew. ExxonMobil, on the other hand, was a key participant throughout. There is thus further evidence, as noted above in the industry survey data from the Carbon Disclosure Project and CERES, of a *pattern in the intra-industry differences – namely, non-US firms were less likely to join and faster to leave if they did join, as compared with US firms.* The GCC was disbanded in 2002, though many of its members continued as individual firms and/or as members of other industry associations to oppose many proposed actions on climate change.

The American Petroleum Institute (API) has been a consistent critic of climate science and opponent of many mitigation measures. For instance, the API (2001) challenged the analysis by the US National Academy of Sciences (2001) of the Third Assessment Report of the UN Intergovernmental Panel on Climate Change (2001), because both scientific organizations concluded that there is a scientific consensus on the existence of a global warming problem and that it results in substantial part from human activities. The George W. Bush administration adopted the critical position of the API, despite the fact that the US National Academy of Sciences report was an official government response by an independent scientific panel to a request by the administration itself to evaluate the IPCC work.

Further, one of the oil industry's Washington lobbyists became Chief of Staff of the White House Council on Environmental Quality in the George W. Bush administration. When his role was exposed in revising an EPA report on the science of climate change, even though he was not a scientist, he resigned his government position and went to work for ExxonMobil (Pooley 2010, 46–48; US House of Representatives 2007, 26–27). In the 2012 presidential election campaign, the head of the API was a close advisor to candidate Romney (*Financial Times* 2012b; 2012c). The political influence of the US oil industry once prompted an executive of a British insurance firm to refer to the *insurance industry in Europe* as "... often being the only large commercial body able or willing to challenge the propaganda and complacency of US business and the oil lobby" (South 2000, 16).

In recent years, however, the policy positions of the API in the energy sector have been challenged as well by a variety of US-based industry associations representing the energy efficiency and renewable energy segments of the broader energy industry. For instance, the American Council for an Energy Efficient Economy (2007) and the US Council for Energy Alternatives have supported climate change mitigation measures that would help their industries. Also, a group of electric utilities has publicly endorsed *mandatory* CO_2 emissions caps on their industry; the firms include Calpine, Con Edison, Keyspan, Northeast Utilities, PG&E Corporation, PPL Corporation, Public Service Enterprise Group, and Wisconsin Energy Corporation. An *ad hoc* international industry organization, "E-55," which includes US-based corporations in its memberships, has supported a variety of measures to address climate change. In addition to firms in the electric power industry, a coalition of high-profile firms and environmental organizations in the form of the US Climate Action Partnership (USCAP) was centrally involved in the legislative process concerning cap-and-trade bills in the 111th Congress in 2009–1010, as discussed in Chapter 5.

2.5 Lobbying and campaign contributions

All of the types of organizations noted above – individual firms, industry associations, environmental organizations, and labor unions – are frequently involved in a wide range of political activities, though not necessarily directly as legally registered lobbyists. Because of the reporting requirements for lobbying organizations and individuals, as well as the data collecting and reporting activities of monitoring and journalistic

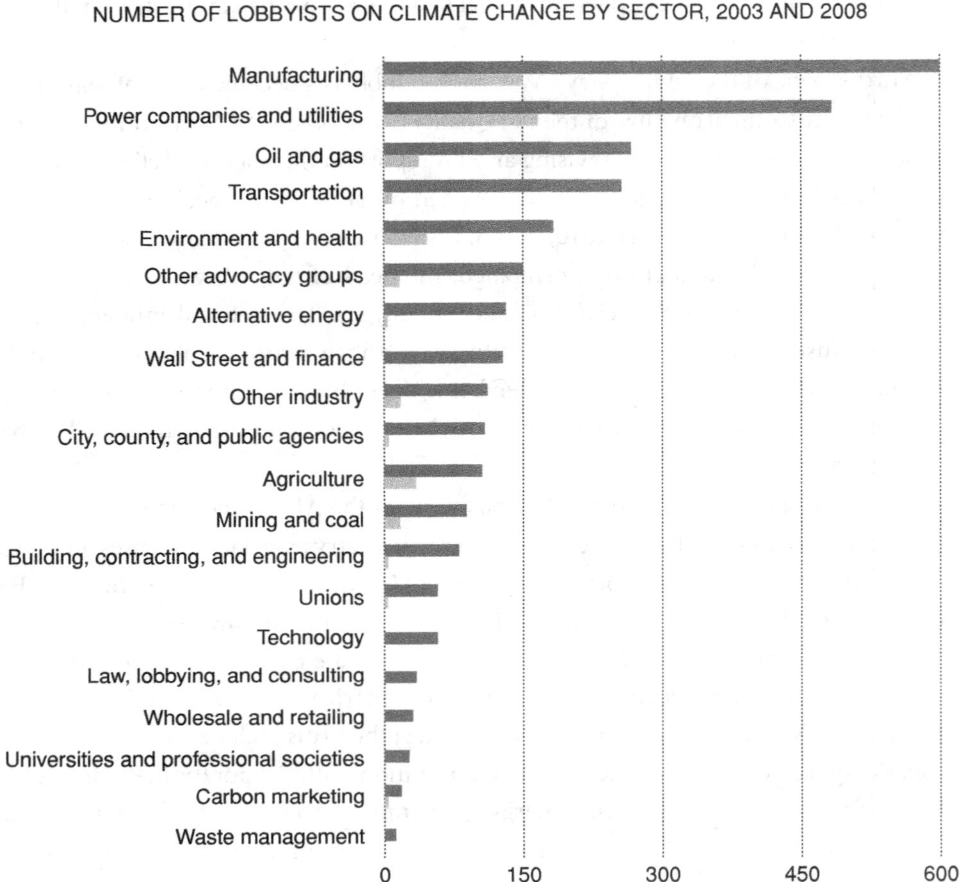

Figure 2.1 Number of individual lobbyists on congressional climate change legislation
Source: Center for Public Integrity (2009b). Used with permission.

organizations, it is possible to get a detailed factual understanding of the nature and extent of those lobbying activities.[8] Several patterns and trends are apparent. First, the numbers of firms, organizations, and individuals lobbying Congress on climate change issues have increased dramatically. The increase is evident in Figure 2.1, which compares

8 For a discussion of limitations in the lobbying reporting system, see Center for Public Integrity (2009a).

2003 and 2008; these were years when there were major climate bills introduced in the Congress – a bill sponsored by Senators McCain and Lieberman in 2003 and a bill sponsored by Senators Warner and Lieberman in 2008. Whereas there were only a few hundred firms and other organizations lobbying Congress in 2003, there were more than 770 in 2008 – and they hired 2340 individual lobbyists, which was more than four lobbyists per member of Congress. By the second quarter of 2009 when the Waxman–Markey bill was under active consideration in the House, there were 1150 firms and other organizations with lobbyists representing their interests (Center for Public Integrity 2009c). During 2009 the oil and gas industry spent more than $150 million on lobbying for climate change and *other energy-related issues*, up from $132 million the year before and a new record for spending on lobbying by one industry (Greenwire 2010b).

The distributions of lobbyists by industry sector in Figure 2.1 reveal that manufacturing and electric power were dominant sectors, with about 500 lobbyists each, and the oil and gas and transportation sectors had about 250 each. Meanwhile environmental and health organizations were represented by about 200 lobbyists; "other advocacy groups" on both/all sides of the issues were represented by about 150 lobbyists. Further, many groups, which were not represented at all or very little in 2003, were able to fund a hundred or so lobbyists each by 2008; these included alternative energy, financial services, and state and local governments.

The alternative energy industry was much more active by 2009, as compared with earlier years. From 1998 to 2009 there was a ten-fold increase from 20 to 200 in the number of alternative energy firms with lobbyists actively involved in the congressional legislative process (Center for Public Integrity 2009b). They spent $30 million, including $5 million by the American Wind Energy Association and nearly $2 million by the Solar Energy Industries Association. Though these amounts represented significant increases for these industries, they were still less than the $7 million spent by the American Petroleum Industry.

Another salient trend has been the increase in campaign contributions. In Figure 2.2 the contributions indicate that the amounts spent per bi-annual election cycle by energy/natural resources firms and other organizations increased more than four-fold in real inflation-adjusted terms in little more than a decade, reaching more than $120 million in the 2011–2012 election cycle. In addition to the overall secular increase, there is naturally a strong pattern of increases in each presidential election cycle. In most years, two-thirds to three-fourths of the contributions have gone to Republican candidates (80 percent in 2012).

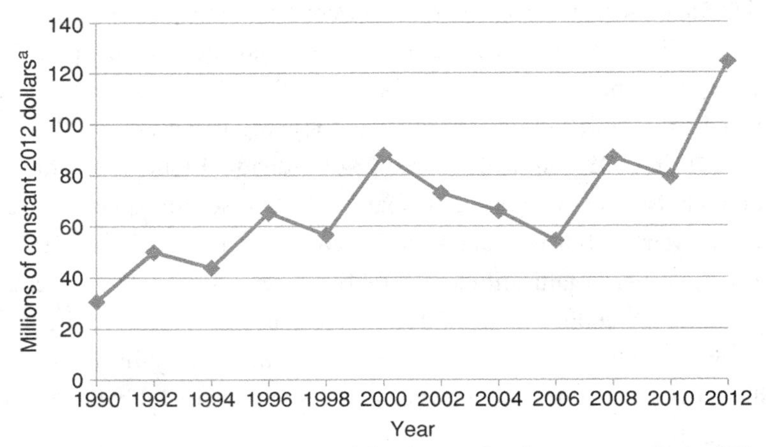

[a]Constant 2012 dollars computed using GDP deflator from US Office of Management and Budget 2013, Table 10-1; bi-annual data are the two years of election cycle, e.g. 2011 and 2012 for 2012.

Figure 2.2 Campaign contributions by energy/natural resources firms and organizations during bi-annual election cycles

Source: Computed by the author from data in Center for Responsive Politics (2011; 2012).

The nature and extent of specific lobbying activities can be illustrated by the coal industry, which faces the prospect of increasing pressures over the long term as measures are undertaken to reduce the greenhouse gas emissions in the electric power sector. In 2008 a new American Coalition for Clean Coal Electricity (ACCCE) was formed by forty-eight firms in the mining, rail, manufacturing, and electric power-generating industries (Center for Public Integrity 2009a). With an annual budget of $45 million, nearly half of which was spent on television advertising, ACCCE has focused attention on the current importance of coal in the US economy, since much of the economy's electricity is generated in coal-fired power plants, and on the potential for carbon capture and storage technologies to reduce the plants' greenhouse gas emissions. At the same time, ACCCE took the position on specific bills in the Congress that it would be willing to support caps on emissions as long as the legislation included provisions that supported coal's continued importance in the economy and ample government funding of research, development, and deployment of carbon capture and storage technologies. In the 2008 election campaigns ACCCE contributed over $300,000 to McCain's campaign, over $200,000 to Obama's campaign, and nearly $100,000 or more to each of eighteen candidates for House or Senate seats, including many members of the House Energy and Commerce Committee. However, Representatives Waxman and Markey, co-sponsors of the cap-and-trade bill, did not accept ACCCE contributions.

In the 2010 congressional election cycle, winning candidates who subsequently became members of the House Energy and Commerce Committee received similarly substantial campaign contributions from the oil and gas industry as well as the coal industry.

2.6 Implications: variations in interests, attitudes, and actions

The economic interests of business – and labor – in climate change issues are highly diverse, not monolithic. As we saw in Chapter 1, the interests vary across industries according to the greenhouse gas intensities of their production processes and products, and according to their vulnerabilities to the impacts of climate change. In the present chapter, the theme of diversity of interests has been further developed by noting in more detail that *there are business opportunities as well as threats to business interests involved in climate change issues.* In fact, for any one industry or even a single firm within an industry, there is a mixture of potential costs and benefits at issue; and the mixes are of course different for mitigation and adaptation measures. There is evidence of such variations at the level of individual firms as well as at the level of industries.

The major politically active and influential umbrella organizations – the Chamber of Commerce, National Association of Manufacturers, Business Roundtable, and US Council for International Business – have consistently opposed proposals to use cap-and-trade or other government regulatory measures to address climate change. Furthermore, many industry-specific associations representing coal, oil, and other fossil-fuel intensive industries have also opposed such measures. However, especially in recent years, new coalitions representing a wide range of service and manufacturing industries, including many information and communications technology firms, have been actively supportive of efforts to pass significant climate change legislation.

Splits among industries and among firms within industries, as well as among industry associations, have often exhibited patterns that could be explained at least partly in terms of differences across issue clusters. This was particularly evident in the positions taken by major industry associations since they represent diverse interests. During the George W. Bush administration, there was a clear preference among the major associations for technological solutions and opposition to regulatory solutions – of whatever form, whether taxes, cap-and-trade or otherwise. But differences among them began to appear when the Business Roundtable created its RESOLVE program of setting emission targets at the industry and firm levels. Although they were strictly voluntary and were not verified – or even in many instances not made public by individual firms – they were still viewed by some industry representatives as a possible step toward eventual mandatory targets in the form of caps in a cap-and-trade system.

Meanwhile, the National Association of Manufacturers (NAM) and the US Chamber of Commerce – with periodic protectionist tendencies on international trade issues and more generally unilateralist positions in international affairs – have increasingly focused on what they have perceived as international competitive threats, especially from China, in prospective cap-and-trade legislation. Thus, they have not only opposed the establishment of a cap-and-trade system, they have done so partly on the grounds of international trade arguments, as discussed further in Chapter 5. The US Council for International Business, on the other hand, with a membership of large multinational firms including some with parent headquarters outside the United States, has been less inclined to focus on international competitiveness arguments; rather, it has been more supportive of multilateral efforts, especially concerning international technology innovation and diffusion issues, and increasingly so in recent years.

In sum, a combination of variable interests among firms and industries, plus changes in positions on issues over time, has created a complex pattern of business attitudes and actions on a wide range of climate change and related energy issues. Nevertheless, although there have been countervailing groups representing alternative energy firms, financial services firms, and other firms supporting more aggressive measures to address climate change, they have not spent as much on lobbying or campaign contributions and have not exercised as much influence as have the traditional fossil fuel industries. Whether and how the positions and relative weights of these industry groups in government policymaking processes will shift in the next few years will be key variables affecting any government policy initiatives.

A central issue about the future of business responses to climate change issues is whether the large, well-established business associations will mostly continue to oppose cap-and-trade legislation and other government policies to mitigate greenhouse gas emissions or whether they will alter their positions to reflect the growing interest in some sectors and firms in taking more action at the federal, state, and local levels of government.

In any case, many individual firms have unilaterally been making significant changes in their production processes, purchasing practices, product design, and product lines as well as their marketing practices – and they are likely to continue to do so. The gaps between leaders and laggards within industries are likely to persist and in some cases become more pronounced. The US-based firms in the automotive and oil industries have been laggards compared with their foreign competitors, and the US insurance firms, in particular, seem especially prone to be mostly laggards compared with their foreign competitors, with only a few US firms taking leadership positions.

Suggestions for further reading and research

The lobbying activities of industries and firms on climate change are covered in Falke (2011) and Pooley (2010), especially during the George W. Bush administration and in relation to cap-and-trade legislation proposed in Congress during 2003–2009. The Center for Public Integrity (www.publicintegrity.org) and the Center for Responsive Politics (www.www.opensecrets.org) compile data, respectively, on lobbying and campaign contributions. A detailed, wide-ranging and critical analysis of business responses to climate change issues is provided by Newell and Patterson (2010).

An early introduction to climate change issues for firms, with a strong emphasis on the Kyoto Protocol, is available in O'Neill and Reinhardt (2000). More recently, the burgeoning literature on business perspectives, practices, and policy-related activities concerning climate change includes Meckling (2012), Glicksman (2010), Hoffman (2005), Jones and Levy (2009), Kamieniecki (2006), Kolk (2000), Kolk and Pinkse (2005), Kolk and Levy (2001), Kraft and Kamieniecki (2007a; 2007b), Layzer (2007), Levy (2005), Levy and Egan (1998), Levy and Newell (2005), Pew Center on Global Climate Change (2010a; 2010c), Pinkse and Kolk (2010), as well as Brewer (2005a; 2005b; 2009a). Goldman Sachs (2009) provides an investment banking perspective on the issues for business. For more broadly based analyses, see Esty and Caves (1983), Hall and Deardorff (2006), and Wright (1996). A classic political science work that emphasizes the varied and conflicting interests of business – even within a single firm – is Bauer, Pool, and Dexter (1963); see also Lowi's (1964) review of it and Salamon and Siegfried (1977). Salorio (1994) analyzes conflicting firm interests and strategies and their implications for their positions on government trade policies.

Two organizations provide regular reports with extensive data and analyses on firms' responses to climate change. The annual reports of the Carbon Disclosure Project provide comprehensive and up-to-date data on firms' responses to climate change issues; the reports include numerous cross-national and cross-industry comparisons as well as access to the responses of individual firms at www.cdproject.int. The reports are based on surveys of the *Financial Times* Global 500 firms. The Coalition for Environmentally Responsible Economics (CERES) has conducted in-depth studies of selected industries

(CERES 2003; 2012a) as well as reports on governance issues related to climate change (CERES 2003; 2006) and surveys of investors' perceptions and actions (CERES 2011b).

Detailed empirical studies of many industries are reported by Climate Strategies (2010). A special issue of *Greener Management International* (2002) includes an overview of the strategic issues by Dunn (2002) and a series of industry-specific studies. Kolk and Levy (2001) and Pulver (2007) have focused on the major oil and auto firms. As for other industry-specific studies, Schmidheiny (1992) addresses climate change issues in the context of the energy industry; Schiller (2012) focuses on the insurance industry, as do Dlugolecki and Keykhah (2002). Hofman (2002) analyzes electricity production and distribution issues; and Nordquist, Boyd, and Klee (2002) report on the cement industry; also see Andersen and Zaelke (2003). See World Business Council for Sustainable Development (2010) and ClimateBiz.com (2010) for a variety of cases. There are numerous brief anecdotal and case study materials from periodicals and collections of articles; these include yearbooks and collections of cases by Cutter publishing (Arris 1997; 1998; 2000; DiPaola and Arris 2001; South 2000). An extensive analysis of issues and responses among Japanese firms is provided by the Development Bank of Japan (2003).

There is a vast literature on emissions trading – which is discussed in Chapters 4 and 5 in the present volume. For context and concepts, see especially Kopp and Toman (2000), International Emissions Trading Association (2003), and Peace and Stavins (2010). Reinhardt (2000) presents a case study of BP's development of an internal emissions trading scheme.

The monthly periodicals *Environmental Finance* and *Carbon Finance* contain information about emissions trading and other corporate practices concerning emissions reductions as well as updates on recent government policy developments. Another monthly, *The Emissions Trader*, is published by the Emissions Marketing Association. The electronic news services of Point Carbon, including *Carbon Market News*, are focused on carbon markets but are also useful sources of information about a variety of climate change issues. *Bloomberg New Energy Finance* and www.Greenbiz.com are both useful for tracking industry developments. The website of the Association of Climate Change Officers at www.acco.org includes updates on government regulations and corporate practices.

APPENDIX 2.1

Umbrella business organizations' positions on climate change issues

The Business Roundtable is an association of approximately 150 large, well-known corporations from diverse industries, including financial services and other services industries, in addition to manufacturing, which has emphasized technological solutions and support for increased government funding of R&D programs, plus a voluntary greenhouse gas emission reduction program called RESOLVE (Responsible Environmental Steps, Opportunities to Lead by Voluntary Efforts). Created in 2003 at the behest of the George W. Bush administration, the program had 107 firms participating by 2004 (Business Roundtable 2004; also see 2001). By 2009 most of the firms that had signed up to participate in the program reported that they had adopted energy efficiency measures in production processes and/or their buildings (Business Roundtable 2009). Information about the precise amounts of reductions by individual firms has not been made readily available, however; nor have the reported measures been subject to independent third-party verification. Nonetheless, it is not surprising that the Business Roundtable was the only large business umbrella organization that cooperated with the US administration in this endeavor; its membership is disproportionately comprised of large, multinational financial services firms that feel less threatened by regulatory measures to address climate change. In fact, such firms face the prospect of new business opportunities in cap-and-trade brokerage and other product lines, as well as investment banking business in assessing firms' "carbon exposure" and advising clients on potential new regulatory risks and other climate-change-related risks and opportunities.

The National Association of Manufacturers (NAM) has a membership including many firms from traditional heavy industries such as steel and automobiles. It has been strongly opposed to US participation in the Kyoto Protocol or the establishment of a national cap-and-trade program. However, it has supported technological programs. It stated in a June 2001 press release (National Association of Manufacturers 2001a) that "the White House's focus on science, research and technology [was] the right way to approach global warming and greenhouse gases." Its Executive Vice President observed of the Kyoto Protocol, "As the facts about the pact come in, it's becoming clear that the

Bush Administration was right and that we are far better off not being a party to this onerous agreement" (National Association of Manufacturers 2001b). The NAM also opposed proposals for a mandatory domestic cap-and-trade emissions trading system when they were being considered in Congress in 2009–2010 (National Association of Manufacturers 2009).

The US Chamber of Commerce represents "more than 3 million businesses, nearly 3,000 state and local chambers of commerce, 830 associations, and over 90 American Chambers of Commerce abroad" (US Chamber of Commerce 2009). It has opposed US participation in the Kyoto Protocol (and other measures to address the problem)

> because [the treaty] does not require the fastest growing sources of greenhouse gases, such as China and India, to take steps to control their emissions. Thus US ratification of the Kyoto Protocol would likely do serious damage to the US economy while doing little or nothing to actually address climate change in the long term. The US Chamber believes that other approaches, such as offering incentives for development of high-efficiency technology and transfers of efficient technology to developing nations, holds the best promise of sustainable progress towards controlling climate change (US Chamber of Commerce 2001; also see 2010).

During congressional consideration of cap-and-trade legislation, the Chamber indicated that it opposed the cap-and-trade bills in the House and the Senate, a position that prompted several prominent firms to leave the organization. At that time, Republican Senator Lindsay Graham, who was for a time one of the leaders of a "tri-partisan" effort with Democratic Senator Kerry and Independent-Democrat Lieberman to establish a national cap-and-trade system, was asked what could increase the number of Republican senators supporting the legislation. He answered that "The Chamber of Commerce and the National Association of Manufacturers need to tell my colleagues it is O.K. to price carbon, if you do it smartly" (quoted in Friedman 2010). The Chamber filed a petition in 2010 to prevent the EPA from implementing a program based on its "endangerment finding," following a Supreme Court case concerning the EPA's legal grounds for doing so, as discussed further in Chapters 4 and 5.

The US Council for International Business (USCIB) has tended to be the most "internationalist" of the major US business umbrella organizations, as it represents mostly large US-based multinational corporations having widespread international interests, and to a lesser extent firms with headquarters in other countries but with substantial business interests in the USA. As an association of large firms with globalized business interests, it tends to favor international cooperation generally and multilateral arrangements specifically for many issues. Though it has opposed establishing a US cap-and-trade system and US participation in the Kyoto Protocol, it has nevertheless been positive about the potential to use the multilateral UN Framework Convention on Climate Change as a basis for greater international cooperation (USCIB 2001). By 2009 the USCIB was yet more supportive of international efforts. In a "Statement on the UN Global Leadership Forum on Climate Change," it "welcome[d] the initiative of United Nations Secretary General Ban Ki-moon to energize political and societal will towards a balanced, strengthened and effective global post-2012 climate agreement" (USCIB 2009).

Another business association, the World Business Council for Sustainable Development (WBCSD), has been strongly supportive of the Kyoto Protocol and a wide range of other mitigation measures. It has also developed an extensive greenhouse gas monitoring and reporting protocol with the US-based NGO World Resources Institute (WRI). The WBCSD (2010) notes that "Business has recognized the need to move from dialogue to action . . . . [T]he WBCSD aims to provide tools and practices for businesses to encourage the implementation of practical solutions to address the climate change challenge." It is based in Geneva, Switzerland, and fewer than 40 of its 150 member firms have been headquartered in the USA. Its impact in the USA has thus been marginal and not publicly visible.

APPENDIX 2.2

The business partnership programs of environmental organizations

The Business Environmental Leadership Council of C2ES (2013) is based on "the belief that business engagement is critical for developing efficient, effective solutions to the climate problem. [They] also believe that companies taking early action on climate strategies and policy will gain sustained competitive advantage over their peers."

The *Environmental Defense Fund*[9] (2010) "engages with companies on many levels, from information sharing via [their] Innovation Exchange, to public policy advocacy, to formal partnership projects that extend over a year in length and result in a public announcement. . . ."

The Natural Resources Defense Council (2010) "works to harness the power of markets to create positive and profitable environmental change. The marketplace has already demonstrated that environmentally sound practices and economic vitality can go hand-in-hand, from hybrid cars to organic food, from green buildings to green jobs. . . . The climate crisis will not be solved by legislation alone. . . ."

The Coalition for Environmentally Responsible Economics (CERES 2010) is a "coalition of investor groups, environmental organizations, and investment funds [that] engages directly with companies on environmental and social issues. CERES companies seek to attain long-term business value and to improve management quality through stakeholder engagement, public disclosure and performance improvements. . . ."

The World Resources Institute (WRI) has developed the Greenhouse Gas Protocol (2010), which is "a decade-long partnership between the World Resources Institute and the World Business Council for Sustainable Development, [and] is working with businesses, governments, and environmental groups around the world to build a new generation of credible and effective programs for tackling climate change. . . . It provides the accounting framework for nearly every GHG standard and program in the world. . . ."

9 EDF has changed its name twice – from Environmental Defense Fund to ED (Environmental Defense) and then back to EDF (Environmental Defense Fund).

CHAPTER 3

Public perceptions and preferences

We have to take climate change out of the atmosphere, bring it down to earth and show how it matters in people's everyday lives.

Glenn Prickett, Nature Conservancy (2010)

Americans Consider Global Warming Real, but Not Alarming.

Gallup Poll Headline (2001a)

Americans' Concerns About Global Warming on the Rise.

Gallup Poll Headline (2013)

The influence of public opinion in government policymaking and the extent to which policy reflects public preferences are clearly important issues, and they are taken up in more detail in Chapters 4–8 in the context of discussions of policy issues. In this chapter the focus is on the substance of public perceptions of the problem of climate change (Section 3.1) and public preferences for government policies (Section 3.2): How much of a problem does the public perceive climate change to be? How have perceptions of the problem changed over time? What kinds of government policies does the public want and not want? The chapter also considers – in Section 3.3 – the factors that are related to those perceptions and preferences: What are the correlates that account for variations in public perceptions of the problem and preferences for solutions? The chapter explores – in Section 3.4 – public opinion as revealed in consumer behavior: Have people changed their behavior as consumers of climate-relevant products, especially concerning energy consumption and energy efficiency?

Section 3.5 concludes that individuals' differences in ideology and party preference are key factors accounting for variations in their attitudes on a wide range of climate change questions. The extent to which these factors reflect underlying differences in economic interests – and perceptions of how climate change and responses to it will affect those interests – are open questions about the political economy of climate change.

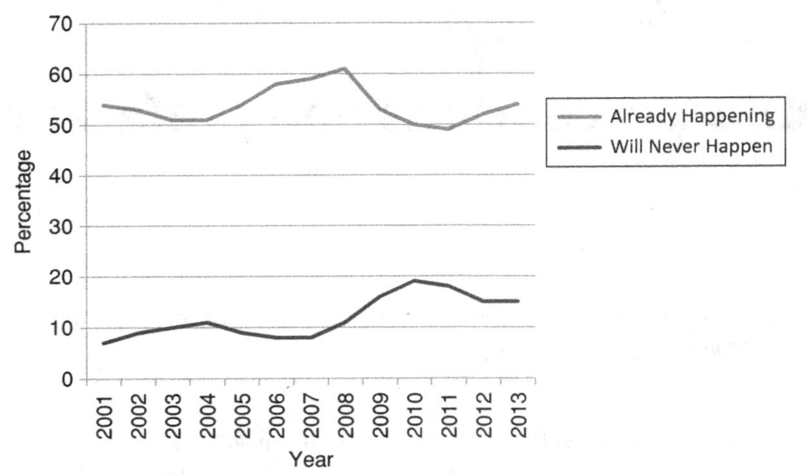

Question: "Which of the following statements reflects your view of when the effects of global warming will begin to happen – [they have already begun to happen, they will start happening within your lifetime, but they will affect future generations, (or) they will never happen]?"

Figure 3.1 Trend in belief about when the effects of global warming will happen

Sources: Gallup (2012; 2013).

At the regional level, there are patterns in opinions that can be related to variations in regional economic profiles. Thus, a political economy approach to understanding responses to climate change issues is helpful at the regional level of analysis, even though its value at the level of individual perceptions and preferences is not yet clear.

Methodological issues about sample sizes, statistical significance levels, question wording, and sponsoring organizations are considered in Appendix 3.1. International comparisons of opinions in the USA with those in other countries are presented in Appendix 3.2.

3.1 Perceptions of the problem

As Figure 3.1 indicates, for seven years (2001–2008) the proportion of the population that believed that the effects of global warming were already happening increased slightly, if irregularly, from about 50 percent to about 60 percent of the population; and the proportion that thought it would never happen was relatively stable at only 7 to11 percent. But between March 2008 and March 2010 there were notable shifts: the percentage who said it was already happening decreased from 61 to 50, and the percentage who said it would never happen increased from 11 to 19 percent

(Gallup 2012). By March 2013, however, 54 percent thought it was already happening (Gallup 2013).

There was thus a paradox: during the period 2008–2010, while the science was accumulating *increasing evidence* of the seriousness of the problem, the proportion of the public believing it was a problem was *decreasing*. By early 2010, furthermore, there was another paradox: while the percentage of the public perceiving a problem was *decreasing*, the percentage preferring greater government action to address the problem was *increasing*. How to explain such patterns? Though these were often portrayed as contradictory developments in superficial interpretations, in fact they were occurring simultaneously among different segments of the population. One segment – about 10–15 percent of the adult population – shifted to a more skeptical position about the existence of the problem, indeed even to the point of denying its existence.

There seem to have been three factors at work in this development: one was a tendency for responses to questions about the existence of the problem to reflect perceptions about the *relative* importance of the problem during a major global recession – as in the well-known expression from a past presidential campaign, "It's the economy, stupid" (Leiserowitz, Maibach, and Roser-Renouf 2010). A second factor was that some of the 10–15 percent of the population were probably prompted into denial by the increasingly serious prospects, including possible "tipping points" of no return, being portrayed by the latest scientific evidence – as in the frequent response to unusually bad news, "It can't be true."

Third, there was probably a tendency, which has been noted on other issues, for some members of the public to harbor a distinctive causal connection between opinions on issues and opinions about political leaders, especially presidents and presidential candidates, namely, "If he or she believes it, it must not be true." Of course, that is the opposite of the traditional notion of a rational citizen in a democratic political system. That is to say, instead of first forming opinions about issues and then opinions about leaders on the basis of the relationship between the citizen's opinion and the leaders' opinions on the issue, many citizens form their opinions about the issue based on their opinions of the leaders.

Thus, in this instance, because candidate Obama and then President Obama declared his acceptance of the scientific evidence about the existence and seriousness of climate change and because he advocated policies to address the problem more aggressively, some citizens "inferred" that climate change must *not* be occurring or must *not* be a serious problem because Obama said the opposite. Such an analytic posture is a way to avoid cognitive dissonance, or maintain cognitive consistency; each of these

Table 3.1 Correlates of shifts in perceptions of the problem (percentages)

Effects of global warming are already happening	Liberals			Moderates			Conservatives		
	2008	2010	Change	2008	2010	Change	2008	2010	Change
	72	74	+2	66	60	−6	50	34	−16

Question: "Which of the following statements reflects your view of when the effects of global warming will begin to happen?"
Source: Computed by the author from Gallup (2010b).

reactions could thus be construed as a kind of irrationality since the views expressed are at odds with the evidence.[1] Yet, these tendencies also represent a kind of cognitive coping mechanism in the face of limited information about a complex topic, about which the public perhaps cannot reasonably be expected to have highly informed, multi-dimensional opinions.

Whatever one's take on these analytic issues, the evidence about some key patterns is clear: the people who shifted their beliefs in the opposite direction of the mounting scientific evidence were disproportionately self-identified "conservatives" (see Table 3.1).

A more detailed snapshot of the "deniers" can be summarized as follows from a Gallup poll (2009): 16 percent said "the effects of global warming" ... "will never happen." Of these "deniers," more than half were male (63 percent), white (77 percent), and conservative (71 percent); and each of these groups was represented disproportionately, compared with their numbers in the general population. In addition, though less than half of the total group of deniers, they were disproportionately upper-income, Republicans, 65 years old or older, from the Midwest, and weekly churchgoers (see Box 3.1).

Regardless, only a minority of the total population are climate deniers. Half to two-thirds, depending on the wording of the question, believe it is happening – and most of them want a variety of actions taken to address it.

What about Superstorm Sandy – did it affect people's perceptions of the problem? The effects on public opinion of Superstorm Sandy in the New York City area in October 2012 have been explored in detail by Krosnick (2013). These are important results because they help to answer a fundamental question about public opinion: Do extreme weather events affect public opinion? Surveys that were conducted just before

1 There was a period beginning in late 2009 when errors in some climate science reports became public and created doubts in some people's minds about the body of climate change evidence more broadly. However, the shift in public opinion had already appeared in national survey data before the controversies about the science emerged. This scientific issue is explored further in the first chapter's Appendix 1.1.

Box 3.1 **Who are the deniers?**

In a 2009 Gallup poll, of the 1012 interviewees, 16 percent said "the effects of global warming . . . will never happen." Of these "deniers," more than half were male (63 percent), white (77 percent), and conservative (71 percent); and each of these groups was represented disproportionately, compared with their numbers in the general population. In addition, though less than half of the total group of deniers, they were disproportionately: upper-income, Republicans, 65 years old or older, from the Midwest, and weekly churchgoers. See the following table.

Percentage of each group believing climate change will "never happen"

Ideology	*Liberal*	*Moderate*	*Conservative*	
	4	8	30	
Party	*Democrat*	*Independent*	*Republican*	
	4	17	30	
Gender	*Female*	*Male*		
	11	21		
Race	*Black*	*Non-white*	*White*	
	7	10	18	
Age	*18–29*	*30–49*	*50–64*	*65+*
	8	17	19	20
Region	*East*	*Midwest*	*South*	*West*
	14	20	14	16
Income (000)	*<30*	*30–74.9*	*75+*	
	6	16	22	
Education	*HS or less*	*Some college*	*College grad*	*Postgraduate*
	18	14	19	13
Religious preference	*Protestant*	*Catholic*	*Non-Christian*	*None*
	17	18	9	12
Church attendance	*Less than monthly*	*Monthly*	*Weekly*	
	10	14	26	

Question: "Which of the following statements reflects your view of when the effects of global warming will begin to happen?"
Source: Computed by the author from Gallup (2009). I am indebted to the Gallup staff for providing the cross-tabs.

and just after the storm offer an unusual opportunity for before–after comparisons. A series of questions tapped a wide range of perceptions – and generally found no statistically significant difference in the before and after results. In particular, there was apparently not a change in the proportions of the respondents thinking global warming is happening or that it is a serious problem. There was a statistically significant shift, however, among a sub-group of the full sample – namely, among the approximately one-sixth who were skeptics about the occurrence of global warming, about one-sixth of those were no longer skeptics; yet, this represented less than 4 percent of the total sample.

Beyond these fundamental questions about believers and deniers – and changes in their perceptions of the problem – there are other ways to identify people's orientations to climate change issues. For example, Leiserowitz, Maibach, and Roser-Renouf (2010); Maibach, Leiserowitz, Roser-Renouf, Akerlof, and Nisbet (2010); and Leiserowitz, Maibach, Roser-Renouf, Feinberg, and Howe (2013) have developed a scale with six categories that provide useful descriptive summaries of Americans' perceptions of the problem of climate change. The six categories representing segments of the population, with the percentages in each group reported sequentially based on national surveys in 2008 and 2010 and with brief descriptions of their features, are presented in Box 3.2.

The data in Box 3.2 are consistent with the short-term trend noted above: between 2008 and 2010, the "alarmed" and "concerned" portions both declined, while the "doubtful" and "dismissive" both increased. Nevertheless in 2010, the "alarmed" plus "concerned" were still a greater share of the population at 39 percent, as compared with the "doubtful" plus "dismissive" groups at 29 percent. By 2012, the combined "alarmed"-or-"concerned" group was 45 percent, while the "doubtful"-or-"dismissive" group was 21 percent.

Yet, despite the occasional shifts in basic perceptions of the problem of climate change, for the most part over the period from 2001 to 2011 there was much continuity. Several survey organizations asking a wide range of questions have found that close to half or more of the population believes climate change has already been happening, about one in three think it will happen in the future, while only about one in seven denies that it will ever happen.

As for the level of worry, a question asked most years in the Gallup polls from 1998 to 2013 revealed some fluctuations, of course, but the percentage of the respondents who indicated that they were worried "a great deal" or "fair amount" has consistently

Box 3.2 Six segments of the US public

The pie charts below present the proportions of the population in 2012 in each of the six segments identified by Leiserowitz, Maibach, and Roser-Renouf (2010); Maibach, Leiserowitz, Roser-Renouf, Akerlof, and Nisbet (2010), and Leiserowitz, Maibach, Roser-Renouf, Feinberg, and Howe (2013). Prior to 2012, although there were some shifts, the basic patterns were similar: the proportions of "alarmed" plus "concerned" citizens exceeded those of the "doubtful" plus "dismissive" – by 51 percent to 18 percent in 2008 and by 39 percent to 29 percent in 2010. Again in 2012, although there had been a shift back toward the distribution in 2008, there were roughly similar numbers of "alarmed" plus "concerned" – i.e. 45 percent – and similar numbers of "doubtful" plus "dismissive" – i.e. 18 percent. In sum, in the years 2008, 2010 and 2012:

- Two-fifths to half of the population were "alarmed" or "concerned";
- One-fifth to one-fourth were "doubtful" or "dismissive";
- The balance in the middle – amounting to approximately one-third of the population – were either "cautious" or "disengaged."

Proportions of six segments, 2012

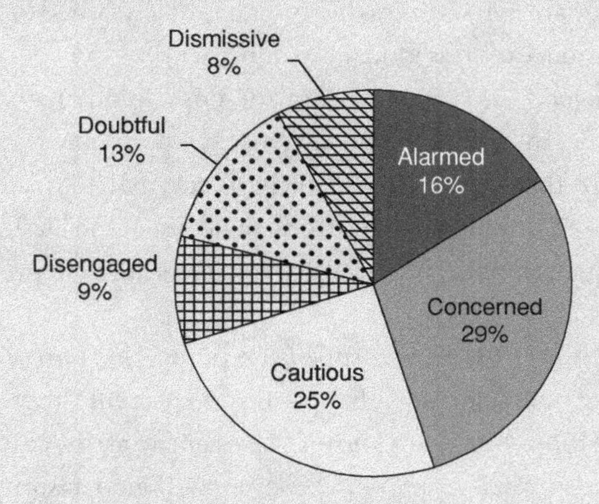

The segments, with their 2008, 2010, and 2012 percentages of the total added in brackets, are defined as follows:

"The ***Alarmed*** [18, 10, 16 percent of the population] are the segment most engaged on the issue. They are very convinced global warming is happening…, is human-caused…, and is a serious and urgent threat. Over two-thirds believe people in the United States are being harmed now. Most of the Alarmed are taking personal action on global warming, and they intend to take further actions in the future.... The majority of the Alarmed support a full range of policies that would reduce carbon emissions.

"The ***Concerned*** [33, 29, 29 percent of the population] – the largest of the Six Americas – are also convinced that global warming is real, human caused, and a serious problem, but they are somewhat less certain in their convictions than are the Alarmed. Moreover, compared to the Alarmed, they are less likely to perceive global warming as a direct threat to themselves or their family, or to future generations of people. While they support a vigorous national response... they are distinctly less involved in the issue than are the Alarmed, and less likely to be taking personal action.

"The ***Cautious*** [19, 27, 25 percent of the population]... also believe that global warming is real; however, they are even less certain than are the Concerned:... Additionally, the Cautious are less likely than the Concerned to believe that climate change is particularly dangerous or threatening. They are less likely than the Concerned to rate global warming as a top national issue priority... [N]evertheless a strong majority of the Cautious... support... CO_2 reduction policies. They are markedly less likely than the Concerned to be taking personal actions of any kind to address global warming per se.

"The ***Disengaged*** [12, 6, 9 percent of the population] haven't thought much about the issue.... By their own admission, they don't know much about global warming and are the segment most likely to say that they could easily change their minds about it. Interestingly, a strong majority of the Disengaged... support [many] CO_2 reduction policies, but they themselves are doing very little to address global warming.

"The ***Doubtful*** [11, 13, 13 percent of the population] are evenly split among those who think global warming is happening (33 percent), those who think it isn't (32 percent) and those who don't know (34 percent), and they are the segment with the highest proportion of people who believe that if global warming is happening, it is caused by natural changes in the environment (81 percent). They tend to say that they have thought about the issue "only a little," yet they also indicate they are somewhat unlikely to change their minds about the issue. Despite their clear doubts about global warming, a strong majority... support [several] CO_2 reduction policies: funding renewable energy research; providing tax rebates for purchases of efficient cars or solar panels; and building more nuclear power plants.

"... [T]he ***Dismissive*** [7, 16, 8 percent of the population] – like the Alarmed – are actively engaged in the issue, but as opponents of a national effort to reduce greenhouse gas emissions. The majority of the Dismissive believe that warming is not happening... and is not a threat to either people or the environment. Ninety-four percent report global warming will not harm people in the United States.... The Dismissive believe global warming should be a low priority for the government.... Most members of the Dismissive segment report that they have thought "some" or "a lot" about global warming, and virtually all... say they are "very unlikely" to change their minds about the issue."

In methodological terms, the six groups were derived *inductively* using "latent class analysis" based on thirty-six variables and four constructs: "global warming beliefs, issue involvement, policy preferences and behaviors."

Sources: Excerpted and compiled by the author from Leiserowitz, Maibach, and Roser-Renouf (2010), Maibach, Leiserowitz, Roser-Renouf, Akerlof, and Nisbet (2010), and Leiserowitz, Maibach, Roser-Renouf, Feinberg, and Howe (2013). Used with permission.

been between about 50 percent and about 70 percent (Gallup 2013). It peaked in 2000; it was 58 percent in 2013 (see Figure 3.2).

3.2 Preferences for government policies

A majority of the public has wanted the national government to take more action to address climate change for more than two decades. This was true in the 1990s as the problem became more salient, it was true during the George W. Bush administration, and it has been true since the advent of the Obama administration. There has been a consensus among Democrats, independents, and Republicans in their support for a wide range of government actions – though certainly not all – at both the state and national levels. There has been essentially a nation-wide consensus favoring more US government action on climate change. For instance, Krosnick (2013) found that a majority in forty-nine of fifty states (except Utah) has supported more national government action.

Before delving further into the empirical evidence to support these generalizations, it is important to recognize that there are caveats to keep in mind in order to avoid making

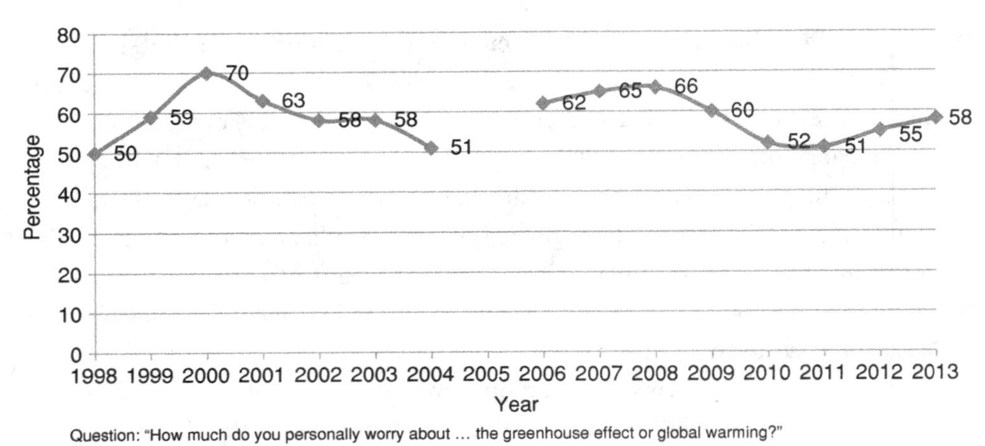

Figure 3.2 Trend in the level of worry (percentage indicating "great deal" or "fair amount")
Source: Gallup (2010a; 2011; 2013).

several erroneous assumptions on the basis of the generalizations. First, the patterns of distributions of *opinions about government policies as solutions* are different from the patterns of *perceptions about the problem.* Second, the majority consensus of course does not support all policy options. Third, there are some policy issues on which there are significant partisan and ideological divisions, for instance support for nuclear power.

Finally, and most importantly from the standpoint of the influence of public opinion in the government policymaking process, the distributions of policy preferences among the public diverge substantially from those of members of Congress and other political leaders. This divergence has been especially evident between Republican party leaders and Republican party adherents; the latter are more widely supportive of a variety of measures, compared with Republican party leaders – a disparity that became greater following the 2010 congressional elections. Such a gap between political leaders and their partisan adherents may persist in part because there is a tendency for the public's assumptions about its own perceptions and preferences to underestimate actual public support for action on climate change (Krosnick 2013). (Whether and to what extent such erroneous assumptions about public opinion are also held by members of Congress and their staff is an important empirical question needing further research. It is also of course a central normative issue in the functioning of a representative political system.[2])

2 Public perceptions of public opinion on climate change underestimate the extent to which the public understands climate change, has knowledge of specific issues, and wants government action on the problem (Rasmussen 2011a; Krosnick 2013).

Keeping in mind these important caveats, we can note in Table 3.2 the high levels of support in 2008 during the presidential elections for a wide array of policies. Except for certain kinds of taxes or other charges and except for nuclear power, there were substantial majorities in favor of numerous policy options. Indeed, the levels of support are in the 80–90 percent or higher range for several of the possibilities that were proffered in the interviews. Although the consensus extended across diverse issues concerning domestic regulatory issues, technology issues, and international issues, the strongest support tended to be for government programs that directly encouraged technological innovation and diffusion.

Data from the 2008 election also make it possible to ascertain the extent to which there was a bipartisan consensus on specific government policy issues. In Table 3.3 there is evidence that there was majority endorsement among both the Obama and McCain supporters for most of the issues. Opinions about building more nuclear power plants or imposing charges on inefficient appliances and cars, however, were split – with a majority of only McCain supporters favoring the former and a majority of only Obama supporters favoring the latter. It should also be noted, of course, that there were some differences of opinion between the two groups – differences that ranged from 3 percent to 31 percent. In sum, then, although there was a bipartisan majority consensus in favor of most of the proposals mentioned, there were also nevertheless partisan differences in the sizes of the majorities.

By the 2012 election, the partisan differences were greater: Although there was strong support for the expansion of renewable energy among both Obama and Romney supporters, there was otherwise much disagreement about a series of fundamental climate change issues (Yale Project on Climate Change Communication and George Mason University Center for Climate Change Communication 2012). A survey focused on international affairs issues during the 2012 campaign (Chicago Council on Global Affairs 2012) found that whereas there was much bipartisan agreement on most issues, on climate change there were sharp differences; for instance, whereas 82 percent of Democrats and 67 percent of independents supported a new international treaty on climate change, only 48 percent of Republicans did so.

Prior and subsequent surveys make it possible to probe more deeply into the nuances of the policy options and opinions about them. For instance, in a Gallup survey in March 2001, a month after the George W. Bush administration announced it would not support mandatory domestic restrictions on carbon dioxide emissions, 48 percent of the public disapproved of the decision and 41 percent approved (Gallup 2002d). In a July 2002 poll, 76 percent of the respondents preferred that the "government set

Table 3.2 Policy preferences during the 2008 election (percentages agreeing)

Issue	WPO.org Survey August 19–20, 2008	Yale–George Mason Survey October 8–14, 2008
Regulatory issues, including taxes		
The government should regulate carbon dioxide as a pollutant		80
[The government should impose] 45 mpg fuel efficiency standard for cars, trucks, and SUVs if it meant increased cost of up to $1000		79
Renewable Portfolio Standard requiring electric utilities to produce at least 20 percent of electricity from wind, solar, or other renewable sources if it meant an average extra cost per household of $100/year		72
Requiring utilities to use more alternative energy, such as wind and solar, even if this increases the cost of energy in the short run	66	
Requiring businesses to use energy more efficiently, even if this might make some products more expensive	61	
Having an extra charge for the purchase of models of appliances and cars that are NOT energy-efficient	43	
Increase gasoline tax by 25 cents per gallon and returning revenues to taxpayers by reducing federal income taxes		33
Energy technology issues		
The government should increase funding for research into renewable energy sources, such as solar and wind power		92
Emphasize more: installing solar or wind energy systems	87	
Support tax rebates for people buying energy-efficient vehicles or solar panels		85
Emphasize more: modifying buildings to make them more energy-efficient	83	
With the rising cost of energy, [shifting to alternative energy sources, such as wind and solar] would save money in the long run	79	
Government subsidy to to replace old water heaters, air conditioners, light bulbs, and insulation, with average additional cost of $5/month in higher taxes		72
Establishment of government fund to make buildings more energy-efficient and teach Americans how to reduce their energy use, at an average additional electricity bill cost of $2.50/month		63
Emphasize more: building nuclear power plants	42	
Build more nuclear power plants		61
International issues		
The USA should be willing to commit to reduce its GHG emissions as part of [a treaty to address climate change] by reducing GHG emissions such as those caused by using oil and coal	78	
The USA should sign an international treaty that requires it to cut emissions of carbon dioxide 90 percent by 2050		90

Sources: Compiled by the author from WorldPublicOpinion.org (2008); Leiserowitz, Maibach, and Roser-Renouf (2009).

Table 3.3 Consensus and conflict about government policy preferences during the 2008 presidential campaign

Issue	Obama supporters	McCain supporters	Difference
Emphasize more: installing solar or wind energy systems	89	86	+3
Emphasize more: modifying buildings to make them more energy-efficient	89	80	+9
Requiring utilities to use more alternative energy, such as wind and solar, even if this increases the cost of energy in the short run	75	60	+15
Requiring businesses to use energy more efficiently, even if this might make some products more expensive	71	55	+16
The USA should be willing to commit to reduce its GHG emissions as part of [a treaty to address climate change by reducing GHG emissions such as those caused by using oil and coal]	94	63	+31
With the rising cost of energy, [shifting to alternative energy sources, such as wind and solar] would save money in the long run	83	73	+10
Emphasize more: building nuclear power plants	33	54	−21
Having an extra charge for the purchase of models of appliances and cars that are NOT energy-efficient	52	39	+13
Less-developed countries should NOT be required to limit their emissions UNTIL the more-developed countries reduce theirs	20	20	0
Less-developed countries should be required to CUT their emissions	19	34	−15
Less-developed countries should be required to MINIMIZE the increase of their emissions through greater energy efficiency	49	42	−7
... accept less-developed country position and proceed to reduce US emissions	46	34	+12
... pressure less-developed countries to limit their emissions but proceed to reduce US emissions together with other developed countries	44	51	−7
... refuse to reduce US emissions if less-developed countries do not agree to limit their emissions	4	10	−6

Source: Compiled by the author from WorldPublicOpinion.org (2008).

standards that require industries to reduce" greenhouse gas emissions; only 16 percent preferred a "voluntary approach to global warming" (Reuters 2002; Union of Concerned Scientists 2002a; 2002b). In the same survey, only 21 percent agreed with the statement that "President Bush's voluntary approach to reducing global warming pollution is enough," or that "Americans will simply adapt to the inevitable changes." A majority of the Republicans (58 to 67 percent, depending on the precise question) who were interviewed in July 2002 preferred mandatory reductions, while only 23 to 33 percent preferred a voluntary approach and supported the administration's policy (Reuters 2002; Union of Concerned Scientists 2002a). The self-identified Republicans and the respondents who voted for Mr Bush in the 2000 election both conformed to the overall majority support for mandatory emissions reductions, which Mr Bush opposed.

In another poll, in the spring of 2001, when asked whether they would be "willing to support tough government actions to help reduce global warming even if each of the following happened as a result," the percentages of respondents saying "yes" were as follows: 47 percent, if "your utility bill went up"; 38 percent, if "unemployment increased"; 54 percent, if "a mild increase in inflation" resulted (Yankelovich/Harris Survey 2001). A differently targeted survey (Aldy, Kotchen, and Leiserowitz 2012), which focused on proposals for a national clean energy standard (NCES), found an average willingness to pay an additional $162 per year for a NCES requiring 80 percent clean energy by 2035.

As for taxes, although large bipartisan majorities have opposed the general concept of increasing taxes in order to discourage greenhouse gas emissions in the transportation and electric power sectors, a plurality or majority of the public would accept taxes if they are revenue-neutral. Further, a plurality or majority would favor increased taxes if the proceeds are used to fund research on alternative energy sources and energy efficiency. In 1990, 59 percent said "yes" they would be "willing to pay an extra 25 [cents] per gallon of gas to reduce pollution and global warming"; in 2001, 48 percent said they would be willing to do so (Yankelovich/Harris Survey 2001). These results bracket a similar finding of another question that was not explicitly about global warming, but is nevertheless germane – "Should the government require improvements in fuel efficiency for cars and trucks even if this means higher prices and smaller vehicles?" – to which 55 percent of the respondents replied "yes."

A national survey during late 2002 and early 2003 asked about three different types of taxes: a gas-guzzler sales tax, a business energy tax, and a gasoline tax (Leiserowitz 2003). In each instance, there was a lengthy preface that included precise amounts of costs to consumers. In sum, whereas a majority said they would support a gas-guzzler tax along the lines indicated, majorities opposed both the business energy tax and the gasoline tax. Of course, if the amounts of money specified in the preface as the costs to consumers were varied, the distributions of responses would also vary. But only in-depth polling focused on these and other features of such taxes could reveal the extent of sensitivity of opinion to the specifics of the proposed taxes; despite this, the available data suggest that a sales tax on gas-guzzlers is one type of tax that would be politically viable.

Opinions have been divided and evolving on cap-and-trade as legislation has been considered by the Congress – which is not surprising since the concept was unfamiliar and a bit difficult to grasp initially even for well-educated people. In any case, the basic patterns and trends have been as follows: As has been the case with opinions about taxes, there has been a clear pattern of greater support for cap-and-trade, if the government proceeds are used for research and development into energy efficiency and alternative energy programs.

As for an international treaty that would require the USA "to cut its emissions of carbon dioxide 90% by the year 2050," 65 percent of the respondents in a 2011 national survey thought the USA should sign such a treaty (Leiserowitz, Maibach, Roser-Renouf, and Hmielowski (2011, 14). It was supported by majorities of Democrats (81 percent), independents (67 percent), and Republicans (54 percent), though among the 12 percent of the self-identifying Tea Party members in the survey, 74 percent were opposed.

Climate change mitigation, energy technology changes, and economic growth

The linkages among climate change mitigation, energy technology changes, and economic growth became particularly pronounced during the development and enactment of the government's response in 2009 to the financial crisis and recession. In that context, climate change policy was to some extent recast as a combination of macro-economic management during a crisis, industrial restructuring, and transportation infrastructure development: Government subsidies for residential solar installations were quick employment-generating expenditures – and climate-friendly energy conservation measures. Subsidies for new battery technologies and electric automobile development were ways to reinvigorate the US auto industry – and climate-friendly industrial restructuring programs. The development of high-speed rail networks was a way to improve the deteriorating transportation infrastructure – and a climate-friendly way to increase public modes of transportation in the national transportation system.

Such reframings of the issues were consonant with patterns that have been revealed in public opinion surveys. The first pattern concerns the relationship between environmental protection and economic growth, while the second concerns the public's priorities for national government policies for addressing climate change and clean energy. As to the first pattern, there has been a continuing, widespread tendency to make environment–economy issues into a simple dichotomous choice – a tendency which has been reflected in a standard Gallup poll question: "With which of these statements about the environment and the economy do you most agree [sequence rotated] – protection of the environment should be given priority, even at the risk of curbing economic growth (or) economic growth should be given priority, even if the environment suffers to some extent?" (Gallup 2011). In an early 2010 survey, the environment received priority by 53 to 38 percent (reflecting a pattern that had held since the mid-1980s, with an exception only in 2009). A year later in early 2011, though, the priorities were reversed with economic growth favored by 54 percent and environmental protection by 36 percent (Gallup 2011).

A more nuanced approach to the issue has been used by Leiserowitz, Maibach, Roser-Renouf, and Smith (2010), and by Krosnick (2010) and Krosnick and Villar (2010). Thus, in a 2010 survey, two key questions were asked (Leiserowitz, Maibach, Roser-Renouf, and Smith 2010). The first question was – with the percentages of the responses in percentages: "Overall, do you think that protecting the environment...

- "Improves economic growth and provides new jobs (56 percent agreed)?"
- "Has no effect on economic growth or jobs (25 percent agreed)?"
- "Reduces economic growth and costs jobs (18 percent agreed)?"

The second question, with the response percentages again in parentheses, was: "When there is a conflict between environmental protection and economic growth, which do you think is more important?

- "Economic growth, even if it leads to environmental problems (35 percent)."
- "Protecting the environment, even if it reduces economic growth (65 percent)."

The parallels in the two distributions in the 2010 surveys are striking. Priority was given to economic growth by only 38 percent and 35 percent of the respondents in the two surveys, while priority was given to protecting the environment by 53 percent and 65 percent. Whether the public's priorities will return to this earlier pattern after the economy has fully recovered from the financial crisis and recession – and as it continues to undergo significant restructuring in some sectors and regions – is clearly an important question about the future domestic political economy context of US climate change policymaking.

As to patterns in public preferences about government policy priorities concerning "global warming" and "clean energy": As the data of Table 3.4 reveal, in a late 2009 to early 2010 national survey (Leiserowitz, Maibach, Roser-Renouf, and Smith 2010), most of the respondents gave only "medium" or "low" priority to "global warming," while 60 percent gave "very high" or "high" priority to "clean energy." In light of such a pattern, it is not surprising that US "climate change" policy has to a significant

Table 3.4 Preferred policy priorities for global warming and clean energy

	Global warming[a]	Clean energy[b]
Very high	13	24
High	25	36
Medium	31	29
Low	31	11

[a] Question: "Do you think global warming should be a low, medium, high or very high priority for the president and Congress?"
[b] Question: "Do you think that developing sources of clean energy should be a low, medium, high or very high priority for the president and Congress?"
Source: Compiled by the author from Leiserowitz, Maibach, and Roser-Renouf (2010).

degree become "clean-energy" policy. Part of the appeal of the latter is the associated distributions of benefits and costs.

The distributions of the benefits and costs of mitigation

There has also been a heightened interest in trying to shift in particular the perceived and actual distributions of the benefits and costs of climate change mitigation measures. These distributions (as introduced in Chapter 1) refer to the concentration/dispersion of effects among the *groups affected* and the distribution in terms of the *timing of the effects* between the short term and long term. For instance, government subsidies of research, development, and deployment of carbon capture and sequestration (CCS) technologies – especially for coal-fired power plants – create short-term benefits for producers of CCS equipment and services, and reduce the costs for the electric power firms that install the equipment. More generally, subsidies for energy efficiency and renewable energy innovation and diffusion create short-term benefits – or reduced costs – for producers and consumers. Similarly, the distributions of revenues from sales of emission credits in a cap-and-trade system can also create short-term benefits for selected groups of producers and/or consumers (see Chapter 5). Such a concentration of increased benefits (or reduced costs) for politically significant *groups* can fundamentally change the political dynamics of climate change policymaking. Changing the distribution of costs over *time* can also change the political dynamics. By shifting the costs of climate change mitigation efforts from specific groups of producers or consumers to the general public via government programs that are funded at least in part by deficit spending, the costs are being deferred to the future.

Such redistributions of benefits and costs along the two dimensions are summarized diagrammatically in Figure 3.3. The diagram reflects the shift (i) *from* a traditional focus

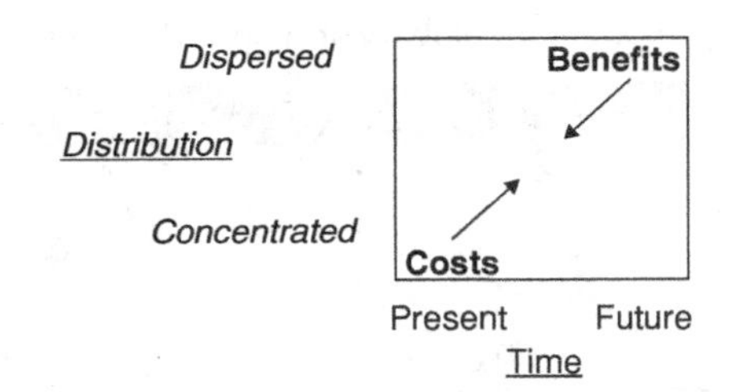

Figure 3.3 Redistributions of benefits and costs of mitigating climate change

on the long-term, widely dispersed worldwide benefits of climate change mitigation in combination with the concentrated short-term costs (ii) *toward* a new focus on the concentrated, short-term benefits in combination with the widely distributed costs that are deferred to the long term. In total, the rich mixture of shifts represented by the two arrows of Figure 3.3 would constitute a significant reframing of the issues.[3]

The costs of inaction

Finally, there has been a reframing of the issues by giving increased attention to the *costs of inaction.* The Stern review of the economics of climate change (Stern 2007) was important in shifting discussion of climate change issues in many countries, including the USA, in this regard. In the USA, as elsewhere, the distributions of the costs of inaction *among groups* and the distributions *over time* are again two key dimensions. As with the analysis above of mitigation measures, the political consequences of the costs of inaction depend on their distributions along the two dimensions. This is why scientific explanations – and public interpretations – of extreme weather events are so important politically: they can change fundamental perceptions about the effects of climate change among groups and over time. For instance, the increased incidence of hurricanes in the most severe category increases interest in the short-term costs of climate change and in the concentration of costs in particular regions of the country; the occurrence of severe, long-term droughts has similar effects (see Chapter 1).[4]

The shift of benefits from the upper right corner to the lower left corner occurs through:

3 Note that the worldwide and long-term benefits of greenhouse gas mitigation actions remain.

4 Whether there is or is not a relationship between climate change and the frequency and/or severity of tornadoes and floods is not yet entirely clear.

- Increased concentration of benefits through subsidies for key producer and consumer groups. N.B. the long-term benefits of greenhouse mitigation remain for people worldwide.
- Creation of benefits in the present and short term through subsidies for key producer and consumer groups.

The shift of costs from the lower left corner to the upper right corner occurs through:

- The shift of costs from concentrated producer and consumer groups to widely dispersed costs for the public generally, through distributions of cap-and-trade revenues to affected producers and consumers.
- The shift of costs from the present to the long term through government deficit spending.

3.3 Correlates of variations in perceptions and preferences

Of course, there are potentially many variables that could explain each of the distributions in the perceptions of the problem and in the preferences for government policies. Most explanatory variables are familiar and widely incorporated in survey research on most issues and thus can be analyzed for climate change in particular – gender and age, for example. Others are of special interest in the present study, which is focused on political economy – for instance ideology, partisanship, and income.[5]

Box 3.1 above on the correlates of denial contains a long list of such correlates. Beyond those and other results above about deniers, here we consider additional evidence. We begin with those characteristics that consistently reveal strong patterns, namely ideology and party, then others that also reveal patterns but not such strong ones, namely income, region, gender, and age.

Ideology and party

The origins of the "affliction of partisanship" as it has been called by Abbasi (2006) have been traced to the 1997 debate over the Kyoto Protocol (Krosnick, Holbrook, and

5 There is an inevitable methodological issue: Do these variables truly "explain" attitudes – or are they merely statistical correlates? Because there are typically multiple causes of variations in individuals' perceptions and preferences, I have chosen to use the term "correlates" in the section title and in much of the discussion. They can be alternatively considered "plausible explanations" rather than reflective of direct cause–effect relationships. This methodological challenge is especially daunting at the regional level of analysis, as we shall see below. In any case, true causal analysis in the most demanding epistemological sense necessarily requires more in-depth survey questions and more advanced multivariate analysis than is feasible in a secondary summary analysis of the data such as this one.

Table 3.5 Origins of partisan gap at the time of Kyoto Protocol debate (percentages)

		Sept–Oct 1997 (before debate)	Dec–Feb 1998 (after debate)	Change
Said global warming was already happening	Strong Democrats	73	87	+14
	Strong Republicans	68	69	+1
	Gap	5	18	+13

Source: Compiled by the author from data in Krosnick, Holbrook, and Visser (2000), 253; also see Abbasi (2006).

Visser 2000). During that debate, there was an increase in the party differences (see Table 3.5).

Since then, variations in individuals' perceptions of the problem of climate change and their preferences for government policies have been consistently found to be strongly related to partisan and ideological tendencies.

Keeping in mind the underlying consensus on many questions, beyond that, the strength of ideology and party as correlates of a wide range of opinions about climate change issues can hardly be overstated. Numerous surveys over many years have consistently found the same pattern: large differences according to ideology and party in perceptions of the problem and in preferences about solutions. Conservatives and Republicans have consistently been less convinced of the seriousness or even existence of the problem (as we saw in the previous section), and they have similarly consistently been less supportive of government measures to address the problem, as compared with liberals and Democrats. Self-identified moderates and independents have been in between, with their distributions typically closer to the Democrats than the Republicans. Although there are strong overlaps between self-classification according to ideology and self-classification according to party, the overlaps are far from perfect. It is thus important to recognize that data that display relationships between party identification, on the one hand, and opinions about climate change issues, on the other, reflect ideological patterns as well, but only quite imperfectly. There are moderate Republicans who consider climate change a serious problem and want serious actions taken to address it, just as there are conservative Democrats who do not. In general, a greater number of self-identified independents are revealed to be closer to the Democratic party adherents than they are to Republican party adherents, when they are asked a follow-up question about whether they "lean toward" one party more than the other.

In an attempt to determine the strength and potential political potency of opinions, the Gallup survey further divided the respondents according to whether each held

Table 3.6 Differences according to party affiliation in perceptions of climate change as a problem, 2011 (percentages)

Question	Democrats	Independents	Republicans	Difference: Dems – Reps
Worry about global warming "great deal" or "fair amount"	72	51	31	+41
Effects of global warming have already begun to happen	62	53	32	+30
Seriousness of global warming is exaggerated in the news	22	43	67	−45
Rise in earth's temperatures is due to pollution from human activities	71	51	36	+35

Source: Compiled by the author from Gallup (2011).

what it termed "directive" or "permissive" opinions based on a follow-up question. That question asked each respondent if they would be "upset" if the USA took the opposite position from the one the respondent had just previously expressed. Those who said they would be "very" or "somewhat" upset were considered to have "directive" opinions because "they *want* their opinions to prevail" (Gallup 2004c). Those who said they would not be upset were considered to be "permissive" – as were those with no opinion – because "they essentially 'permit' the country's leaders to do whatever the leaders deem best" (Gallup 2004c). Well over half of all respondents and of each of the party affiliation groups were in the "permissive" category. Yet, there were also substantial numbers of Democrats (40 percent) and independents (36 percent) who held strong "directive" opinions in favor of US support of the Protocol. On the other side, there were few people with strong opposing views. However, only 13 percent of the Republicans held a similarly strong positive view; at the same time, only 19 percent of the Republicans held "directive" views opposing the Protocol. Thus, as with recognizing whether global warming has already been occurring, there were significant partisan differences. Data from a 2011 survey (Gallup 2011) reported in Table 3.6 found strong partisan patterns.

Regions

Regional variations and income variations overlap with each other and with patterns in both ideological and partisan identifications, and disentangling these is not feasible in the present study. However, we can determine, separately, whether there are notable regional patterns in opinions at the aggregate level and whether there are patterns according to income at the individual level. It is essential to be clear about the difference between patterns at the *aggregated* level for regions versus the *individual* level for income. Otherwise, one can commit the error in statistical analysis known as the "ecological fallacy," that is making inferences about individuals on the basis of aggregated regional data.[6]

A national survey conducted by the Survey Research Laboratory at the University of Oregon makes it possible to compare opinions across regions – West, Midwest, South, Northeast (Leiserowitz 2003). The sub-sample sizes are rather small by the normal standards of survey research, and as a consequence the confidence intervals are relatively large (+/−7 percent for two of the groups and +/−9 percent at the 95 percent level, for the other two). Yet, there are several patterns that are discernible. First, there is a consensus across regions on several issues. There were substantial majorities in all four regions favoring US participation in the Kyoto Protocol; US reductions of greenhouse gas emissions, regardless of whether other countries do so; and US regulation of carbon dioxide in particular. Substantial majorities in all regions also supported renewable energy subsidies. At the same time, there was a national consensus cutting across these four regions against an increase in gasoline taxes. Otherwise, though, there were some substantial regional differences. The Northeast has been consistently the most supportive of action to address climate change, and the South has typically been the least supportive/most opposed. Thus, 42 percent of the respondents in the Northeast favored a business energy tax, while only 22 percent in the South favored such a measure. Similarly, 65 percent of the Northeasterners favored a gas-guzzler sales tax, while only 50 percent of the Southerners did so.

In sum, there is evidence here of the kinds of regional differences identified in Chapters 1 and 2 about economic geography. In particular, the Northeast – which does not have oil, gas, or coal deposits (except in Pennsylvania) or much automotive manufacturing – is the most supportive of several types of specific measures to mitigate climate change. The South, on the other hand, tends to be the least supportive – in part because of the importance of the oil and gas industries in several states (Texas, Louisiana, Oklahoma) and the importance of the coal industry in other states (Kentucky, Tennessee, West Virginia).

Putnam (2000) finds much variance among states in their "social capital" (i.e. "connections among individuals – social networks and the norms of reciprocity and

6 Note that the term "ecological fallacy" is not about ecology as concerns the natural environment; rather, it is about differences in the level or unit of analysis in statistics.

trustworthiness that arise from them") and thus their engagement in civic affairs and their attitudes and behaviors concerning a wide range of issues. Although differences among states in attitudes and policies concerning climate change are not included in his analysis (which was mostly about a time period preceding the emergence of climate change as a salient issue), the pattern of variations among states in their "social capital" is (imperfectly) related to the pattern of state climate change policies: The three West Coast states (Washington, Oregon, and to a lesser extent California) and the six New England states of the Northeast are above average in "social capital," while the Southern states are generally below average. The patterns elsewhere are more complex. An empirical analysis of the relationship could add to both the literature on US responses to climate change and the literature on collective decision-making about public goods, such as the greenhouse-gas-absorbing atmosphere, as well as the literature on civic engagement and community life.

Gender, age, and race

Gender consistently exhibits moderately strong relationships to opinions on climate change issues, with females being typically more concerned about it and more supportive of actions to address it. For instance, a national survey in late 2010 (Borick, Lachapelle, and Rabe 2011) found that 63 percent of the females interviewed accepted the existence of global warming, compared with 53 percent of the males.[7]

The evidence about age as a correlate is not so consistent. In fact, the most extensive analysis available (Feldman, Nisbet, Leiserowitz, and Maibach 2010) on age-related differences in opinions concluded:

> Overall, the survey data, collected between December 24, 2009 and January 3, 2010, offer no predictable portrait of young people when it comes to global warming: While less concerned about and preoccupied with global warming than older generations, they are slightly more likely to believe that global warming is caused by human factors and that there is scientific consensus that it is occurring. They are also somewhat more optimistic than their elders about the effectiveness of taking action to reduce global warming. And, while they are less open to new information about global warming than older generations, they are much more trusting of scientists and President Obama on the issue....

7 A parallel national survey in Canada found higher proportions for both females and males at 80 percent and 79 percent, respectively.

> Nationwide, liberals and conservatives exhibit wide differences in their beliefs about global warming, with conservatives more skeptical and less engaged than liberals, and this ideological divide is no different among young Americans. . . .

As for race, although there are some inconsistencies in the findings among surveys, there is a common tendency for whites to be less concerned about climate change and less supportive of government policies to address it, as compared with other groups.

3.4 Consumer attitudes and behavior

Surveys have revealed that the adoption of energy efficiency measures, particularly at home, has been a popular tendency for many years. Thus, in Gallup surveys in 2000, 2003, 2007, and 2010, 80 to 85 percent of the respondents indicated that they had reduced their household's use of energy during the previous year (Gallup 2010c); more specifically, 81 percent reported in 2010 that they had replaced standard light bulbs with compact fluorescent bulbs during the year.

Beyond these few indicators of consumer behavior, however, the data are not so consistent. In 2009, 68 percent of the respondents to another national survey (USA Today/Gallup 2009) said they had taken steps during the year to make their homes more energy-efficient. As for their motives, 71 percent said they did so "mostly to save money" while 26 percent said they had done so "mostly to improve the environment."

In a 2008 survey, 28 percent said they had made "major changes" and 55 percent said they had made "minor" changes in their "shopping and living habits over the last five years," including in energy efficiency measures in their homes and transportation (Gallup 2008). In another survey about consumers' purchasing behavior, 71 percent of the respondents in a 2008 survey of ABC News, Planet Green, and Stanford University (2008), reported that they were doing something to reduce their carbon footprint, but without an indication of the nature or magnitude of their efforts. In yet another survey – by CBS News/*New York Times* (2008) – 41 percent indicated that they bought products made from recycled materials "regularly" and 49 percent indicated that they did so "occasionally." But the proportions that car-pooled "regularly" or used mass transit "regularly" and those who used solar, wind, or geothermal for hot water, heat, or electricity were all less than 10 percent.

3.5 Implications: shifting consensus – with ideological and partisan differences

Overall, the preponderance of the evidence from scores of surveys over two decades by many organizations representing diverse interests indicates that there has been a persistent public consensus that climate change is happening and that the government should be doing something about it. Majorities have favored more regulation of business's emissions and products, and especially more government subsidies of the development, production, and use of energy-efficient and renewable energy goods and services. Furthermore, the public consensus on many of the specific issues has been bipartisan in support of a wide range of government policies.

Yet, there have also been significant variations along partisan and ideological lines – variations in fundamental perceptions of the problem and in preferences about policies for addressing it. Though these partisan differences have generally become greater over time, they remain less pronounced among the public than among members of Congress and other party leaders.

As for regional variations, publics in Northeastern and West Coast states have generally been more favorable toward a range of government measures at the national level, as compared with publics in Southern, Midwestern, and Mountain states.

There have also been substantial variations over time and among groups in many specific dimensions of perceptions of the problem and how to try to solve it. As for perceptions of the problem, though more than half the public has consistently "worried a great deal" or "a fair amount" about it, the percentage has fluctuated over time.

There have been strong and consistent patterns in some correlates of these and other variations in perceptions of the problem: Self-identified liberals and Democrats are more likely than self-identified conservatives or Republicans to believe that climate change is happening and thus also naturally more likely to be worried about it. Other patterns are less strong or consistent but nevertheless evident: males, whites, and Christians are generally less likely than their counterparts to believe climate change is happening.

The evidence about the relationship of income and other economic variables at the level of individual perceptions and preferences is not so clear. It remains to be seen from further research whether and how strongly individuals' economic circumstances are independent causal factors affecting perceptions and preferences about climate change issues – or whether they are part of a mélange of interacting factors including partisan and ideological variables.

Finally, it should be noted again, the evidence of these patterns in this chapter pertains only to the public. The extent and nature of any consensus or differences or correlates among business leaders and political leaders are a different matter. Variations in business leaders' opinions and the importance of industry variations among them – and also variations among firms within industries – are related to their firms' economic interests and their firms' organizational culture as well as their own personal sense of responsibility. In the next several chapters, the evidence about government policymaking reveals that political leaders' opinions and government policies have frequently not been consistent with public perceptions and preferences.

Suggestions for further reading and research

Reports by Krosnick (2010; 2013) and by Krosnick and Villar (2010) are essential reading, as are reports of the Yale Project on Climate Change (Leiserowitz 2007a; 2007b; 2007c), many of which have been undertaken together with the George Mason University Center for Climate Change Communication (Leiserowitz, Maibach, and Roser-Renouf 2009; 2010; Maibach, Leiserowitz, Roser-Renouf, Akerlof, and Nisbet 2009; Leiserowitz, Maibach, Roser-Renouf, and Hmielowski 2011). Also see Brewer (2005c).

The Gallup Poll and numerous other major survey organizations routinely ask questions about climate change, especially each year in March to April about the time of the annual Earth Day. The Gallup surveys are especially useful for tracking long-term trends (see especially Gallup 2001a; 2004a; 2004b; 2008; 2010a; 2011; 2012; 2013); other surveys (e.g. Harris 2001a; 2001b; 2002) are often focused on detailed analyses of complexities and nuances in opinions about specific elements of perceptions and preferences.

Bi-annual election year national surveys sponsored by the Chicago Council on Global Affairs (formerly Foreign Relations) – often with the German Marshall Fund of the United States – provide snapshots that can be used to trace attitudes over time during the congressional and presidential campaign seasons; see especially Chicago Council on Foreign Relations (1999; 2004; 2012) and Chicago Council on Foreign Relations and the German Marshall Fund of the United States (2002a; 2002b; 2002c; 2002d; 2002e).

Multi-country surveys and compilations of the results of studies over time by World-PublicOpinion.org (for instance, 2007a; 2007b; 2008; 2012) are excellent sources of data on patterns and trends in the USA in a comparative international perspective.

Opinions on climate change issues in California are regularly assessed by the Public Policy Institute of California (e.g. 2005; 2010a; 2012). The survey sample sizes are typically about 2000 and thus enable detailed comparisons among many subgroups. See Chapter 4's Appendix 4.2 in the present volume for illustrative results.

For additional details about these and other resources, see the individual references to these organizations and the specific sources listed for the tables, figures, and boxes in the chapter.

APPENDIX 3.1

Methodological issues about the surveys

Throughout the chapter the analysis is based on polls conducted by major survey organizations during the period from January 1989 to April 2013.

Sample size

The national sample sizes are typically about 1000 or more and thus usually have sampling errors of approximately +/−3 percent at the 95 percent confidence level; in some instances, however, the samples are smaller and have sampling errors of approximately +/−5 percent.

Question wording

Of course, the precise wording of questions in any survey can make a difference in the distribution of responses and interpretations of them; in the case of opinions about the topic of this book, in particular, an obvious manifestation of this question-phrasing issue concerns the use of the terms "climate change" or "global warming" in survey questions. Fortunately, a survey (German Marshall Fund of the United States 2009) addressed this issue by asking identical questions – except for the use of the terms "climate change" and "global warming" – in randomly split subsamples of the same survey. The results appear in the table below. There were differences of 0 to 3 percent in most response categories but a difference of 5 percent in the proportions who thought it was a "critical" problem, with the term "global warming" thus evoking an only marginally greater concern. In any case, most climatologists use the term "climate change" because it conveys the broad array of changes, including the increased frequency of extreme weather events of diverse types and such impacts as sea level rise and droughts, as well as the increase in the earth's mean surface temperature associated with the term "global warming." In this book, I therefore generally use the term "climate change," but in this chapter in particular, of course, whatever term has been used in a survey question has been retained in the presentations of the data and the associated discussion.

Table A3.1a "Climate change" or "global warming"? How much do differences in question wording matter?

	Percentage of responses			
Wording of possible threat[a]	Critical	Important but not critical	Not important	Not sure/ Decline
"Global warming"	44	37	18	1
"Climate change"	39	40	20	1
Difference	+5	−3	−2	0

Question: "Below is a list of possible threats to the vital interest of the United States in the next ten years. For each one, please select whether you see this as a critical threat, an important but not critical threat, or not an important threat at all."
[a] The question was posed to respondents in randomly split samples in the same survey.
Source: German Marshall Fund of the United States (2009).

Table A3.1b Partisan differences in acceptance of climate change as a fact – a comparison of results from two survey sponsors (percentages)

Sponsors	Dates	Totals	Democrats	Republicans	Difference: Democrats – Republicans
CNN	12/2–3/2009	68	86	52	+34
Fox	12/8–9/2009	63	83	46	+37
Difference		+5	+3	+6	−3

N.B. The surveys were one week apart and used different wording. Also note that one defined the population of the survey as registered voters, while the other defined it as the adult population.
Questions:
CNN: "Which of the following statements comes closest to your view of global warming? Global warming is a proven fact and is mostly caused by emissions from cars and industrial facilities such as power plants and factories. Global warming is a proven fact and is mostly caused by natural changes that have nothing to do with emissions from cars and industrial facilities. Global warming is a theory that has not yet been proven."
Fox: "Do you believe global warming exists?"
Sources: CNN/Opinion Research Corporation (2009); FOX News/Opinion Dynamics Poll (2009); also see Krosnick (2013).

Also, from time to time I have used the term "global warming" for stylistic variety in the text.

There are other survey question-wording issues that cannot be so easily resolved – for instance, whether a question asks about the respondents' acceptance of the "theory" that greenhouse gas emissions from human activities cause climate change, or asks them whether they believe there is "solid scientific evidence" that climate change has been occurring, or asks them whether they agree with the "consensus among scientists" that there is such a causal connection. These and other specific question-wording issues are addressed in the context of the interpretations of specific survey results. Also see Krosnick (2013).

Survey sponsorship

Yet another issue about the consistency of survey results concerns the possible effect of survey sponsorship on the phrasing of questions and the results. It is fortunately possible to compare two surveys – one sponsored by CNN (CNN/Opinion Research Corporation 2009) and the other sponsored by Fox News (Fox News/Opinion Dynamics 2009) – taken a week apart and asking about respondents' acceptance of climate change as a fact. As the data of Table A3.1b above indicate, the differences were 3 to 6 percent, again only marginally different.

APPENDIX 3.2

International comparisons of public opinion in the USA and other countries

This appendix reports the results of multi-country surveys that included comparisons between the United States and other countries during the period 2006–2012.

In a late 2009 survey of fourteen countries, including many developing countries, the USA had the smallest proportion of its population (except for Russia) that believed climate change was already happening, the smallest percentage (except China) that thought it was a "very serious problem," and nearly the smallest percentage on several other indicators of understanding of the effects of climate change (see Table A3.2a).

Table A3.2b focuses on comparisons with opinions in Europe – in particular, the question of what the priority issue should be among a list of eight that were offered to the respondents. The percentage of Americans putting climate change at the top (9 percent) was similar to the percentage in Bulgaria, Romania, Slovakia, and Turkey (all of them below 10 percent). At the other extreme, in Germany 42 percent said it should be the top priority, with nearly 30 percent saying so in France, Spain, Portugal, and the Netherlands.

In Table A3.2c, the results are presented for 2007, 2008, and 2009 of surveys in the USA and sixteen other countries in nearly all regions of the world. The question posed was whether the respondents considered global warming "a very serious problem, somewhat serious, not too serious, or not a problem." Although neither the regional patterns nor the time trends were entirely consistent, it was apparent that in 2009 the proportion of Americans responding "very serious" was close to the bottom of the list, with only China and Poland lower. In five of the other sixteen countries, as well as the USA, the percentage in 2009 was lower than in 2007, while in most of the countries the percentage increased.

Finally, in Table A3.2d there is evidence of widespread disapproval of US responses to climate change as of 2006 in twenty-four highly diverse countries; in sixteen of them more than half of the respondents "disapproved."

Table A3.2a Perceptions of the problem: the gap between the USA and the rest of the world (percentages)

	Climate change is "happening now"[a]	Climate change is a "very serious problem"[b]	Unchecked climate change will affect the "likelihood of natural disasters, like droughts or floods . . . a lot"[c]	Unchecked climate change will affect the "price of food and other essential goods . . . a lot"[d]	Unchecked climate change will affect the "need to move their homes to different locations . . . a lot"[e]
USA	34	31	34	32	16
France	47	43	73	59	39
Japan	61	38	59	44	12
Russia	27	30	48	24	27
Turkey	58	79	68	67	56
Bangladesh	67	85	68	59	54
China	71	28	61	34	26
India	59	62	54	54	41
Indonesia	40	61	57	31	9
Vietnam	86	69	84	53	42
Egypt	35	60	41	31	31
Kenya	88	75	62	68	50
Senegal	75	72	58	43	45
Mexico	83	90	84	77	64

[a] Q7: "When do you think climate change will start to substantially harm people in [name of respondent's country]?"
[b] Q1: "In your view, is climate change, also known as global warming, a very serious problem, somewhat serious, not too serious, or not a problem?"
[c] Q4e: "If climate change is left unchecked worldwide, how much do you think climate change will affect each of the following in our country? . . . The likelihood of natural disasters, like droughts or floods."
[d] Q4d: "If climate change is left unchecked worldwide, how much do you think climate change will affect each of the following in our country? . . . The price of food and other essential goods."
[e] Q4g: "If climate change is left unchecked worldwide, how much do you think climate change will affect each of the following in our country? . . . People's need to move their homes to different locations."
Source: World Bank (2009a; 2009b).

Table A3.2b Policy priorities: the transatlantic gap (percentages)

	Climate change	International terrorism	International economic problems	Easing tensions in Middle East	Others
USA	8	26	21	15	15
Germany	42	13	12	18	19
France	30	18	19	17	12
Spain	30	28	12	13	7
Portugal	29	17	19	13	17
Netherlands	28	21	14	19	16
Italy	25	25	19	14	14
UK	21	22	16	19	23
Romania	9	35	27	13	19
Slovakia	9	33	29	10	18
Turkey	8	47	22	9	8
Bulgaria	7	38	33	14	13

Question: "Which of these [eight issues] should be the top priority for the next American President and European leaders?"
Source: German Marshall Fund of the United States (2009).

Table A3.2c The US gap with the rest of the world (percentage answering "very serious")

	2007	2008	2009	Gap with USA 2009
USA	47	42	44	–
UK	45	56	50	+6
France	68	72	68	+24
Germany	60	61	60	+16
Spain	70	61	67	+23
Poland	40	51	36	−8
Russia	40	49	44	0
Turkey	70	82	65	+21
Egypt	32	38	54	+10
China	42	24	30	−14
India	57	66	67	+23
Indonesia	43	46	46	+2
Japan	78	73	65	+21
Argentina	69	70	69	+25
Brazil	88	92	90	+46
Mexico	57	70	65	+21
Nigeria	NA	45	57	+13

Question: "In your view, is global warming a very serious problem, somewhat serious, not too serious, or not a problem?"
Source: Pew Global Attitudes Project (2009).

Table A3.2d Foreigners' opinions of US responses to climate change (percentage who "disapprove")

	2006
Argentina	79
Australia	68
Brazil	73
Chile	63
China	35
Egypt	59
France	86
Germany	84
Great Britain	79
Hungary	53
India	23
Indonesia	52
Italy	74
Kenya	21
Lebanon	68
Mexico	67
Nigeria	25
Philippines	22
Poland	31
Portugal	79
Russia	36
South Korea	45
Turkey	65
UAE	55

Question: "Thinking about the past year, please tell me if you approve or disapprove of how the United States government has dealt with each of the following: . . . The US handling of global warming or climate change."
Source: Council on Foreign Relations (2009a; 2009b).

CHAPTER 4

State and local governments – and the courts – in a federal system

The science is there. It's time to stop debating it and to start dealing with it.

New York Mayor Michael Bloomberg (2007)

I don't accept the premise that man is the cause of global warming, if global warming even exists.

Candidate for Governor of Illinois, Kirk Dillard (2009)

There are people who . . . don't believe there is such a thing as global warming. They're still living in the Stone Age . . . .

California Governor Arnold Schwarzenegger (2009)

Because the United States is a federal political system, many local and state governments have been able to take their own initiatives on climate change issues, while others have of course lagged behind. When Congress failed to pass a national cap-and-trade bill in 2010, the climate-related activities of state and local governments took on renewed significance, and their potential for collectively making important contributions to reductions of greenhouse gas emissions became more widely recognized. California receives special attention in the chapter, because it is not only the biggest state in population and economic terms, but also one of the most advanced in responding to climate change issues. Throughout the chapter, the context is a federal political system, in which sub-national governmental entities have much leeway, but in which at the same time there are constant conflicts about who can do what. (On federalism, see Goulder and Stavins (2011); Derthick (2010); and Litz (2008) about climate change issues; more generally, on issues in federal systems, see Rodden (2008).)

The chapter addresses the following questions: What kinds of policies have cities, states, and regions adopted (Sections 4.1, 4.2, and 4.3)? Which have been leaders – and which have been laggards? What have been the regional patterns in state and city policies? In Section 4.4: To what extent do city, state, and regional policies collectively make up for lagging national policies? How much difference could they make in the future? In Section 4.5: What conflicts have emerged in relations between states and cities, on the one hand, and the national government, on the other? How has the judicial system become involved in climate change issues, and what have the key issues and decisions been? What are the implications for climate change policy of the institutional features of a federal political system? In Section 4.6: Why do "swing states" play a disproportionate role in presidential elections, and how does this affect climate change policymaking? More generally, what is the role of climate change in elections? The concluding Section 4.7 discusses the implications of the facts and patterns of political geography that have been presented in the chapter.

4.1 Local government policies

It is useful to recall that four-fifths of the US population and slightly over half of the world population live in urban areas[1] and that issues about the sources of GHG emissions and impacts of climate change are often different for urban and rural populations – for instance, the relative importance of buildings as sources of emissions, the vulnerability of urban water and electrical infrastructure to sea level rise, and the contributions of public transportation systems to GHG mitigation. Also of special interest is the exposure of urban residents to extreme heat conditions as a result of the combination of local heat-island effects and global warming effects on temperatures (Stone 2012). Figure 4.1 depicts the difference in urban and rural temperatures in the USA over a period of a half century. While both were increasing over the period, it is evident that urban temperatures tended to be about 1.5 degrees Fahrenheit greater. There was also a regional pattern; in Figure 4.2 the relatively greater increases over half a century in the Southwest and Southeast in the numbers of US cities that have experienced extreme heat waves is evident.

Urban-specific issues about the sources and impacts of climate change have prompted many US cities to undertake a variety of countermeasures. Gore and Robinson (2009) have identified a broad range of governmental functions performed by local governments in the USA that affect greenhouse gas emissions; local governments thus offer many opportunities for emission reductions, especially in the residential and commercial building sector and the transportation sector of local economies, as the following list of local governments' activities suggests:

1 At the beginning of 2013, 82 percent of the US population of 315 million lived in urban areas, as did slightly over 50 percent of the world population of 7 billion (US Census Bureau 2013; UN 2013).

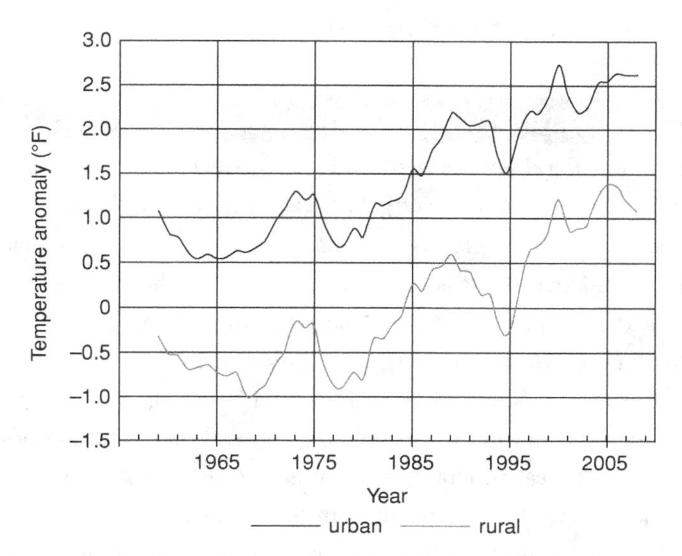

Figure 4.1 Urban versus rural temperature trends in the USA

Source: Stone (2012), 85, Fig. 3.7. Used with permission.

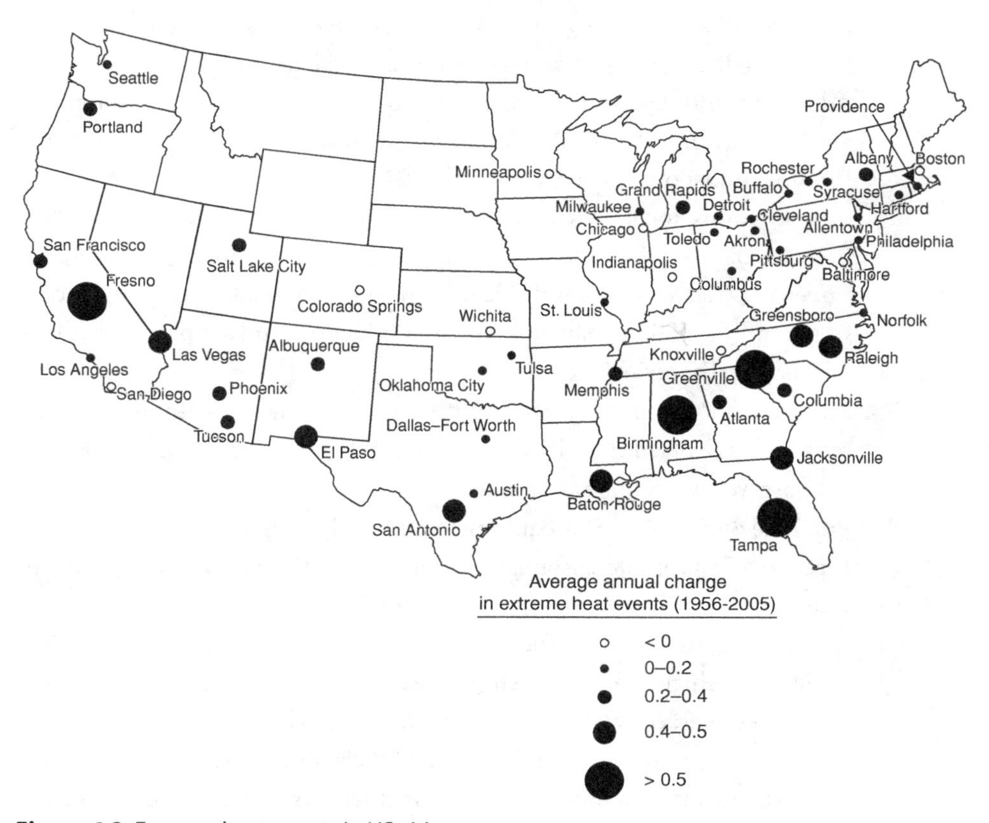

Figure 4.2 Extreme heat events in US cities

Source: Stone (2012), 72, Fig. 3.2. Used with permission.

Box 4.1 The US Mayors' Climate Protection Agreement

A. We urge the federal government and state governments to enact policies and programs to meet or beat the target of reducing global warming pollution levels to 7 percent below 1990 levels by 2012, including efforts to: reduce the United States' dependence on fossil fuels and accelerate the development of clean, economical energy resources and fuel-efficient technologies such as conservation, methane recovery for energy generation, waste-to-energy, wind and solar energy, fuel cells, efficient motor vehicles, and biofuels;
B. We urge the US Congress to pass bipartisan greenhouse gas reduction legislation that 1) includes clear timetables and emissions limits and 2) a flexible, market-based system of tradable allowances among emitting industries; and
C. We will strive to meet or exceed Kyoto Protocol targets for reducing global warming pollution by taking actions in our own operations and communities such as:
   1. Inventory global warming emissions in City operations and in the community, set reduction targets and create an action plan;
   2. Adopt and enforce land-use policies that reduce sprawl, preserve open space, and create compact, walkable urban communities;
   3. Promote transportation options such as bicycle trails, commute trip reduction programs, incentives for car pooling and public transit;
   4. Increase the use of clean, alternative energy by, for example, investing in "green tags", advocating for the development of renewable energy resources, recovering landfill methane for energy production, and supporting the use of waste-to-energy technology;
   5. Make energy efficiency a priority through building code improvements, retrofitting city facilities with energy efficient lighting and urging employees to conserve energy and save money;
   6. Purchase only Energy Star equipment and appliances for City use;
   7. Practice and promote sustainable building practices using the US Green Building Council's LEED program or a similar system;
   8. Increase the average fuel efficiency of municipal fleet vehicles; reduce the number of vehicles; launch an employee education program including anti-idling messages; convert diesel vehicles to biodiesel;
   9. Evaluate opportunities to increase pump efficiency in water and wastewater systems; recover wastewater treatment methane for energy production;

10. Increase recycling rates in City operations and in the community;
11. Maintain healthy urban forests; promote tree planting to increase shading and to absorb CO_2; and
12. Help educate the public, schools, other jurisdictions, professional associations, business and industry about reducing global warming pollution.

Source: US Conference of Mayors (2005).

- Promulgating and enforcing building regulations that affect energy efficiency and renewable energy projects.
- Conducting energy audits and installing equipment and structural retrofits in city buildings.
- Deploying and maintaining infrastructure such as street lights that affect energy consumption/efficiency.
- Purchasing and operating fleets of vehicles – buses for public transport, trucks for trash collection, cars for police forces.
- Developing and/or operating district heating and cooling systems.
- Developing and maintaining transportation infrastructure such as subway systems, bus and tram systems, and bicycle lanes.
- Managing landfill sites.
- Planning land use through zoning.

These and other local governmental activities have acquired direct climate change relevance as more than a thousand cities participate in the Conference of Mayors' Climate Protection Agreement (see Box 4.1).

Map 4.1 indicates the geographic distribution of cities that have signed up to the Agreement. Of course, to some extent the concentrations in California and the Northeast reflect the concentrations of population conurbations in those areas; furthermore, there are important exceptions in the large and varied regions between the coasts – for instance, in Chicago, Illinois, and nearby cities, and in the states of Iowa and Minnesota in the Midwest as well as North Carolina and Florida in the South. Yet, overall, the disproportionate numbers of cities on or near the West Coast and East Coast, with a relative paucity in between, are indications of basic regional patterns.[2] A study by Brown, Sarzynski, and Southworth (2008) of the 100 largest metropolitan areas in the

2 Some US cities have become involved in an *international* effort to address climate change (see Bonn Center for Local Climate Action 2011).

Map 4.1 Cities in the US Conference of Mayors' Climate Protection Agreement
Source: US Conference of Mayors (2010); also see Center for Climate Strategies (2010).

USA found that seven of the ten with the smallest per-capita carbon footprints in their transportation and residential energy use were on the West Coast, while nine of the ten with the largest were in the South or Midwest. Among all cities regardless of size, Lancaster, a mid-sized city in southern California, was the first in the country to require that solar electric systems be installed in new homes beginning in January 2014 (Mulkern 2010).

The West Coast city of Portland, Oregon, offers a good example of an early adopter of a climate action plan (Gore and Robinson 2009). After formulating a plan in the early 1990s, its *per capita* emissions had declined about 13 percent by 2007, and it has kept *total* emissions to 1990 levels, even with population increases and economic growth.

Why have Portland and more than a thousand other US cities taken climate change seriously, while many others have not? Put in the context of a political economy perspective: "Independent municipal climate action challenges orthodox assumptions about collective action problems. Why in the face of a problem replete with free-riding are [some] municipalities taking actions to reduce GHG emissions?" (Gore and Robinson 2009, 153). There are many reasons for the variations among cities' responses, but among them are a series of ideological factors, including commitment to environmental protection goals and actions, as well as economic interests, including the presence or absence of greenhouse-gas-intensive industries. Also, some local leaders understand more clearly and/or value more highly the *co-benefits* of climate action – co-benefits such as reductions of more localized air pollution, or less congestion resulting from improved public transport systems. Similar factors are at work at the state level in some states.

4.2 State government policies

State government climate change policies and related energy policies are important because some individual states are among the world's major sources of greenhouse gases. There are thirteen states which – if they were countries – would rank among the forty highest-ranking countries in the world in terms of greenhouse gas emissions (Rabe 2009). Texas, which ranks first among the states as an emitter of greenhouse gases, by itself contributes a little over one-tenth of total US emissions; in fact, the carbon dioxide emissions of Texas are greater than those of either France or the United Kingdom. Collectively, the ten highest states in terms of their carbon dioxide emissions from fossil fuel combustion contribute about one half of the US total (see Table 4.1).[3]

The patterns of industry concentration in those states are not surprising: Two states in the South – Texas and Louisiana – have major oil and gas and petrochemical sectors. Three are Midwestern automotive industry states – that is, Ohio, Illinois, and Indiana.[4] Four are coal-dependent states: Pennsylvania, as well as, again, Ohio, Illinois, and Indiana.

How have states responded to the problem of greenhouse gas emissions from these and other industries and to other climate change issues? As with cities, there is much variety in the types of measures that states have taken. Some involve direct regulation of producers, such as emissions caps that create carbon prices; others are designed to increase the uptake of energy-efficiency and renewable-energy technologies.

3 An important distinction between production and consumption – or in more personal terms, producers and consumers – as the "sources" of emissions is addressed in Chapter 7 in an international context. Similar conceptual, methodological, and ethical issues pertain in sub-national inter-state and international contexts.

4 Michigan dropped out of the top ten as its auto sector shrank. It was replaced by Georgia.

Table 4.1 Ten largest-emitting US states of carbon dioxide emissions from fossil fuel combustion[a]

State	Percentage of US total	Metric tonnes (mil)[b]
Texas	12.1	653
California	6.9	371
Pennsylvania	4.7	254
Ohio	4.6	248
Florida	4.6	245
Illinois	4.3	231
Indiana	4.0	216
Louisiana	3.9	211
New York	3.2	174
Georgia[c]	3.2	173
Total of 10 states	52.2	2776
National total	100.0	5383

[a] Data are for 2010.
[b] These are *absolute amounts* of emissions of carbon dioxide. For state-level data and rankings on *per capita* emissions of all greenhouse gases, see Appendix 1.1 at the end of Chapter 1.
[c] Georgia replaced Michigan in the top ten as manufacturing in the auto industry in Michigan declined.
Sources: Compiled by the author from US Environmental Protection Agency (2013a), Table 2-5 (2013b).

States are able to undertake a wide variety of measures to mitigate greenhouse gas emissions – some of them decided independently but as complements to national government measures, and some of them with the encouragement of the national government. At the same time, there are also some state actions that are legally prohibited and that can prompt court cases. The federal nature of the constitutionally prescribed political system thus entails a mélange of cooperative and conflictive relationships. Within the broad framework of constitutional provisions, plus laws passed by Congress and regulations promulgated by national regulatory agencies, states have adopted many GHG mitigation policies in the energy, transportation, and building sectors, in addition to climate-specific actions.

Altogether, thirty-nine states have some kind of Climate Action Plan (see Table 4.2 below). Electricity production and distribution is the sector with the most significant sources of GHG emissions subject to state regulations. In some states, electricity providers are required to give consumers choices among electricity sources such as hydro or nuclear or wind or solar, as opposed to fossil fuel plants. The construction of new coal-fired power plants is subject to state approvals, or disapprovals, as is the switching from coal to natural gas. Among the many types of measures that are evident in the table, Renewable Portfolio Standards (RPSs) have been especially prominent

Table 4.2 Elements of state governments' climate change policies[a]

Program element	Number of states
Climate action	
Climate Action Plan	39
GHG Targets	24
GHG Registry	43
Adaptation Plan	26
Carbon Cap/Offset for Power Plants	15
Energy sector	
Renewable Portfolio Standard	39
Net Metering	46
Energy Efficiency Resource Standard	34
Carbon Capture and Storage Incentives	12
Transportation	
Vehicle GHG standards	16
Biofuel Mandates and Incentives	44
Low-Carbon Fuel Standard	14
Medium and Heavy-Duty Vehicle Standards	40
Plug-in Electric Vehicle Incentives	35
Building sector	
Residential Building Energy Codes	40
Commercial Building Energy Codes	40
Appliance Efficiency Standards	15
Green Standards for State Government Buildings	44

[a] For the sake of brevity, not all types in the original source are included.
Source: Compiled by the author from Center for Climate and Energy Solutions (2013a); also see Climate Strategies (2010); Litz (2008); and Northrop and Sassoon (2008).

(Rowlands 2010), with thirty-nine states, producing more than half the country's electricity, having adopted them. Because electric utilities are still largely regulated at the state level, the states that have not adopted them are mostly coal-intensive states in the Southeast plus some oil and coal states in other Southern and Western Mountain states. It is relatively easy for state governments to impose RPSs, while the national government has not yet done so despite repeated attempts over many years within the Congress to do so.

Within the transportation sector, in addition to national renewable fuel policies, which require increasing amounts of biofuels to be blended into motor vehicle fuels nationally by producers, many states have tax incentives for blending biofuels and/or establishing biofuel infrastructure; some have state-level advanced biofuel research and development programs. California has been a leader in establishing GHG emission limits from motor vehicles – a precedent being adopted by other states. On the other hand, national-level policies concerning aviation, off-road vehicles, and other areas have pre-empted state policies in those areas.

Many energy efficiency measures are the result of local and state government policies, especially in the building sector, where building codes are overwhelmingly locally formulated and enforced and where subsidized building insulation programs are state-run, with substantial national government subsidies. The energy efficiency programs of electric utilities – such as offering home energy audits and providing detailed data about energy consumption – are encouraged and regulated by many state authorities.

Each state's total number of programs can be summed as an indicator of the extent of its commitment to action on climate change issues, and again there are regional patterns in such programs: of the ten states with the highest scores, nine are on the West Coast or in the Northeast; of the ten states with the lowest scores, six are in the South.

Many of these state policies are "demand–pull" policies, which complement the "supply–push" research and development policies of the national government (discussed in Chapter 6) as ways to address the market failures associated with technology innovation and diffusion. Such policies are thus particularly important in the development of solar, wind, and other renewable energy sectors. The RPS program in Texas, for instance, places much emphasis on wind; and as a consequence, Texas is the leading wind-power state in terms of installed capacity (Database of State Incentives for Renewables and Efficiency 2011). Texas thus represents an unusual state: it is both a leading state in renewable energy production and consumption, but at the same time a leading state in its per-capita emissions and in its strong opposition to many national-level programs designed to mitigate climate change.

California

Of course, some states have attracted special attention, not only because of their large populations and thus their economic, political, and environmental significance, but also because of their climate policies – California being the most obvious example (see Farrell and Hanemann 2009, who have examined the panoply of California programs in detail). The importance of California is evident in many statistics: It is the largest US state, with a population of more than 38 million. Its gross state product of about $2 trillion is approximately 14 percent of the US national GNP – and larger than the

GNPs of all but seven countries. Its greenhouse gas emissions of 371 million metric tonnes CO_2e in 2010 were the second-largest among US states – after Texas.

The importance of California in the USA, however, goes beyond these elementary demographic, economic, and environmental statistics. California is a leader on environmental issues, and its policies are often emulated by other states; for instance, seventeen other states indicated they would adopt similar motor vehicle tailpipe GHG emissions standards to those in California – actions which would extend the standards from California's one-tenth share of the national new car market to almost half the national market.[5] California has adopted – or attempted to adopt – other measures, which are pending, such as emissions restrictions on ships entering its ports. Indeed, the California government has expanded its international linkages on climate change and related energy issues to cities, regional governments, and national governments in Asia, South America, and Europe as well as North America, through a variety of agreements (Fickling 2010). Such actions sometimes threaten traditional national government policymaking prerogatives and can pose constitutional issues about the relationship between states and the national government in a federal political system, and thus often require court decisions for clarification.

Because of its many initiatives, California is often portrayed as an exemplar for paths to high-efficiency/low-carbon economies: As a result of a variety of energy efficiency, renewable energy, and other measures, between 2000 and 2008 California's per-capita GHG emissions on a CO_2e basis declined 8.3 percent, from 13.4 to 12.5 tonnes; and the GHG intensity of its economy declined 27.3 percent, from 355.8 metric tonnes per million dollars of state gross product to 258.7 metric tonnes per million dollars. During the same period, its population increased 11.8 percent, and its gross state product increased 20.2 percent in real terms. Whatever the verdict may be on its efforts to date, the paths of such indicators will be tracked carefully over the next many years as additional programs become fully implemented.

The state Global Warming Solutions Act of 2006 (which is also known as Assembly Bill 32 or simply AB32) is central to California's climate change policies. The Act and subsequent administrative actions have established a series of programs whose goal is the reduction of GHG emissions to 1990 levels by 2020 – i.e. 427 million metric tonnes of carbon dioxide equivalent. The programs can be conveniently considered in two groups – those involving "early action" beginning in 2010 and those concerning

5 California Assembly Bill 1493, which was adopted in 2002, mandates GHG emissions of new vehicles to be 30 percent less by 2016.

a cap-and-trade system beginning in 2012 (see Appendix 4.1 for summaries of the program).

The entire program was the subject of a state-wide vote known as Proposition 23 in the November 2010 election; the proposition would have suspended implementation of AB32 until the state's unemployment level was below 5.5 percent for four consecutive quarters. Because the unemployment rate was more than 12 percent at the time of the election and had been below 5.5 percent for four consecutive quarters only three times in several decades, the proposition represented a virtual nullification of the program, or at least a likely deferral of it for many years. The proposal was defeated – and thus the program was supported – by a vote of 61 percent to 39 percent.

A state-wide survey in July 2010, which had found that 61 percent of "likely voters" favored "the state law that requires California to reduce its greenhouse gas emissions back to 1990 levels by the year 2020," also found that while 80 percent of Democrats and 73 percent of independents favored the measure, only 39 percent of Republicans were so inclined. It also found that the percentages who believed that "the effects of global warming would never happen" were as follows: Democrats, 5 percent; independents, 14 percent; Republicans, 40 percent – the latter an increase from 18 percent in 2007 (Public Policy Institute of California 2010a; 2010b). Two years later, while 88 percent of Democrats and 76 percent of independents thought it was "necessary to take steps to reduce the effects of global warming right away," only 38 percent of Republicans thought so (Public Policy Institute of California 2012). See Appendix 4.2 for additional data on public opinion in California.

A broad-based coalition of supporters of AB32 was co-chaired by a Republican, George Shultz (who was Director of the Office of Management and Budget, Secretary of Defense, and Secretary of State during the presidency of Ronald Reagan), and by a San Francisco hedge fund manager, Tom Steyer. Also Microsoft's Bill Gates, Silicon Valley venture capitalist John Doerr, and others contributed millions of dollars to defeat Proposal 23 and were thus in support of AB32; on the other side, BP, Chevron, and ExxonMobil were among the contributors (Greenwire 2010a).

State insurance regulations

State and local governments are not only addressing issues about reducing greenhouse gas emissions, they are also addressing issues about *adaptation to the impacts of climate change.* Among the many adaptation issues are those concerning flood and storm damage to communications and transportation infrastructure and to residential and commercial property. Because of the importance of state governments' function as insurance regulators, there has been increasing interest in the impact of climate change on the insurance industry and its underwriting and damage claims policies.

As a result of concern among state insurance regulators about the possible impact of climate change on insurance company solvency as well as the availability and costs of insurance, which are traditional concerns of insurance regulators, a task force of the National Association of Insurance Commissioners (NAIC) developed a mandatory disclosure survey to be completed by insurance firms each year. Adopted by a unanimous vote at a 2009 NAIC meeting, the Insurer Climate Risk Disclosure Survey covers many issues about firms' risk management practices associated with climate change. A year later, about a month before the insurers were required to submit their responses, at an NAIC meeting in March 2010, there was an unexpected reconsideration of the plans for the survey – in particular whether its results would be made public or remain confidential. It was decided that public disclosure would be voluntary, after a handful of states, led by Indiana and joined by Alabama, South Carolina, Utah, and others, withdrew their earlier support for mandatory disclosure (*Financial Times* 2010a; ClimateWire 2010a). Because of its own state law requirements, Pennsylvania nevertheless made the results public (Pennsylvania Insurance Department 2010). The questions posed in the survey, as well as an excerpt from the introductory discussion of its context, are presented in Appendix 4.3.

4.3 Regional programs

The Regional Greenhouse Gas Initiative (RGGI), which began operation in 2009, includes the nine Northeastern states of Connecticut, Delaware, Maine, Maryland, Massachusetts, New Hampshire, New York, Rhode Island, and Vermont. Though it is limited to the carbon dioxide emissions of electric power plants, it is nonetheless significant as the first mandatory cap-and-trade system in the USA prior to the creation of California's (see Box 4.2 and also Selin and VanDeveer 2010). A study by Hibbard and Tierney (2011) found that in its initial three years, customers saved approximately $1.1 billion on electric bills and $174 million on natural gas and heating bills, and 16,000 jobs were created. The power plants' revenues were $1.6 billion lower, though their revenues still exceeded their costs. Meanwhile, the Midwest Greenhouse Gas Reduction Accord (MGGRA) is far behind in its institutional development and operational implementation (see Box 4.3).

Box 4.2 Regional Greenhouse Gas Initiative (RGGI)

Year of initial agreement: 2005

Year of initial operation: 2009

Targets and timetables: 2009–2014, cap stabilizes emissions at 188 million tons annually; 2015–2018, cap declines by 2.5 percent per year for total reduction of 10 percent

Coverage: CO_2 emissions from fossil-fuel-fired electric power plants, 25 megawatts or greater in size, 209 facilities region-wide as of December 2010

Members and Observers:

US state Members: Connecticut, Delaware, Maine, Maryland, Massachusetts, New Hampshire, New York, New Jersey, Rhode Island, Vermont

US state Observer: Pennsylvania

Canadian province Observers: New Brunswick,[a] Ontario,[b] Québec[c]

[a] New Brunswick is also an Observer in the WCI.

[b] Ontario is also an Observer in the MGGRA and a participant in the WCI.

[c] Quebec is also an Observer in the MGGRA.

Sources: Compiled by the author from Regional Greenhouse Gas Initiative (2010).

Box 4.3 Midwestern Greenhouse Gas Reduction Accord (MGGRA)

Year of initial agreement: 2007

Year of initial operation: pending

Targets and timetables: pending

Coverage: pending

Members and Observers:

US state Members: Iowa, Illinois, Kansas, Michigan, Minnesota, Wisconsin

US state Observers: Indiana, Ohio, South Dakota

Canadian province Member: Manitoba

Canadian province Observer: Ontario[a]

[a] Ontario is also a participant in the WCI and an Observer in the RGGI.

Source: Compiled by the author from Midwestern Greenhouse Gas Reduction Accord (2010).

In another regional development, California and as many as five other US states and four Canadian provinces had agreed that they would cooperate to create an international cap-and-trade program, beginning in 2012, through the Western Climate Initiative (2010). However, by early 2013 the plan had been largely abandoned, with its principal remnant consisting of a plan for California and Quebec to link their cap-and-trade systems.

4.4 Collective impact of local, state, and regional programs

Of course, from broader national and international perspectives, a key question about these programs at the local, state, and regional levels is how significant their total collective impact on GHG emissions is or could be. In order to assess the past and potential contributions of sub-national governmental actions to US emissions reductions, it is useful to group the emissions sources (Bianco, Litz, Meek, and Gasper 2013) as follows – with the percentage contribution to total national GHG CO_2e in 2010 indicated in parentheses for each: electric power plants (40 percent), transportation (30 percent), residential and commercial heating (7 percent), industry (13 percent), others including non-CO_2 which is mostly industry (10 percent).

Sub-national governmental programs have made and will continue to make substantial contributions to emissions reductions in the first three categories – power plants, transportation, buildings – which are together responsible for about three-fourths of US emissions. In the power plant sector, some cities own and operate electric power plants and thus make a variety of investment and operational decisions about their fuels and other features, such as whether they are combined power and centralized heat producers. At the state level, there are electricity-pricing rules and other regulatory policies such as renewable energy requirements in place. In the transportation sector, many cities of course operate their own public bus and subway systems, and regional and state agencies operate and regulate commuter railroad systems. In the building sector, local and state building codes directly affect the energy efficiency of heating and cooling systems. So there are many ways that sub-national governmental programs can impact greenhouse gas emissions.

But can the cumulative impacts be quantified? The data cited above, for instance, for Seattle, Washington, the state of California, and the Regional Greenhouse Gas Initiative are exceptions and cannot be extrapolated to other cities or states. In fact, so many of the state and local programs are recent or still under development that their total impact is likely to be marginal. There is in fact a paucity of empirical evidence about the impacts to date. However, as for the potential for the future, in all scenarios in the extensive analysis for World Resources Institute by Bianco, Litz, Meek, and Gasper (2013), the biggest impact is in the electric power sector. Altogether they estimate a wide range of potential reductions. For instance, assuming a federal government "middle of the road" scenario, state programs could add 2 percent to 7 percent in reductions by 2020 beyond the estimated 12 percent reductions from federal policies. An analogous set of estimates for 2035 includes state contributions of 5–15 percent reductions in addition to federal reductions of 26 percent. Extrapolating from those estimates for the future, the current reductions are likely to be less than 5 percent. For the future, the estimated reductions within the next decade or so from state-level programs are likely to be quite modest, but after about two decades there are likely to be significant effects.

One particular area of increasing importance and special interest at the state and local levels is limitations on the exploration and/or production of shale gas. Some states and local governments have imposed moratoria pending the clarification of a variety of health, safety, social, and environmental impacts. This is a climate change issue for several reasons (Brewer 2013a; 2014): one is that the GHG emissions from producing electricity with natural gas are about 40–50 percent less than producing it with coal; the second is that during exploration and production the occurrence of "fugitive" methane releases, with its relatively high global warming potential compared with carbon dioxide, may substantially offset – or perhaps even more than offset – the emissions reductions compared with coal; a third concern is that the relatively low cost of shale gas may undermine investment in low-carbon renewable alternatives such as wind and solar over the long run. These and a variety of other issues about shale gas are under close examination – and like many other climate change issues are becoming involved in legal disputes.

4.5 Court cases: state and local governments versus the national government

Since federalism is a key institutional feature of the US political system, it is not surprising that there have been a large number of court cases involving issues about the roles of the national and state governments in addressing climate change and about the interpretation of legislation at the national and state levels (Gerrard 2007; Gerrard and Howe 2011). In addition, there have been cases in which attempts have been made to require industry to reduce emissions and to disclose more about approaches to climate change issues, as discussed in Chapter 2; and there have been cases in which industry has tried to stop governmental actions, as discussed in Chapter 5.

The most consequential legal case pitting states against the national government was the case of *Massachusetts v. EPA*. The state of Massachusetts – joined by eleven other states and three cities – filed a complaint against the national government's EPA during the George W. Bush administration for failing to use the Clean Air Act to regulate greenhouse gas emissions on the grounds that they constituted an "endangerment" to public health and welfare. The states joining Massachusetts were California, Connecticut, Illinois, Maine, New Jersey, New Mexico, New York, Oregon, Rhode Island, Vermont, and Washington, and the cities were New York, Baltimore, and Washington, DC. On the other side, the states siding with the EPA were Michigan, Alaska, Idaho, Kansas, Nebraska, North Dakota, Ohio, South Dakota, Texas, and Utah. There were thus familiar regional patterns in the states' positions.[6]

The case was decided by the Supreme Court in 2007 by a 5–4 vote – along ideological lines with Alito, Roberts, Scalia, and Thomas voting against the majority consisting of Breyer, Ginsburg, Souter, Stevens, and Kennedy (often a swing vote in the middle in Supreme Court cases). The majority found that greenhouse gases could be considered pollutants under the definition of the Clean Air Act and that therefore the EPA could regulate greenhouse gas emissions. However, it also found that the EPA needed to determine that greenhouse gases "endangered" public health and welfare in order to apply the Clean Air Act to such emissions. The majority opinion written by Justice Stevens said in part:

> The harms associated with climate change are serious and well recognized. . . . [The Environmental Protection Agency in 2007 during the George W. Bush administration] does not dispute the existence of a causal connection between man-made greenhouse gas emissions and global warming. . . .
>
> (US Supreme Court 2007, in *Massachusetts v. EPA*)

This Supreme Court decision was a landmark event in a series of developments in the regulatory process centered in the EPA and in the federal court system. After a two-year delay in the EPA's response to the Supreme Court decision during the remainder of the George W. Bush administration, the EPA under the Obama administration initiated an "endangerment finding" process that led in 2009 to the conclusion that greenhouse gases did endanger the public health and welfare. This finding was challenged in the courts by the states of Alabama, Texas, and Virginia, but the US Court of Appeals for

6 In addition, there were many environmental organizations allied with Massachusetts and many industry associations allied with the EPA (i.e. the Alliance of Automobile Manufacturers, National Automobile Dealers Association, Engine Manufacturers Association, and Truck Manufacturers Association).

Table 4.3 Thirty-five states' positions on EPA's "endangerment" finding

States (Census region/division) supporting EPA	States (census region/division) opposing EPA
Arizona (West/Mountain)	Alabama (South)
California (West/Pacific)	Alaska (West/Pacific)
Connecticut (Northeast/New England)	Florida (South)
Delaware (South)[a]	Hawaii (West/Pacific)
Illinois (Midwest)	Indiana (Midwest)
Iowa (Midwest)	Kentucky (South)
Maine (Northeast/New England)	Louisiana (South)
Maryland (South)[a]	Michigan (Midwest)
Massachusetts (Northeast/New England)	Mississippi (South)
Minnesota (Midwest)	Nebraska (Midwest)
New Hampshire (Northeast/New England)	North Dakota (Midwest)
New Mexico (West/Mountain)	Oklahoma (South)
New York (Northeast/Middle Atlantic)	South Carolina (South)
Oregon (West/Pacific)	South Dakota (Midwest)
Pennsylvania (Northeast/Middle Atlantic)	Texas (South)
Rhode Island (Northeast/New England)	Utah (West/Mountain)
Vermont (Northeast/New England)	Virginia (South)
Washington (West/Pacific)	

[a] Often considered "Northeast/Middle Atlantic" states despite their official classification as "South/South Atlantic."
Source: Compiled by the author from *New York Times* (2010).

the District of Columbia sided with the EPA (see Table 4.3 for lists of states on both sides). Yet another series of legal issues that emerged from this process concerned a "tailoring" rule, according to which the EPA changed the threshold size of installations for Clean Air Act coverage under the new greenhouse gas standards. This rule was also challenged in the courts, which again sided with the EPA.

As of the beginning of 2011, the EPA regulations were thus in effect. While the EPA tried to implement its new rules, however, nearly four years after the Supreme Court case that enabled it to proceed under the terms of the Clean Air Act, it encountered opposition in some states, notably Texas. As the account in Box 4.4 indicates, the relationship between the EPA and the state of Texas was problematic.

Many court cases have involved California, partly because of its relatively activist approach to climate change issues and partly because the Clean Air Act granted California a privileged position in the sense that it was allowed to have policies that were more stringent than national laws because the air pollution problem in the Los

Box 4.4 The Environmental Protection Agency (EPA) and the state of Texas at odds

. . . The [EPA] is embroiled in a tooth-and-claw legal fight over its attempt to take over greenhouse gas permitting authority in the state [of Texas] in the face of Texan intransigence. Most legal experts think EPA is in the stronger position, although Texas may have a couple of cards up its sleeve. In the past two weeks, [Texas] has launched a new two-pronged assault to try and counter EPA. First, it filed a petition in the 5th US Circuit Court of Appeals seeking review of EPA's finding that Texas' permitting process was not in compliance with the agency's rules. . . .

Then, on Dec. 30 it filed a new petition in the US Circuit Court of Appeals for the District of Columbia taking issue with EPA's decision to effectively take over greenhouse gas permitting in Texas with almost immediate effect. . . . The 5th Circuit denied Texas' request for an immediate stay but has not yet taken any other action. The D.C. Circuit issued a temporary stay preventing the EPA takeover until after both sides have filed briefs. . . . Texas is also involved in the wider litigation attacking the various EPA rules upon which the agency is basing its plan to regulate greenhouse gases. . . . There are 13 states in total challenging the regulations, but Texas . . . is the only one that has refused point blank to play ball with EPA while the legality of the rules is debated in court. That is what prompted EPA to announce its plans to take over greenhouse gas permitting for new sources. Lawyers familiar with the case . . . mostly agree that if EPA sticks to its guns it will ultimately win out. But Texas could notch up a short-term victory of its own over EPA's failure to follow the usual procedures when it announced its takeover plans on [December] 23. It did not open up the process to notice and comment, Texas noted in its most recent petition. EPA says it did not give Texas a year to prepare, as outlined in the Clean Air Act, because Texas had failed to respond with a proposal for how it would modify its permitting process so it would be in line with the new rules. [One lawyer] noted that there is the possibility of greenhouse gas permitting authority switching back and forth between EPA and Texas before the litigation is over. . . . [Another lawyer] noted that the 12 other states that objected to the EPA rules have all been able to reach an interim agreement.[a]

[a] A federal court ruled in favor of the EPA and against the state of Texas in January 2011.

Source: Reprinted from *Greenwire* with permission from Environment & Energy Publishing, LLC, www.eenews.net, 202/628-6500. Paragraphing altered by the author.

Angeles area had been so severe at the time the original Clean Air Act was passed. In a case where the central issue was whether California could impose restrictions on the tailpipe greenhouse gas emissions of motor vehicles, the George W. Bush administration joined with the auto manufacturers and unions in an attempt to prevent the California law from being implemented. California argued that it had the prerogative under the Clean Air Act to adopt more stringent standards than the national governments' standards; the national government, auto makers, and auto union argued that California was prohibited from doing so by the interstate commerce clause of the Constitution, which reserves the right to regulate interstate commerce to the federal government. The case was pending in a US District Court when the Obama administration entered office. In the context of the administration's bailout of the US auto firms GM and Chrysler during the financial crisis in 2009, the case was withdrawn by the auto firms and union. Thus, the constitutional issue about the application of the interstate commerce clause was left unresolved.[7]

In sum, in the US federal political and legal system, both the federal government and the states can and do use legal procedures to try to advance their policies over the opposition of the other. In some instances, groups of states have been on opposing sides. The outcomes in terms of winners and losers have been mixed: Sometimes the national government has won and sometimes it has lost.

4.6 "Swing" states and other electoral system issues

Because of the federal nature of the US political system, states play a central role in national presidential elections. A particular feature of those elections of special relevance to climate change issues is the disproportionate impact of a relatively small number of "swing" states, sometimes referred to as "battleground" states. They are the typically dozen or so states where the margin of victory tends to be relatively small and where in any given election either major party candidate has a decent chance of winning the state's electoral college votes (Gimpel, Kaufmann, and Pearson-Merkowitz 2007; cf. Shaw 1999). Much of the attention, political activity, campaign expenditures, and fund-raising in a presidential election, therefore, tend to be concentrated in those

7 Some cities have also sued the federal government because of the failure of the government to address climate change and thereby reduce its damages – Oakland, California, because of the damage done to its port area by sea level rise, and Boulder, Colorado, because of the damage done to its agricultural and snow tourism industries by drought.

relatively few states where the outcome is in doubt, from early in the campaign until election day.

The precise list of the particular battleground states varies with each election, though some states such as Ohio and Pennsylvania are regularly included because of their large size and relatively balanced party distributions. Swing states of special interest in regard to climate change issues are three coal states – not only Ohio and Pennsylvania, but also West Virginia. Pennsylvania and West Virginia are major coal-producing states. In Ohio, coal is a significant factor in the state economy because it is the principal fuel for producing electricity, though most of it comes in from other states. In fact, the combination of Ohio's location adjacent to the country's major coal region and its coal-dependent electric power industry make it a de facto coal state. Its neighbor, Indiana, is also a highly dependent coal state, with substantial coal production of its own; however, except for the 2010 election, Indiana has had a long tradition of voting Republican in presidential elections and is therefore not generally considered a "swing" state.

In an unusually close presidential election, West Virginia by itself – with a population of 1.8 million, or only 0.6 percent of the national population – can shift the outcome. In the 2000 presidential election, because a shift from George W. Bush to Gore of West Virginia's five electoral votes would have changed the outcome of the presidential election, it has been suggested that Gore's identity as a strong environmentalist, with a special interest and record of support for climate change mitigation, cost him the election at the margin (Agrawala and Andresen 2001; Rosenkranz 2002, 232, n23).[8] At an August 2000 meeting in Charleston, West Virginia, during the presidential election campaign, Mr. Bush met with officials of a coal industry association and a labor union to hear their complaints about government regulations on the industry. Later that day, Donald Evans, Mr. Bush's campaign chairman and subsequently Secretary of Commerce, called the industry and union officials to discuss government regulations. The two officials then created the Balanced Energy Coalition, which contributed to the Bush campaign and encouraged mine workers to vote for him.

Four years later, in the 2004 presidential election, a shift of 60,000 votes out of 5.6 million cast in Ohio would have reversed the outcome of that election. A shift of about 1 percent of Ohio's votes would have meant its 20 electoral votes would have been for Kerry instead of Bush, enough to give Kerry the majority of the electoral votes and thus victory in the national election. During the presidential campaign,

8 In such a close election, of course, other issues in other states could also be considered to have made a difference in the overall election outcome.

government subsidies for the development of alternative coal technologies to reduce greenhouse gas emissions became an issue when Senator Kerry proposed a multi-year $10 billion program, which exceeded President's George W. Bush's more modest $2 billion program.

Another feature of the presidential election system has also affected climate change policymaking – namely the "electoral college" which over-represents sparsely populated states that tend to be more conservative and more opposed to climate change mitigation measures, while under-representing densely populated states that tend to be more liberal and in favor of climate change mitigation measures. This particular feature was of course especially prominent in the 2000 election, in which candidate Gore won 48.4 percent of the popular vote, which was 0.5 percent more than George W. Bush's 47.9 percent (with other candidates receiving the balance). Bush nevertheless won the election with 50.5 percent of the electoral college vote, while Gore received 1.0 percent less at 49.5 percent (US Federal Election Commission 2010).

A related phenomenon of the institutionalized over-representation of small population states and the under-representation of large population states in the Senate is considered in Chapter 5 in relation to cap-and-trade legislation. In that chapter there is also a discussion of the distortion of representation in the House of Representatives resulting from the practice whereby partisan considerations yield voting district boundaries that also distort popular vote outcomes in such a way that advocates of action on climate change are under-represented and opponents are over-represented.[9]

4.7 Implications: political geography

The US federal political system allows state and local governments much autonomy in a wide range of climate-relevant policy domains, and thus many of them have significant programs with direct implications for climate change mitigation and/or adaptation; these governmental units are especially important in electricity production and consumption, transportation systems, and building practices. Each state of course has its own distinctive mix of industry interests and climatic features, which are addressed

9 Because an understanding of the distortions of representation in both the Senate and the House requires an understanding of their distinctive institutional features, a more detailed explanation is deferred to Chapter 5. However, in as much as the distortions are also the result of a federal political system in which states play distinctive roles in elections for the president and both of the houses of the national legislature, a brief discussion in the present chapter about the federal political system is also appropriate.

according to each state's distinctive political institutions and political culture. It is not an exaggeration to say that the US response to climate change is occurring in substantial part in distinctive ways in fifty states, the District of Columbia, five offshore territories,[10] and 566 tribal nations in 35 states.[11]

While climate change is a salient item on the agendas of many of these sub-national governments, there are significant regional patterns in the responses: local and state governments on the West Coast and in the Northeast have generally been the most active in adopting their own climate and energy programs. It is in these areas of the country where traditions of environmental protection are strong and where heavy manufacturing industries and extractive industries are generally absent (oil in California being an important exception). The tendencies in these parts of the country are in contrast to many parts of the Midwest and South, where there has traditionally been less concern about environmental issues and where the auto, coal, oil, and petrochemical industries are important.

Empirical evidence about the actual impacts on emission reductions of these many programs is generally sparse. However, econometric modeling of prospective impacts indicates reductions by 2020 and 2035 from state programs, which though certainly not negligible, are likely to be well below the impacts of federal programs. Of course, these are only roughly indicative and are based on a wide range of economic and policy assumptions that may turn out to be off the mark. Yet, the relative size of California and its extensive programs which are still evolving give it a special significance not only nationally but even internationally.

A further implication of the analysis of the chapter is that the complex mix of separate and shared responsibilities of the national government with state and local governments inevitably leads to much conflict as well as cooperation. State governments (and occasionally local governments) have consequently become embroiled in important court cases with the national government, as the constitutionally prescribed rights and responsibilities of governmental units are addressed at all levels of the federal judicial system. In some instances, the national government finds itself aligned with some states and in opposition to others, as the states themselves form coalitions to support or oppose particular policies. The entire federal-level court system – including

10 There are five territories with a Delegate or Commissioner in the House of Representatives: American Samoa, Guam, Northern Mariana Islands, Puerto Rico, and US Virgin Islands. Like the Delegate from the District of Columbia, they are not allowed to vote on bills and other matters "on the floor," though they have some other limited voting rights in committees.

11 In terms of area, they are collectively equivalent to being the fourth-largest state.

the Supreme Court, Appeals Courts, and District Courts – is now centrally involved in many key questions about climate change issues. The courts' involvement in climate change issues is not restricted to national–state government relations, however; it extends to central issues about the constitutionality of national government policies, as we shall see in the next chapter.

The next chapter also provides additional details about how the regional patterns of political geography in this chapter are reflected at the national level in Congress as it addresses cap-and-trade and other policy options.

Suggestions for further reading and research

For tracking government policy developments at the state and local levels, the Georgetown Climate Center at www.georgetownclimate.org is particularly useful.

Much of the rapidly increasing literature on the policies of state and local governments has been spawned by Rabe (2002; 2009; 2010c). Other important contributions include Betsill and Bulkeley (2006), Byrne, Hughes, Rickerson, and Kurdgelashvili (2007), Goulder and Stavins (2011), Pew Center on Global Climate Change (2009a; 2010a), Posner (2010), and Selin and VanDeveer (2009). An extensive compilation of state policies is available in Fickling (2010) and in Natural Resources Defense Council (2012).

The principal sources of information about legal issues are a collection of readings published by the American Bar Association and edited by Gerrard (2007). Gerrard and Howe (2011) have developed a comprehensive compilation of cases with descriptions of the issues and outcomes; the compilation is updated periodically and is available at www.climatecasechart.com. Also see Engel (2010) for a discussion of the role of the courts in climate change policymaking.

On climate change issues in relationship to federalism, see Derthick (2010) and Litz (2008); on federalism in general see Rodden (2008). On "swing" states (or "battleground" states) in presidential elections, see especially Gimpel, Kaufmann, and Pearson-Merkowitz (2007) and Shaw (1999).

Stone (2012) provides detailed evidence of the significance of heat-islands in US cities and the implications for climate change impacts and policies.

Information about California policies is available at the website of the state Air Resources Board at www.arb.ca.gov, while extensive data from state-wide annual surveys is available at www.ppic.org, the website of the Public Policy Institute of California, which is an independent survey organization.

APPENDIX 4.1

California's early action and emissions trading programs

The following summaries are based on the decisions and information provided by the California Air Resources Board or ARB (2010), which is part of the state Environmental Protection Agency. The wide range of *early action* items include the following:

- Low-carbon fuel standard – The LCFS calls for a reduction of at least 10 percent in the carbon intensity of California's transportation fuels by 2020. The affected fuels include gasoline, E85 ethanol, diesel, biodiesel, liquefied natural gas, hydrogen, and electricity.
- Landfill methane capture – An ARB regulation "requires owners and operators of certain . . . landfills to install gas collection and control systems, and requires existing and newly installed gas and control systems to operate in an optimal manner."
- Mobile air conditioning – A series of measures have been adopted to "reduce hydrofluorocarbon (HFC) emissions associated with mobile air conditioning . . . and the recovery of refrigerants from decommissioned refrigerated shipping containers."
- Consumer products with high global warming potentials – Restrictions are placed on "pressurized containers that utilize nitrous oxide (N_2O) including aerosol cheese and dessert toppings, as well as hydrofluorocarbon (HFC) propellant products such as boat horns, pressurized gas dusters, and tire inflators."
- Shore power for ocean-going vessels – Operators are required to reduce "emissions from diesel auxiliary engines on container ships, passenger ships, and refrigerated-cargo ships while berthing at" any of six California ports.

As for the *cap-and-trade program*: By a vote of 9–1 in December 2010, the ARB passed a series of regulations for the commencement of operations in January 2012. The first auction was held in November 2012. The cap on annual emissions, beginning at the business-as-usual level forecast for the year in 2012, is reduced by 2 percent per year during 2012–2014, and then 3 percent per year during 2015–2020. Its coverage includes about 360 firms with 600 facilities responsible for 80 percent of the state's GHG emissions. Initially, major industrial sources and utilities are covered; then beginning in 2015, the program covers distributors of transport fuels, natural gas, and propane. Other specific provisions include:

- Free allowances during 2012–2014.
- Auctions held quarterly by the ARB, with the first auction in late 2012.
- Offsets from agriculture and forestry projects (including in Brazil and Mexico and perhaps other countries) allowed up to 8 percent of a firm's emissions.
- Banking of allowances is permitted.
- An Allowance Price Containment Reserve (equaling 4 percent of total allowances issued) enables firms to buy allowances from one-third tranches at $40/$45/$50 per metric tonne in 2012 – prices which increase by 5 percent plus inflation each year thereafter.
- Linkage to other programs, including through the Western Climate Initiative, is anticipated.

APPENDIX 4.2

Opinions in California

In California there has been a bipartisan consensus on many issues, including support for government subsidies and regulations for energy efficiency and renewable energy, and for reducing automotive and factory GHG emissions. But there has also been partisan conflict over carbon taxes and cap-and-trade (Table A4.2). The proportion favoring a cap-and-trade regime for companies was only about half, while the proportion favoring a carbon tax was slightly over half. Otherwise, about three-fourths or more of the respondents favored each of the other five options. Again, however,

Table A4.2 Partisan differences in Californians' policy preferences (percentage in favor)

Issue	All adults	Democrats	Independents	Republicans	Difference: Democrats – Republicans
Regulate GHG emissions from power plants, cars, factories[a]	76	86	79	54	32
Carbon tax on companies[a]	56	73	52	33	40
Cap-and-trade for companies[a/b]	49/53	57/55	47/44	36/32	21/23
Require automakers to reduce GHG emissions from new cars[a/b]	78/78	90/89	81/79	55/52	35/37
Require increase in use of renewable energy sources such as solar and wind power by utilities[a]	85	91	85	71	20
Require an increase in energy efficiency for residential and commercial buildings and appliances[a/b]	76/77	86/88	77/77	63/56	23/32
Require industrial plants, oil refineries, and commercial facilities to reduce their emissions[a/b]	80/82	91/92	81/79	63/57	28/35

[a] = 2009
[b] = 2012
Source: Compiled by the author from Baldassare, Bonner, Paluch and Petek (2009); and Public Policy Institute of California (2012).

there were substantial gaps in the distributions for registered Democrats and registered Republicans, with the possibility of requiring utilities to use more solar and wind revealing the smallest partisan gap of 20 percentage points and cap-and-trade revealing the largest gap at 40 points. Slight shifts are evident in some of the data for 2009 and 2012.

APPENDIX 4.3

Insurer Climate Risk Disclosure Survey by state insurance commissioners

[Introduction to the survey:]

The goal of the *Insurer Climate Risk Disclosure Survey* is to provide regulators with substantive information about the risks posed by climate change to insurers and the actions insurers are taking in response to their understanding of climate change risks. Disclosure of climate change risks is important because of the potential impact of climate change on insurer solvency and insurance availability and affordability across all major categories of insurance: property casualty, life and health. . . .

Questions:

1. Does the company have a plan to assess, reduce or mitigate its emissions in its operations or organizations? If yes, please summarize.
2. Does the company have a climate change policy with respect to risk management and investment management? If yes, please summarize. If no, how do you account for climate change in your risk management?
3. Describe your company's process for identifying climate change-related risks and assessing the degree that they could affect your business, including financial implications.
4. Summarize the current or anticipated risks that climate change poses to your company. Explain the ways that these risks could affect your business. Include identification of the geographical areas affected by these risks.
5. Has the company considered the impact of climate change on its investment portfolio? Has it altered its investment strategy in response to these considerations? If so, please summarize steps you have taken.
6. Summarize steps the company has taken to encourage policyholders to reduce the losses caused by climate change-influenced events.
7. Discuss steps, if any, the company has taken to engage key constituencies on the topic of climate change.

8. Describe actions your company is taking to manage the risks climate change poses to your business including, in general terms, the use of computer modeling.

Source: National Association of Insurance Commissioners (2010); as reported by National Association of Mutual Insurance Companies (2011). Used with permission.

PART III

NATIONAL GOVERNMENT POLICIES

CHAPTER 5

Regulating carbon dioxide and other greenhouse gases

. . . [Y]ou will never have energy independence without pricing carbon. . . . The technology doesn't make sense until you price carbon.

Senator Lindsey Graham (2010)

A federal cap-and-trade program is perhaps the most significant endeavour undertaken by Congress in over 70 years. . . .

Senator Debbie Stabenow and nine other senators (2009)

This is the most complicated issue I have addressed in my [half century] in Congress.

Representative John Dingell (2008)

This chapter addresses a variety of questions focused on congressional consideration of climate change legislation over a period of more than a decade: What killed the BTU tax in 1993 (Section 5.1), and what killed cap-and-trade legislation in 2010 (Section 5.2)? What were the patterns of voting behaviour in the House and the Senate when cap-and-trade legislation was considered? How much influence has business lobbying had on the outcomes of the effort to establish a national cap-and-trade system? How do institutional distortions of representation patterns affect the influence of carbon-intensive areas versus low-carbon areas? In Section 5.3, the questions concern the regulatory activities of the Environmental Protection Agency (EPA) and the Security and Exchange Commission (SEC): In the aftermath of the abandonment of cap-and-trade legislation in the Senate in 2010, what has been happening to EPA rules – and legal actions to oppose them? What has the SEC been doing about climate change issues?

The chapter builds on the analysis of business interests and lobbying activities in Chapter 2 by examining their role in the legislative process for specific bills in the House and Senate. It also compares congressional attitudes and actions with public perceptions of the problem and preferences for solutions in Chapter 3. It furthermore applies the patterns of economic and political geography in Chapter 4 to the issues of whether and how to develop a national mandatory cap-and-trade system.

As for the title of the chapter, I have opted for the relatively cumbersome but descriptively accurate expression "regulating carbon dioxide and other greenhouse gases" rather than the commonly used short-hand expressions referring to "pricing carbon" or "regulating carbon." Short-hand references only to "carbon," though, inevitably enter into the discussion in quotations, as for instance in the initial chapter header quote above.

5.1 Congressional action and inaction, 1988–2008

Climate change on the Senate agenda, 1988–1992

More than two decades of periodic attention to climate change in the Senate and occasionally in the House began in the summer of 1988, with hearings before the Senate Committee on Energy and Natural Resources.[1] At that time, James Hansen, Director of the Institute for Space Studies of the National Aeronautics and Space Administration (NASA), reported to the committee that global warming was mostly attributable to human activities that released carbon dioxide and other greenhouse gases. As the hearings happened to be held during an unusually hot summer in the USA involving heat-related deaths in Chicago, the testimony received considerable attention in the Congress and elsewhere. Within the Congress, the reactions were, to a significant degree, already following partisan lines.[2] However, there were no moves at that time towards putting a price on greenhouse gas emissions, either through the establishment of a cap-and-trade system or the imposition of taxes. After a brief lapse, climate change re-entered the formal Senate agenda after the George H.W. Bush administration signed the UN Framework Convention on Climate Change at an international conference in Rio de Janeiro in 1992 (see Chapter 7).

1 Also, in 1988, an international conference in Toronto on climate science attracted some congressional interest. Of course, it had been the subject of much concern among climatologists, including scientists in the United States government, since the 1950s (see the book's Introduction for a chronology).

2 Among the Democrats on the committee expressing concern about the problem were Senators Wirth and Gore, while Republican Inhofe was highly critical of Hansen's analysis. More than two decades later, Wirth and Gore were still actively involved in climate change issues, though no longer as senators, while Inhofe was still a prominent climate change denier in the Congress.

Table 5.1 House of Representatives vote in 1993 on the Omnibus Budget Reconciliation Act (including a BTU tax provision)

	Democrats[a]	Republicans	Totals
Supported	219 (86%)	0 (0%)	219 (51%)
Opposed	38 (14%)	175 (100%)	213 (49%)
Number voting	257	175	432

[a] Includes Sanders, independent of Vermont.
Source: Compiled by the author from US House of Representatives (1993).

The BTU tax in the House, 1993

It was in 1993 that a formal proposal to price carbon entered the legislative process, when the Clinton administration proposed a BTU tax based on the heat content of fossil fuels.[3] In the economic and political context of the time, it was proposed as a way to increase government revenues and thus reduce the budget deficit, which was a priority of the administration. It was also presented as a tax on pollution (i.e. a "bad") because it would increase energy efficiency, as opposed to an income tax on labor or wealth (i.e. "goods"); as a result, it had some business support as well as widespread support among environmental organizations (Erlandson 1994). Because a tax on energy tends to be regressive, there were provisions to reduce its impact on low-income consumers. As it progressed through the House of Representatives, opposition to the bill increased among agricultural interests and some industries, where increased energy costs were of particular concern; it nevertheless passed the House by a narrow majority with a strongly partisan division (see Table 5.1).

In the Senate, as lobbying activity against the proposal became more intense from the American Petroleum Industry, the National Association of Manufacturers, and other organized business lobbies, it was opposed by senators from fossil-fuel-intensive states such as Texas, Louisiana, and West Virginia. It was abandoned in the Senate without a formal floor vote, and its defeat became a symbol of the difficulty of addressing climate change issues through energy taxes or any other kind of tax.[4]

3 A BTU (British Thermal Unit) is a unit of energy indicating the amount of heat needed to raise the temperature of a pound of water by 1 degree Fahrenheit. It is commonly used in English-speaking non-metric countries such as the USA and some English-speaking metric countries such as Canada. The use of a term of foreign origin was not lost on the critics of the tax when it was being debated in Congress.

4 The widespread assumption about political obstacles to taxes, especially in regard to public opinion, may be overstated. There has been majority public support for a sales tax on gas-guzzler

Cap-and-trade in the Senate, 2003, 2005, and 2008

The best known "early" proposal for a mandatory national cap-and-trade system was introduced in the Congress by Senators McCain and Lieberman in 2003. It drew upon many years of experience with cap-and-trade arrangements for mitigating air pollution causing acid rain (see Chapter 8). Known as the Climate Stewardship Act (S. 139), the McCain–Lieberman proposal in 2003 would have established a nation-wide system covering all greenhouse gases and most sectors of the economy. Despite its defeat, the formal introduction of this proposal into the congressional legislative process and the vote on it in the full Senate were politically significant for several reasons: It had bipartisan sponsorship in the two senators who introduced it – Republican McCain of Arizona, and then-Democrat but later-Independent Democrat Lieberman of Connecticut. Among the ten co-sponsors, furthermore, there were two Republicans – both of them from New England states, i.e. Chafee of Rhode Island and Snowe of Maine.[5] The outcome of the vote (43–55) symbolically superseded the unanimous vote in 1997 on the Byrd–Hagel amendment (see Chapter 7); as a more recent vote, the 2003 vote on the McCain–Lieberman bill became the more indicative and politically relevant vote in the Senate, as observers and participants assessed the changing domestic political context of climate change issues. However, the distribution of the vote was highly partisan, with three-fourths of Democrats voting in favor and nearly nine-tenths of Republicans voting against (see Table 5.2, Panel A).

The 2003 consideration of a bill to establish a national cap-and-trade system made clear the wide-ranging design issues that need to be addressed when developing such a system (see Box 5.1).

Such issues were again at the center of attention in 2005 when a new Congress was being organized following the 2004 elections and proposals to create a mandatory national cap-and-trade program received support from the non-governmental, but politically prominent and bipartisan, National Commission on Energy Policy (2004).[6]

automobiles, and in any case there is more support for any tax that is made revenue-neutral by an offsetting reduction in a different tax, as compared with one that is not revenue-neutral. See Chapter 3 for details.

5 As a senator from Connecticut, Lieberman was also from a New England state – one with significant financial services interests in the form of residents who commute to Wall Street or work in banking and brokerage offices in Connecticut.

6 One of its co-chairs was Republican Boyden Gray, White House Chief of Staff in the George H.W. Bush administration and subsequently US Ambassador to the European Union in the George W. Bush administration. The Commission was disbanded in December 2010, in the aftermath of the defeat of cap-and-trade legislation in the Senate.

Table 5.2 Senate votes on cap-and-trade legislation

	Democrats	Republicans	Totals
A. Vote on McCain–Lieberman, Climate Stewardship Act, 2003			
Supported	37 (79%)[a]	6 (12%)	43 (44%)
Opposed	10 (21%)	45 (88%)	55 (56%)
Number voting	47	51	98

[a] Senator Jeffords, independent of Vermont, is included with the Democrats.
Source: Compiled by the author from data in US Senate (2003).

	Democrats	Republicans	Totals
B. Vote on McCain–Lieberman Amendment, Energy Policy Act, 2005			
Supported	32 (74%)[a]	6 (11%)	38 (39%)
Opposed	11 (26%)	49 (89%)	60 (61%)
Number voting	43	55	98

[a] Senator Jeffords, independent of Vermont, is included with the Democrats.
Source: Compiled by the author from data in US Senate (2005).

	Democrats	Republicans	Totals
C. Vote on Boxer Manager's Amendment to Lieberman–Warner, Climate Security Act, 2008			
Supported	44 (92%)[a]	4 (11%)	48 (57%)
Opposed	4 (8%)	32 (89%)	36 (43%)
Number voting	48	36	84

[a] Senator Sanders, independent of Vermont, is included with the Democrats.
Source: Compiled by the author from data in US Senate (2008).

Its recommendations for a cap-and-trade system included a series of specific features: It endorsed the imposition of mandatory emission limits, with incrementally decreasing limits expressed in macro-economic intensity terms, but with a relatively low "safety valve" price on emission allowances. The Commission also supported government funding for carbon capture and sequestration projects – in particular, $3 billion over ten years "to demonstrate commercial-scale carbon capture and geologic sequestration at various sites," and $4 billion over ten years for "integrated gasification combined cycle (GCC) coal technology and for carbon capture and sequestration."

It furthermore recommended a wide range of energy-efficiency and renewable-energy policies that would directly affect climate change. The Commission's report was thus significant, in part because it demonstrated the extent to which climate change issues had become enmeshed in tangible and specific ways with a wide range of

Box 5.1 Cap-and-trade system design issues[a]

Coverage

- What sectors (industries) should be included – electric power, transportation services, manufacturing (chemicals, aluminum, etc)?
- What greenhouse gas(es) should be covered – carbon dioxide, methane, others?
- What entity emission levels should be established as thresholds for inclusion in the system?
- What percentages of total national emissions are included in the system?

Emission reduction targets

- What should be the amounts of the targets?
- What years should be used as target years?
- What base years should be used for computing the targets?

Allocation of credits

- Should credits be allocated by auction? Or by free allocation? Or a combination?

Inter-temporal issues

- Should there be recognition for "early emissions reductions" – i.e. either those undertaken voluntarily before the advent of the system and/or those taken under mandatory state or regional cap-and-trade systems?
- Should banking of unused allowances for future use be possible?
- Should borrowing of allowances in the present against expected future allocations be possible?

Price limits

- Should there be price floors below which allowance prices cannot go?
- Should there be price ceilings above which allowance prices cannot go?

Revenues

- Should revenues from sales of allowances be designated for specific uses – e.g. climate-friendly R&D programs, government budget deficit reduction?
- Should revenues be offset by tax reductions?

International

- Should purchases of offset credits in non-governmental or United Nations programs involving projects such as reforestation be credited? Which programs?

- Should border adjustment measures be used to offset relatively low prices on imports from countries without comparable measures to reduce greenhouse gas emissions? Which imports, from which countries?

Administration

- What should be the date of entry into operation?
- What verification procedures should be in place to ensure compliance?
- What penalties should there be for non-compliance?

[a] Not an exhaustive list

Sources: Compiled by the author from US Congressional Budget Office (2001; 2008); and US Energy Information Agency (2003). On transparency issues in carbon markets, see Brewer and Mehling (2014).

energy policy issues, including especially energy R&D and subsidy programs. For instance, the Commission's report – and Senators McCain and Lieberman – endorsed increasing support for the nuclear power industry, and there were provisions to that effect in the revised bill that was submitted to the Senate in 2005. (See Chapter 6 for further information on energy programs.)

As a consequence of this blending of climate change and energy issues, the Senate vote in 2005 on a revised version of the McCain–Lieberman bill occurred in the context of a debate in the Senate on the significant and extensive Energy Policy Act of 2005. Despite changes in content and context between 2003 and 2005, however, the distribution of votes was similarly partisan: three-fourths of Democrats in favor, nearly nine-tenths of Republicans against (see Panel B of Table 5.2). Further disaggregation of the vote reveals more about the interactions of region and industry with party differences, as discussed in Box 5.2.

Another change in 2005 marked an expansion of the congressional dialog on climate change issues to include *adapting to its impacts*; adaptation became more salient, including among politicians from areas where there have been disproportionate numbers of climate deniers. For instance, Senator Stevens of Alaska commented about climate change that "it is clearly not the actions of mankind that have brought about the [climate] changes that we are seeing." Yet, he also acknowledged the *impact* of climate change, which was already evident in Alaska, when he said, "We have to devise some way to deal with Alaska's problems" from rising sea levels and increasing storm

Box 5.2 Disaggregation of a Senate vote: the interplay of party, region, and industry patterns

In order to consider whether industry location also had an effect in the Senate vote on McCain–Lieberman, the data can be disaggregated by examining the "deviating" votes within each party – that is, the Democrats who voted against the bill and the Republicans who voted in favor of it. Of the ten "deviating" Democrats, four were from the highly fossil-fuel-dependent states. Of the six Republicans who deviated from their party's majority, four were from New England states. Yet, there were enough exceptions among these small groups to indicate that regional economic interests and ideological differences were among several factors operative in the votes.

The votes can also be further disaggregated by examining the states where the votes were split – that is, where one senator from a state voted in favor and the other voted against the proposed cap-and-trade system. In those ten states where there was a split, 80 percent of the Democrats voted in favor and 80 percent of the Republicans voted against. Thus, party affiliation was clearly a stronger factor than any state-level industry interest represented by the two senators from each state.

In four of the ten "split" state delegations, senators of the same party voted on opposite sides. In two of those states, a Democratic senator voted contrary to the position of the state's dominant industry; Rockefeller in West Virginia (a coal state) and Stabler in Michigan (an auto state) voted for the proposal. The cross pressures on individual senators from party affiliation and local industry interests can be strong and lead to ambivalence or switching votes. For instance, Democratic Senator Landrieu of Louisiana, an oil and gas state, voted against the McCain–Lieberman bill in 2003. However, she indicated that she might change her mind in 2004 when the possibility of another vote arose (E&E Daily 2004). Thus, while she was swayed more by state industry interests in 2003, she indicated she might be swayed more by party affiliation subsequently. Further, Senators Rockefeller and Stabler, noted above, Democrats from respectively West Virginia and Michigan, changed their inclinations by 2010, when the Kerry–Lieberman cap-and-trade bill was under consideration.

damage; and he supported federal government subsidies to relocate Alaskan fishing villages that were vulnerable to coastal erosion (E&E Daily 2004).

The Senate considered cap-and-trade legislation again in 2008 – this time with Republican Senator John Warner of Virginia being a co-sponsor with Senator

Lieberman. (Senator McCain opted out of a Senate leadership role on climate change legislation in 2008, a presidential election year, when he was the Republican candidate.) There were some differences between the 2005 and 2008 bills, including in particular more subsidies for nuclear power. However, this and other changes did not garner any additional Republican votes, and there were still not enough votes for cloture, though 57 percent of those voting did vote for cloture (see Table 5.2, Panel C).

5.2 Cap-and-trade bills in the 111th Congress, 2009–2010

House

In 2009, during the first session of the 111th Congress, the House took the initiative on climate change in the form of the American Clean Energy and Security Act (H.R. 2454) introduced by Representatives Waxman of California and Markey of Massachusetts. Title III, which would have established a cap-and-trade system, was the most contentious portion of the bill and attracted the most attention; other titles were sweeping in their inclusion of a wide range of energy issues. In fact, much of the bill was directly about energy efficiency and renewable energy, and these were important issues in negotiations among members of the House, as well as representatives of business, labor, consumer, environmental, and other organizations. The energy provisions, furthermore, made it possible to gain the support of business sectors that were otherwise opposed to the establishment of a cap-and-trade system. The complexity and scope of the bill are reflected in Appendix 5.1, which lists its titles and subtitles.

The bill addressed both market failures in the form of the negative externalities associated with greenhouse gas emissions and market failures in the form of the positive externalities associated with technological innovation and diffusion. The two were explicitly linked, for instance, by providing for the use of revenues from allowance auctions in the cap-and-trade system to fund energy technology R&D programs, such as electric vehicles in the auto industry and carbon capture and sequestration in the coal industry. Allowance auction revenues would also have been used to facilitate economic adaptation and transition programs, such as job training for workers in directly affected industries and assistance for electricity consumers.

As the bill moved through the committee process and associated negotiations among key members of the House from both parties, the chairs of two committees were particularly important – Representative Peterson, Chair of the Agriculture Committee, and Representative Dingell, Chair of the Energy and Commerce Committee. Although Representative Peterson decided not to hold Agriculture Committee hearings on the bill, he did insist on several significant changes to accommodate the interests of the agricultural sector and rural areas, including the specification of the Department of Agriculture rather than the Environmental Protection Agency as the agency to administer the conservation program of subsidies to farmers for greenhouse gas sequestration, on the assumption that the Department of Agriculture would be more responsive to the desires of the agricultural industry. Dingell, as representative from the state of Michigan, had a long history of active involvement in legislation affecting the auto industry, including air pollution issues. He was thus in a key institutional position, represented an important industry with big stakes in the issue, and he was also among the most knowledgeable members of the House about the substantive issues involved – in short, he was in a strong position to influence the content of the bill. Dingell's committee held extensive hearings, as it had done previously when Dingell himself had proposed climate change legislation, and it made numerous changes in the original proposal by Waxman and Markey – changes that favored the auto industry. Issues about the elements and design of a cap-and-trade system are discussed further in Chapter 8, which considers policy options for future actions.

In addition to auto and agricultural interests that affected the content of the House bill, the interests of the electric power industry, the coal industry, the oil and gas industry, the chemical industry, and other industries also came into play as the bill progressed to a final vote. Examples of changes made to accommodate the interests of a particular industry and congressional district were subsequently discussed by Representative Boucher of Virginia, following his defeat in the November 2010 election after twenty-eight years in the House.[7] He discussed his role in negotiating provisions that were supportive of the coal industry, which is especially important in Virginia, in exchange for his support for the bill (Peck 2010). He noted that he "had to spend the better part of two months in an intense series of negotiations to modify the original draft . . . so as to make it acceptable for the coal-fired utilities" (Peck 2010). He was able to achieve several specific changes that catered to the coal and electric power industries:

7 He attributed his defeat mostly to a wide range of *other* issues, but acknowledged that his support for the climate change bill was a factor in his defeat, despite the favorable provisions for the coal industry in his district that he successfully negotiated into the bill.

- Free allocation instead of auctioning of allowances for firms' GHG emissions.
- The right of firms to purchase offsets up to 2 billion tons of CO_2e per year – a way for firms to gain credit for GHG emission reductions without actually reducing their own.
- About $10 billion in government subsidies over ten years for development and deployment of carbon capture and sequestration (CCS) facilities, which would be especially important for reducing CO_2 emissions from coal-fired electric power plants.
- Up to $150 billion in free allowances for utilities that deployed CCS technologies.

Beyond Boucher's congressional district in Virginia, these items were instrumental in gaining support for the bill from some of the largest (and most coal-intensive) electric utilities in the country, as well as the Edison Electric Institute (EEI), which represents the interests of most of the country's large electric utilities. The EEI was in a difficult position; for its corporate members include not only several large coal-dependent utilities (such as American Electric Power, Duke Energy, and Southern Company), but also utilities (such as Exelon) that are nuclear-intensive. It took some time for the EEI to resolve the conflict of interests among its members in this regard; but it played an important role in the Senate consideration of cap-and-trade in 2010, as it had in the House in 2009.

When the Waxman–Markey bill was passed by the House on June 26, 2009, its provisions for the cap-and-trade system included the following:

- Economy-wide emission reductions, compared with 2005, increasing from 3 percent by 2021, to 17 percent by 2020, to 42 percent by 2030, to 83 percent by 2050.
- Coverage of 85 percent of emissions, including electricity producers, oil refineries, and other energy-intensive industries such as steel, cement, and paper.
- An increasing proportion of auctioned allowances, starting with about 15 percent (and thus 85 percent free allocations), with revenues from auctions distributed to energy consumers, government energy R&D programs, energy efficiency subsidies, and other uses.
- A minimum auction price of $10 initially in 2012, followed by annual increases of 5 percent plus the inflation rate. A strategic reserve of allowances for additional auctioning if allowance prices were greater than 160 percent of the three-year average.
- Provisions allowing the banking of allowances (i.e. saving allowances for future use) and also borrowing allowances (i.e. drawing on future expected allowances).
- Fines of two times the market value of the shortfall in allowances.

Table 5.3 House vote on Waxman–Markey, American Clean Energy and Security Act, 2009

A. Party distributions

	Democrats	Republicans	Totals
Supported	211 (83%)	8 (5%)	219 (51%)
Opposed	44 (17%)	168 (95%)	212 (49%)
Number voting	255	176	431

Source: Compiled by the author from US House of Representatives (2009a).

B. Regional distributions

	West[a]	Midwest	South	Northeast	Totals
Supported	56 (58%)	47 (47%)	50 (33%)	66 (80%)	219 (51%)
Opposed	40 (42%)	53 (53%)	102 (67%)	17 (20%)	212 (49%)
Number voting	96	100	152	83	431

[a] 62 percent of the California representatives' votes were in favor; 67 percent of the votes of the three West Coast states (California, Oregon, Washington) were in favor.
Source: Compiled by the author from data in *Washington Post* (2009); based on US House of Representatives (2009a).

- Up to 2 billion tons of offsets (domestic and international).
- Designation of the Federal Energy Regulatory Commission and the Commodity Futures Trading Commission as the implementing regulatory agencies.
- Pre-emption of existing state and regional cap-and-trade systems, with exchanges of their allowances for national system allowances.

The vote on the floor of the House was 219 in favor of the bill and 212 against; 83 percent of the Democrats voted in favor, while 95 percent of the Republicans voted against (see Table 5.3). As noted above, there had been a remarkably similar distribution of the votes in the House sixteen years earlier in 1993, when the overall division was 219 in favor and 213 against the proposal to establish a BTU tax, with 95 percent of the Democrats in favor and 100 percent of the Republicans opposed.

Not only party, but also regional, factors played an important role in the vote: four-fifths of the votes cast by representatives from Northeastern states and nearly three-fifths of those from Western states were in favor of establishing a cap-and-trade system, while slightly less than half from the Midwest and only one-third from the South were so inclined.

These partisan and regional variations have been exacerbated by the redistricting of the House seats following the 2010 census. As happens in many states after a census, there is a "gerrymandering" of some district lines such that the party in control of a state legislature can draw the lines so that its own supporters gain increased representation from being distributed widely enough while supporters of the opposition party are concentrated geographically so that they have fewer representatives of their party. That process of distorting the pattern of representation has resulted in the over-representation of Republican voters and under-representation of Democratic voters. Thus, in the 113th Congress beginning in January 2013 the number of Republican members was about 7 percent greater than the number of Democratic members, though in the nation-wide popular vote Democratic candidates received about 1 percent more votes than did Republican candidates. There is therefore a net distortion of 8 percent in the difference between the distribution of popular votes by individual voters, on the one hand, and the distribution of seats and thus votes in the House of Representatives, on the other hand. This partisan distortion affects climate change because of the persistent party differences in attitudes towards climate change issues among the members of the House.

Senate

Following passage of the Waxman–Markey bill in the House in 2009, congressional action shifted to the American Power Act in the Senate in 2010. It was initially proposed by Senators Kerry of Massachusetts and again Lieberman of Connecticut; but as part of an effort to gain sufficient Republican votes to achieve the 60 votes needed for cloture, Republican Senator Graham of South Carolina was added as a prospective co-sponsor of the bill during negotiations before it was to be formally introduced.[8] A preliminary unofficial head count (E&E Daily 2010d) in early 2010 indicated that about forty Democrats were likely to vote for a cap-and-trade bill along the lines being developed by Kerry and Lieberman, while about thirty-three Republicans were likely to vote against it. Among the remaining twenty-seven "fence sitters," there were seventeen

8 Senator Graham subsequently withdrew his support because of differences with the Senate Democratic leaders about priorities and procedures in the legislative process on immigration policy reform issues; the leader gave immigration reform priority, while Senator Graham wanted climate change and associated energy issues ahead of immigration reform on the Senate agenda. Except for thus being connected in the politics of gaining support on key Senate votes, the issues were otherwise unrelated; it is thus an important instance of how issues about otherwise unrelated issues can derail climate legislation.

Democrats and ten Republicans, whose support might be forthcoming. The challenge for the supporters in the Senate and for the administration was to find a combination of Democrats and Republicans to achieve sixty votes for cloture.

A letter to Kerry, Graham, and Lieberman sent by ten Democrats outlined their concerns.[9] Because of the importance of the issues and the influence of the senators (mostly from auto and coal states),[10] it is useful to note their reservations in detail:

> It is essential that any clean energy legislation include a package of provisions that strengthens American manufacturing competitiveness, creates new opportunities for clean energy jobs, and defends against the threat of carbon leakage by maintaining a level playing field for domestic manufacturers. Key provisions needed for a manufacturing package include:
>
> 1. Invest in American Manufacturing Competitiveness:
>   Provide Assistance for Retooling and Clean Energy Manufacturing....
>   Support Research, Development, and Deployment of Low-Carbon Industrial Technologies....
> 2. Support American Manufacturers of Clean Energy Technologies....
>   Level the Playing Field and Prevent Carbon Leakage....
>   Keep Energy Costs Low for Manufacturers....
>   Phase-In [coverage of] Manufacturers to Allow for Planning and Transition....
>   Provide Full Funding for Rebates to Energy-Intensive, Trade-Exposed Industries....
>   Apply Border Measures To Prevent Carbon Leakage....
> 3. Ensure a Long-Term Future for American Manufacturing:
>   Clarify Federal and State Greenhouse Gas Standards....
>   Promote Meaningful International Action....
>   Provide Community Economic Adjustment Assistance and Worker Training....

Such concerns were partly responsible for the proliferation of alternative bills; in addition to the Kerry–Lieberman bill, there were many other bills with overlapping coverage of climate change and energy issues. Collectively, they revealed the intensive and extensive attention that the issues had been receiving in the Senate for many years. Some of the bills had bipartisan sponsorship, some were proposed by Democrats, and

9 This followed a similar expression of concerns the previous year, when sixteen Democratic senators had publicly registered their reservations about the Lieberman–Warner cap-and-trade bill (Stabenow et al. 2009).

10 Sometimes referred to as the "gang of ten," the senators and their states were: Bayh (Indiana), Brown (Ohio), Byrd (West Virginia), Specter (Pennsylvania), Casey (Pennsylvania), Hagan (North Carolina), Levin (Michigan), McCaskill (Missouri), Stabenow (Michigan).

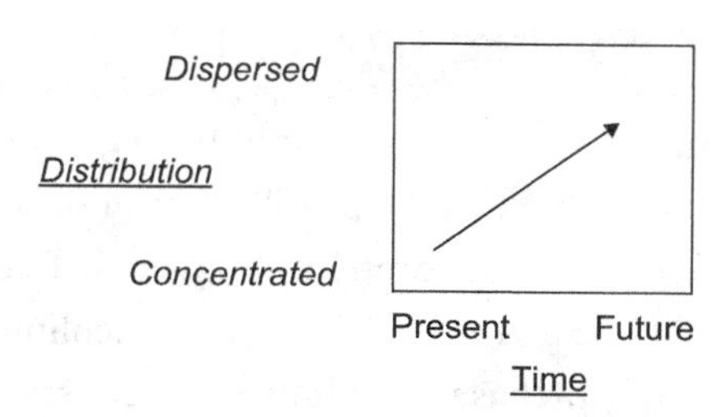

Figure 5.1 Redistributing the costs of climate change legislation: dispersing them and shifting them into the future

some were proposed by Republicans. Their similarities and differences are evident in the summaries of the bills presented in Appendix 5.2.

Like the Waxman–Markey bill that had been passed in the House, this panoply of Senate bills posed a wide range of issues of direct significance to the economic interests of nearly all sectors of the American economy. There was therefore a significant uptick in business lobbying, as well as lobbying by unions, consumer organizations, environmental organizations, and others (see Chapter 2). Although umbrella business organizations, such as the Chamber of Commerce and the National Association of Manufacturers, as well as industry-specific organizations such as the American Petroleum Institute, continued to oppose cap-and-trade proposals, there were new divisions in the ranks of the business organizations that actively lobbied. The divisions were evident in the formation – and eventual split within – the US Climate Action Partnership (USCAP), a coalition of environmental groups and corporations formed specifically to support cap-and-trade legislation. There were also splits within some industries, as we noted above in the role of the Edison Electric Institute in the passage of the Waxman–Markey bill in the House. There was also active lobbying by energy-efficiency and renewable-energy firms – for instance, by ACORE (the American Council for Renewable Energy). Some individual firms also actively supported cap-and-trade legislation, first in the House and then in the Senate, among them GE and American Electric Power (AEP). The diversity of the organizations and their positions are evident in Appendix 5.3, where there is an extensive list of the individual organizations and firms that were actively involved.

In conceptual terms, the intention of such lobbying activities was of course to redistribute the costs and benefits of any legislation – in particular to shift them along both dimensions of Figure 5.1 – that is, the degree of the concentration–dispersion of the costs and the timing of the costs along the present–future continuum. The extent of

Box 5.3 International leakage and competitiveness issues

One of the central issues in the development of the legislation and negotiations in both the House and the Senate was the possible use of offsetting border measures (OBMs) to address international competitiveness and free-rider concerns (Van Asselt and Brewer 2010; Van Asselt, Brewer, and Mehling 2009). OBMs would require US importers to purchase greenhouse gas emission allowances in order for imported goods to be allowed to enter the USA from countries deemed by the US government not to be making "satisfactory" efforts to mitigate greenhouse gas emissions. The requirement for such allowance purchases would be an alternative to offsetting border measures in the form of tariffs. Some firms feared that their international competitive position would be undermined by lower energy prices in non-participating countries. In the USA, these issues have become salient in regard to emerging economy countries, especially Brazil, China, and India. Studies suggested, however, that only a few industries exhibit the combination of energy-intensity and trade-intensity that would make them vulnerable to the problem; the relevant industries include steel, chemicals, and cement (Grubb, Brewer, Sato, Heilmayr, and Fazekas 2010). There have also been concerns that such a tariff might be challenged in a WTO dispute settlement case, and the outcome of such a case would inevitably be uncertain. In any case, such issues became important in the manufacturing sector, especially in the Midwest. The salience of these issues is abundantly clear and explicit in the letter from ten senators excerpted at length above. Whatever the evidence may have been, manufacturing firms in general feared – or at least claimed to have feared – detrimental effects on their international competitive position that could not be overcome.

these redistributive efforts was especially evident in 2009–2010, first in the House and then in the Senate.

Yet, it is nevertheless true that some individual industries more clearly have economic interests at stake as compared with other industries. This fact was apparent, for instance, in analyses of the international competitiveness impacts on a few industries, if a cap-and-trade system were to be put in place. As a result there were calls for the use of "offsetting border measures" and/or other measures to address the concerns of some industries. See Box 5.3 for more discussion of these issues.

It is important not to misinterpret the tangible consequences of some key provisions for affected industries. In particular, as Stavins (2010a) points out, the distributions of the tangible implications of the arrangements depended on the identities of the "ultimate beneficiaries" of the allocations of allowances – whether by auctions or free distribution. As a result, the actual implications for industries can be quite different from first appearances. For instance, Stavins calculated that the majority (about 82 percent) of the benefits of the distribution of free allowances and revenues from auctions from the provisions of the Kerry–Lieberman bill would have accrued to consumers, taxpayers, and government programs; industry would only have received about 18 percent of the benefits (see Table 5.4 for the details).[11]

In any case, there was not sufficient support in the Senate in 2010 to gain the sixty-vote super-majority vote needed for cloture – at least in the judgment of Senator Reid, the Democratic Leader, and the White House. The proposal was thus abandoned without a formal vote, despite the prospects of a near or even bare simple majority of votes in favor of passage and substantial support for it within the electric power industry.

Senate voting rules are skewed against the least GHG-intensive states and in favor of the most GHG-intensive states because of the cloture rule in combination with the built-in over-representation of the least populous states and under-representation of the most populous states. Many of the most GHG-intensive states are relatively sparsely populated, while many of the least GHG-intensive states are relatively large and densely populated. These patterns of economic geography are related to two fundamental institutional features of the Senate, thus causing distortion in representation patterns, compared with a body that would provide equal representation based on population. First is the structural feature that each state, regardless of population, has two senators. Second is the procedural requirement for a super-majority of sixty votes (60 percent) in order to close debate and have a vote on the substance of proposed legislation. The net result is an under-representation of states with low per-capita GHG emission levels and an over-representation of states with high per-capita GHG emission levels.

Figure 5.2 provides evidence of the extent of this phenomenon. In the figure, the fifty states have been arranged into five groups of ten states in each group – i.e. "quintiles" – according to the relative levels of their GHG emissions per capita. Thus, the vertical bar on the left represents the ten states with the lowest levels of per-capita emissions, while the vertical bar on the right represents the ten states with the highest levels of per-capita emissions. The vertical left scale is the proportion of the total national population of each group of ten states. The population of the ten states with the

11 Still, the matter of the *relative* sizes of allocations for the customers, as well as the firms themselves among competitors, within an industry is of enormous interest. For a firm's competitive position within an industry is affected by the relative amounts of allowances that it and its customers receive.

Table 5.4 Potential ultimate beneficiaries of allowances and auctions in Kerry–Lieberman bill

Nature and recipients of benefits	Percentage of total
Assistance to consumers and taxpayers, of which:	*57.7*
Direct assistance, of which:	34.0
Energy and tax refunds for low income households	11.7
Allowances auctioned for universal tax refunds	22.3
Indirect assistance to mitigate impact on energy consumers, of which:	23.7
Electricity local distribution companies	18.6
Natural gas local distribution companies	4.1
State programs for home heating oil, propane, and kerosene customers	0.9
Rural energy savings (consumer loans to implement energy efficiency measures)	0.1
Government programs, of which:	*24.9*
State renewable and energy efficiency programs	0.6
State and local agency programs to reduce emissions through transport projects	1.9
Grants for national surface transportation system	1.9
Auctioned allowances for Highway Trust Fund	1.9
Domestic adaptation	1.0
International adaptation	1.0
Allowances auctioned for deficit reduction	7.4
Remaining allowances auctioned to offset bill's impact on deficit	6.1
Auction from cost containment reserve	3.1
Industry, of which:	*17.6*
Allocations to covered entities, of which:	12.3
Energy-intensive, trade-exposed industries	7.0
Petroleum refiners	2.2
Merchant coal-fired electricity generators	2.2
Generators under long-term contracts without cost recovery	0.9
Technology funding, of which:	5.1
Carbon capture and sequestration incentives	3.8
Clean energy technology R&D	0.7
Low-carbon manufacturing R&D	0.3
Clean vehicle technology incentives	0.3
Other domestic priorities, of which:	0.2
Manufacturing plant energy efficiency retrofits	0.1
Compensation for early-action emissions reductions prior to cap	0.1

Source: Adapted and computed by the author from Stavins (2010a). Used with permission.

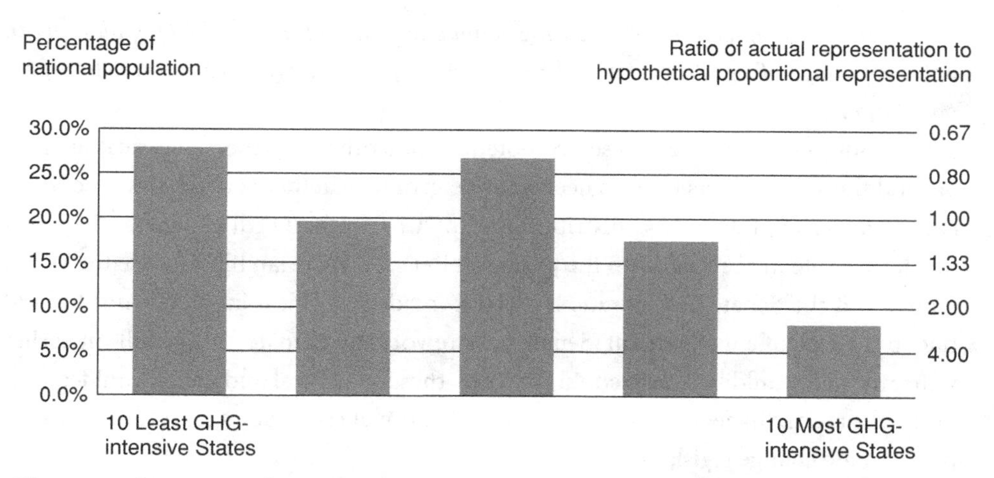

Figure 5.2 Distortions of representation in the Senate according to the greenhouse gas intensity of states' economies

lowest per-capita emissions is 86.1 million, or 27.9 percent of the total population. The population of the ten states with the highest per-capita emissions is only 25.0 million, or 8.1 percent of the total population.

However, because each state is entitled to two senators regardless of population, each group of ten states actually has 20 percent of the representation in the Senate. The 8.1 percent of the population in the highest GHG-intensity states has 20 percent of the representation in the Senate, while the 27.9 percent of the population in the lowest GHG-intensive states also has 20 percent of the representation in the Senate. The ratio of the over-representation of the former to the under-representation of the latter is 27.9/8.1, or 3.4. The same basic pattern also obtains for the second and fourth quintiles, though not for the third in the middle. The ratios of actual representation to the hypothetical proportional representation (i.e. equal representation for each citizen) are displayed in the right vertical scale, indicating that *the least GHG-intensive states have only about three-fourths of the representation that a proportional system would yield while the most GHG-intensive states have about one and a half times the share of the votes that a proportional system would yield.*

Further, the cloture rule is such that forty-one senators (41 percent of the total) can kill a bill that is supported by a simple majority by preventing it from coming to a vote. In principle, in the case of GHG intensity of the states, it may be necessary to get sixty senators from low-GHG-intensity states representing 74.3 percent of the population to outvote forty senators from high-GHG-intensity states representing 25.7 percent of the population. *Hence, senators representing about one-fourth of the national population can prevent passage of legislation favored by senators representing about three-fourths of the population.*

It should be noted that these are potential outcomes. Further empirical analysis beyond the scope of this book is needed to determine whether actual votes on climate change issues approach these distributions. In 2010, in the 111th Congress, there was no cloture vote in the Senate on the proposed Kerry–Lieberman bill because the White House and the Senate Democratic leadership made a political judgment not to hold one in the face of a likely defeat. Senate cloture votes on climate change bills in earlier years, though, could be analyzed. In any case, these structural and procedural features of the Senate are likely to pose significant institutional constraints on future efforts to pass climate change legislation.

Role of the President

The demise of the Kerry–Lieberman bill and thus the collapse of congressional efforts to establish a national cap-and-trade system in 2010 prompted much comment on the President's role in the legislative process on climate change issues.

During the 2008 presidential election campaign, candidate Obama had explicitly indicated his support for a cap-and-trade system. He thus entered the presidency on record for supporting a system that would cover nearly the entire economy and include 100 percent auctioning of allowances. In addition, key members of his administration had already been strong supporters of action to address climate change. In fact, Vice President Biden, when he was a freshman in the Senate in 1988, had sponsored the Senate's first climate legislation. Both Obama and Biden, as well as Secretary of State Clinton, had voted in favor of the McCain–Lieberman bill while they were in the Senate. Further, many of Obama's top advisors were strong supporters of action on climate change – including the White House Science Advisor, John Holdren; Secretary of Energy, Steven Chu; Administrator of the Environmental Protection Agency, Lisa Jackson; and Director of the White House Office of Energy and Climate Change Policy, Carol Browner (commonly known as the "climate czar").

During the period of 2009–2010, when the climate change bills were pending in the House and Senate, there was much controversy about the part played by the president and others in the administration in trying to get the bills passed. The controversy focused on two questions: Was the President involved enough in the congressional policymaking process? Did he and/or his advisers – including especially those who were directly involved in congressional relations and more generally legislative strategy – make tactical mistakes at key points? Among supporters of the legislation, there was considerable sentiment that the answers were, respectively, "no" and "yes." Among other close observers of congressional-presidential relations, there was disagreement.

President Obama faced a domestic political dilemma when cap-and-trade legislation was introduced in the Congress in 2009, the first year of his presidency. If such legislation passed with his support, there was a fear that the Democrats in Congress who would be up for re-election in the fall of 2010 might lose votes over the issue – perhaps enough to lose some seats and thus conceivably lose party control in the Senate and/or House of Representatives; this was a particular concern for Democrats from auto, coal, and oil and gas states in the Midwest and South. On the other hand, if the President had pressed such Democrats for their support, made deals with reluctant Republicans and then lost the vote in Congress, he would have been perceived as ineffective as a leader in domestic politics, and he would have lost status and influence within the context of legislative-executive relations.

Analyses of the possible voting outcomes in the Senate in 2010 (e.g. ClimateWire 2009) indicated that about fifty Democrats – including Independent Democrat Senator Lieberman, who had long been a supporter of cap-and-trade legislation – would vote in favor of cloture of debate and for the legislation; however, there were not enough votes to achieve the sixty votes needed for a cloture vote.[12] There were enough Democrats from oil and gas, coal, and auto states to prevent cloture without some Republican support. As a result of these developments and calculations, it was clear from the beginning that the President needed at least a small group of Republican senators to vote in favor – perhaps as few as three or four but more likely as many as seven or eight. Depending on how many Republican votes he could get, he probably also needed to win some more Democratic votes – perhaps two to five. It was, as was often said by many observers at the time, "an uphill fight" to get to the sixty votes for cloture (see especially Lizza 2010 for details of the legislative process and politics).

In an attempt to gain a few Republican votes, Obama agreed to support (a) increased subsidies for the nuclear power industry and (b) increased offshore drilling for the oil industry.[13] The latter, of course, subsequently put him in an awkward situation because of the Gulf oil spill in 2010. Further, the administration missed a window of

12 The number of Democratic senators was reduced by one to fifty-nine with Senator Kennedy's death and the election of a Republican in a special state election to replace him. Subsequently, the death of Senator Byrd of West Virginia may also have made the prospects for passage worse, when his Democratic successor announced that he would vote against. (Byrd had seemed to be wavering from his long-time opposition to action on climate change, though there was no evidence that he would actually vote in favor of the cap-and-trade bill.)

13 In conjunction with their efforts to gain a few Republican senators' support, the administration was also seeking industry support – or at least less active industry opposition. The US Chamber

opportunity – in the view of many activists on climate change – to use the impact of the oil spill as a "teaching moment" about the risks and costs of inattention to the environmental impacts of continued reliance on oil and the apparent need for more careful regulation of the practice.[14]

In addition, the administration and/or Senate Democratic leadership made three tactical judgments in the legislative process that were criticized by many climate action activists: First, when the House passed the Waxman–Markey bill in June 2009, there could have been a push in the Senate for action at that time, especially since several committees in the Senate had held many hearings on the issue over nearly a decade and because there had been floor votes and committee votes on the issue in the recent past. Moreover, in 2009, there were climate change bills and climate-related energy bills already pending in the Senate or otherwise in the process of being developed. In short, climate change was on the Senate agenda in 2009, as it had been for several years. However, the administration did not want to jeopardize health care reform – which they considered a more immediately pressing issue.

The second tactical issue was the suggestion by Senate Democratic Leader Harry Reid of Nevada of the possibility that immigration reform would be taken up in the Senate before the impending election.[15] This angered Republican Senator Graham, who understood there was an agreement not to do that; and his support was essential to gaining the few Republicans needed to support the cap-and-trade legislation if it were to have a chance of passage. In fact, Graham was a prospective co-sponsor with Senators Kerry and Lieberman of a major climate change bill that had been developed over many months in extensive negotiations with industry and environmental representatives, and he had become an outspoken advocate of action on climate change.

Finally, as on some other issues, there were questions about whether the president had been sufficiently involved in the negotiations on the climate change bill in the Senate; perhaps if he had been willing to expend more political capital, the Kerry–Lieberman bill could have passed the Senate (E&E Daily 2010a; 2010b; 2010c; Luce and Harvey 2010; Revkin 2010).

Whatever opportunities may have been missed to pass climate change legislation during the 111th Congress, the prospects in the 112th Congress were notably less propitious.

of Commerce and the National Association of Manufacturers were especially problematic in this regard.

14 The President did make at least one speech in which he linked the oil spill to climate change and energy issues (Obama 2010).

15 Senator Reid was engaged in a close re-election campaign in Nevada at the time, and there was much opposition to cap-and-trade legislation in his state.

5.3 The new domestic political context starting with the 112th Congress

As a result of the 2010 congressional elections, new incumbents in key House leadership positions in the 112th Congress were hostile to action on climate change issues; indeed, some House leaders were prominent deniers or skeptics of climate science. They included the Speaker of the House, the Chair of the Committee on Energy and Commerce, and the Ranking Majority member of the same committee, as well as the chair of the fifty-seven-member Tea Party Caucus (see Box 5.4).

Early in the 112th Congress they and others in the House supported a series of amendments concerning climate change that were offered in the context of debate over a Continuing Resolution for the fiscal year 2011 budget.[16] The following amendments were passed by the House (Sugarman 2011) – but subsequently rejected by the Senate – as the legislation for the FY2011 budget moved through the congressional budgeting process:

- A motion to prevent the Environmental Protection Agency from undertaking any regulatory actions concerning greenhouse gas emissions – Amendment 466 to the Continuing Resolution (H.AMDT 101).
- A motion to prevent the government from paying the salaries and other expenses for the offices of the Assistant to the President for Energy and Climate Change in the White House or the Special Envoy for Climate Change in the State Department – Amendment 204 to the Continuing Resolution (H.AMDT 89).
- A motion to prevent the creation of a Climate Service in the National Oceanic and Atmospheric Administration (NOAA) – Amendment 495 to the Continuing Resolution (H.AMDT 148).

The new partisan situation and increased hostility toward action on climate change in the House in 2011, following defeat of cap-and-trade legislation in the Senate in 2010, of course made significant direct action on climate change mitigation legislation by Congress virtually impossible. However, the administration was able to continue with efforts to change attitudes by reframing the issues and offering alternative visions of the future.

16 The fiscal year 2011 budget cycle represented an extreme case of delay in congressional decision-making: the FY2011 budget was not finally passed until April 2011, more than six months late and hence more than halfway into the fiscal year, which started the previous September 1. (See Appendix 6.1 for details of the budgeting process.)

Box 5.4 Quotes concerning climate change from leaders of the House of Representatives in the 112th Congress

Speaker of the House, John Boehner: "... the idea that carbon dioxide is a carcinogen that is harmful to our environment is almost comical. Every time we exhale, we exhale carbon dioxide. Every cow in the world, you know, when they do what they do, you've got more carbon dioxide."[a] (ABC News 2009)

Chairman of the Committee on Energy and Commerce, Fred Upton: "Upton has in the past called climate change 'a serious problem' – a phrase he deleted from his website [in 2010] after declaring his intent to run for committee chairman.... He was subsequently asked during an interview after becoming committee chair, 'You believe that the climate is changing but you're not convinced human activity is causing the change, is that your position?' Upton replied 'It is.'" (*National Journal* 2011)

Ranking Majority Member of the Committee on Energy and Commerce, Joe Barton: "There is no dispute that average world temperatures have risen over the past 100 years. While a precise measurement is difficult to pinpoint, most scientists believe the temperature increase to be within the two to three degrees Fahrenheit range. The documented increase in temperature since the 1800s, and the projected increase over the next hundred years, appears well within the range of natural variation. In light of research conducted by numerous scientists on both sides of the debate, the theory that human actions are responsible for changes in global temperatures is a serious one, and worthy of continued research. I have, however, not been convinced the theory is strong enough to warrant the immediate and draconian measures called for by some segments of the environmental community." (US House of Representatives, Office of Congressman Joe Barton 2011)

Chair of the Tea Party Caucus,[b] Michelle Bachmann: "... because of this underlying bill [H.R. 2454, commonly known as the Waxman–Markey cap-and-trade bill], the federal government will virtually have control over every aspect of lives for the American people. It is time to stand up and say: We get to choose. We choose liberty, or we choose tyranny – it's one of the two." (US House of Representatives 2009b)

"I want people in Minnesota armed and dangerous on this issue of the energy tax [i.e. the cap-and-trade bill, H.R. 5424] because we need to fight back. Thomas Jefferson told us 'having a revolution every now and then is a good thing,' and the people – we the people – are going to have to fight back hard if we're not going to lose our country. And I think this has the potential of changing the dynamic of freedom forever in the United States." (University of Minnesota, Center for the Study of Politics and Governance 2009).

[a] There are two mistakes about greenhouse gases in this statement. If you cannot recognize them, you might want to go back to Table 1.2 and Appendix 1.3 in Chapter 1.

[b] The Tea Party Caucus was established as a "caucus" in the House of Representatives in the 112th Congress. It is a particularly significant one because of its influence within the Republican party's decision-making processes, including within the House Republican caucus. Caucuses are not legally prescribed elements of the House structure as established in the Constitution (as the position of Speaker is) or established by official House action every two years at the beginning of each new Congress (as the legislative committees are).

5.4 Regulatory agencies: Environmental Protection Agency (EPA) and Securities and Exchange Commission (SEC)

With the change in administrations in January 2009, both the Environmental Protection Agency and the Security and Exchange Commission undertook new initiatives concerning their respective regulatory responsibilities. The EPA has been focused on the emissions of firms' production processes and products, while the SEC has been focused on the impacts of climate change on firms' strategic and competitive positions – both positive and negative – and the firms' responsibilities under SEC regulations to disclose those impacts to investors.

EPA

It is important to note several features of EPA regulations concerning greenhouse gas emissions: First, they have been promulgated on the basis of legislation passed by Congress such as the Clean Air Act; Congress can subsequently change the legislation, either to prescribe more or less regulation. Congress can also affect the implementation of the regulations through the budget process, as was attempted early in the 112th Congress, particularly in the House of Representatives. Second, EPA regulations are subject to legal challenges in the courts – to push for either more or less regulation. As we saw in Chapter 4, states are often parties to these cases, sometimes with different groups of states on opposite sides. Although the court cases are sometimes resolved relatively quickly at the Federal District Court or Appeals Court levels, they can go to the Supreme Court and take years to resolve. The court cases involve a wide range – from the application of regulations to particular sites or to entire industry sectors, or even the constitutionality or existence of the regulations. Third, changes in administrations can lead to substantial changes in regulatory emphases – as for instance in the change from the George W. Bush administration's tendency to minimize EPA involvement in issues concerning GHG emissions to the Obama administration's more aggressive use of the EPA in addressing climate change. In short, the EPA is constrained by the institutional context in which it develops and administers regulations – an institutional context that includes all three branches of the national government, as well as the states in a federal political system.

EPA regulations concerning GHG emissions have not involved cap-and-trade arrangements; rather they require a mix of specific emissions limits, the use of "best available technology," and other direct controls. The regulations entail several different sets of regulations, which can be highlighted as follows (US Environmental Protection Agency 2011a; 2012a; 2012b; 2012c):

- A mandatory GHG reporting program was established as a result of the Consolidated Appropriations Act passed by the 110th Congress in the context of the fiscal year 2008 budgeting process.
- In December 2009, the EPA issued its final action concerning the endangerment to public health and welfare represented by greenhouse gases – an action that followed from the US Supreme Court case of *Massachusetts v. EPA* in 2007 (see Chapter 4 for details of the case).
- A "tailoring rule" issued in May 2010 limits the application of the Clean Air Act (CAA) permitting process for new and existing industrial facilities to large GHG emitters such as electric power plants, refineries, and cement plants – which collectively account for about 70 percent of US GHG emissions from "stationary sources." The tailoring rule exempts small farms, restaurants, and other small commercial facilities from the CAA permitting requirements.
- A few months later in December 2010 the EPA proposed a schedule of implementing GHG standards under the CAA for fossil-fuel electric power plants and petroleum refineries – facilities that account for about 40 percent of total US GHG emissions.
- In March 2012 the EPA announced a new regulation limiting (only) new electric power plants to 1000 pounds of carbon dioxide emissions per megawatt of electricity

produced – thus in effect requiring new coal-fired power plants to include carbon capture and storage capabilities.

- Also in 2012 the EPA issued a regulation concerning greenhouse gas emissions for hydraulic fracturing natural gas wells, whereby the producers must capture methane in particular (a GHG that can be resold).
- Another regulation that was finalized in 2012 mandated an average for cars and light trucks of 54.5 miles per gallon by 2025 – an increase from 29 miles per gallon in 2012 and a requirement of 35.5 by 2016.
- A series of regulations are directed specifically at motor vehicles. One set concerns the Renewable Fuels Standard (RFS) program mandated by the Energy Policy Act of 2005 and the Energy Independence and Security Act of 2007, a program that establishes minimum renewable fuel content requirements for transportation fuels. Another set – formulated with the National Highway Transportation Safety Administration – concerns fuel efficiency and GHG emissions standards for automobiles and "light-duty" vehicles. Yet another set of regulations concern fuel efficiency and GHG emission standards for medium-duty and heavy-duty vehicles, including pickup trucks, as well as large trucks and buses. Finally, a state-specific EPA decision in June 2009 allowed California an exemption from the federal Clean Air Act so that the state could impose GHG emission standards on motor vehicles on the basis of a previously passed state law. This represented a reversal by the Obama administration of the position taken by the George W. Bush administration, which had opposed the California law.

The totality of these regulations – some of which are at various stages of litigation in federal courts – could have a significant impact on the aggregate level of US GHG emissions, particularly through reductions in the emissions of electric power plants and motor vehicles. It has been estimated by the World Resources Institute (2010e) that strong implementation of all of the existing federal regulations across several sectors of the economy could result in reductions, compared with business-as-usual projections, equivalent to 22 percent in 2020. The actual impacts on emissions will of course depend on the diverse decisions of presidents, court judges, and members of Congress.[17]

17 In an unusually explicit and formalized denial of the occurrence of global warming, the House Committee on Energy and Commerce voted 31 to 20 *against* the following proposed amendment to the Energy Tax Prevention Act (H.R. 910): "Congress accepts the scientific finding of the Environmental Protection Agency that 'Warming of the climate system is unequivocal, as is now evident from observations of increases in global average air and ocean temperatures, widespread melting of snow and ice, and rising global average sea level.'" All thirty-one of the Republicans

SEC

Though regulatory developments in the Securities and Exchange Commission have not entailed the drama or attracted the same level of attention as the developments in the EPA, one SEC development nevertheless could lead to important changes in the regulatory environment for firms – and perhaps be a harbinger of even more significant changes in the future. In particular, the SEC has developed "staff interpretive guidelines" to clarify the implications of climate change for firms' reporting requirements under existing SEC rules (US Securities and Exchange Commission 2010; and Smith, Morreale, and Drexler 2010). It explicitly notes the potential for positive as well as negative impacts on individual firms; i.e. there are business opportunities as well as business risks. In order to capture both the generality of its potential scope as firms apply it to their reporting practices and at the same time its precision as administrative law, excerpts from it are reproduced in Box 5.5.

5.5 Implications: interests, ideologies, and influence in Congress

Congressional deliberations on climate change have been marked by conflicts along a combination of party, region, and industry lines, with strong persistent partisan differences appearing in each critical vote. In the face of opposition, supporters of the legislation tried to gain support by shifting costs into the future through a variety of provisions and by the redistribution of costs more thinly and broadly, as compared with concentrating them more intensively in a few industries.

Industry interests have of course also played an important role. The members of Congress from coal, oil, and auto states were able to gain significant industry-specific concessions in the House and prevent passage in the Senate. In short, traditional fossil-fuel-dependent industries and representatives in Congress from greenhouse-emissions-intensive states in the Midwest and South have been able to prevent the creation of a national cap-and-trade system, despite recorded majority support in the House of Representatives in 2009, a likely majority in the Senate if a vote had been taken in 2010.

Key structural and procedural features of the Senate have over-represented carbon-intensive states and under-represented low-carbon states. Going forward, in the House

on the committee voted against the amendment, and all twenty of the Democrats who voted – of twenty-three on the committee – voted in favor of it (US House of Representatives, Committee on Energy and Commerce 2011).

Box 5.5 Excerpts from Securities and Exchange Commission interpretive release on firms' disclosure of information concerning climate change

IV. Climate Change-related Disclosures

... The following topics are some of the ways climate change may trigger disclosure required by [SEC] rules and regulations.... These topics are examples of climate change-related issues that a registrant may need to consider.

A. Impact of Legislation and Regulation

... [T]here have been significant developments in federal and state legislation and regulation regarding climate change. These developments may trigger disclosure obligations under Commission rules and regulations, such as pursuant to Items 101, 103, 503(c) and 303 of Regulation S–K. With respect to existing federal, state and local provisions which relate to greenhouse gas emissions, Item 101 requires disclosure of any material estimated capital expenditures for environmental control facilities for the remainder of a registrant's current fiscal year and its succeeding fiscal year.... In addition to the Regulation S–K items discussed in this section, registrants must also consider any financial statement implications of climate change issues in accordance with applicable accounting standards, including Financial Accounting Standards Board ("FASB") Accounting Standards Codification Topic 450, Contingencies, and FASB Accounting Standards Codification Topic 275, Risks and Uncertainties....

Item 303 requires registrants to assess whether any enacted climate change legislation or regulation is reasonably likely to have a material effect on the registrant's financial condition or results of operation....

We reiterate that climate change regulation is a rapidly developing area. Registrants need to regularly assess their potential disclosure obligations given new developments.

B. International Accord

Registrants also should consider, and disclose when material, the impact on their business of treaties or international accords relating to climate change....

C. Indirect Consequences of Regulation or Business Trends

Legal, technological, political and scientific developments regarding climate change may create new opportunities or risks for registrants.

These developments may create demand for new products or services, or decrease demand for existing products or services. For example, possible indirect consequences or opportunities may include:

- Decreased demand for goods that produce significant greenhouse gas emissions;
- Increased demand for goods that result in lower emissions than competing products; . . .
- Increased competition to develop innovative new products;
- Increased demand for generation and transmission of energy from alternative energy sources; and
- Decreased demand for services related to carbon based energy sources, such as drilling services or equipment maintenance services. . . .

D. Physical Impacts of Climate Change

Significant physical effects of climate change, such as effects on the severity of weather (for example, floods or hurricanes), sea levels, the arability of farmland, and water availability. . . .

Possible consequences of severe weather could include:

- For registrants with operations concentrated on coastlines, property damage and disruptions to operations, including manufacturing operations or the transport of manufactured products;
- Indirect financial and operational impacts from disruptions to the operations of major customers or suppliers from severe weather, such as hurricanes or floods;
- Increased insurance claims and liabilities for insurance and reinsurance companies; . . .
- Decreased agricultural production capacity in areas affected by drought or other weather-related changes; and
- Increased insurance premiums and deductibles, or a decrease in the availability of coverage, for registrants with plants or operations in areas subject to severe weather. . . .

Source: Excerpted by the author from US Securities and Exchange Commission (2010).

of Representatives, beginning with the 113th Congress in 2013, the distortion of representation resulting from the gerrymandering of districts on the basis of the 2010 census has led to an over-representation of Republican-held seats and an under-representation of Democratic-held seats, compared with the national popular vote; the former held 7 percent more seats in the House while the latter won 1 percent more popular votes nation-wide in the election for the House seats.

As a result of the impasse in Congress, the Environmental Protection Agency has been developing more extensive regulations and the Securities and Exchange Commission has taken small steps in that direction. These developments on the part of federal-level regulatory agencies – in combination with the court cases as well as state- and local-level programs discussed in Chapter 4 – have significantly broadened the number and diversity of venues where climate change is on the active agenda. Furthermore, as we shall see in the next chapter, the expansion of the active policymaking arenas has also included many energy issues and thus many programs of the Department of Energy.

Suggestions for further reading and research

For tracking climate change issues in the Congress and in executive agencies, several electronic newsletters are essential reading: Climate Change News at www.eesi.org, Climate Wire at www.eenews.net, and InsideEPA.com at www.environmentalnewsstand.com.

The politics of cap-and-trade bills in the Congress are examined in detail in Falke (2011), Lizza (2010), and Pooley (2010). There is a detailed analysis by Richards and Richards (2009) of the evolution of bills in the Senate. The following are illustrative of sources of information and assessments of the Waxman–Markey bill at key stages of the legislative process: An analysis of an informal discussion draft released on March 31, 2009, is available from the Pew Center on Global Climate Change (2009b), which has been renamed Center for Climate and Energy Solutions with a website at www.c2es.org. A copy of the bill as introduced is available at US Library of Congress (2009a). An analysis of the formal bill after action by the House Committee on Energy and Commerce is available from the US Congressional Research Service (2009). A summary of it, as passed by the House, is available from World Resources Institute (2009). Comparisons of different versions as the bill progressed through the legislative process, and analyses of changes that were made to it, are available at Natural Resources Defense Council (2009).

For more on the economics of cap-and-trade in general and specific proposals in the Congress, see Stavins (2010a; and more generally www.belfercenter.ksg.harvard.edu); also see reports by the US Congressional Budget Office (2001) and the US Congressional Research Service (1999a; 1999b; 2000a; 2000b; 2000c; 2000d). Betsill (2009) analyzes the possibility of a NAFTA cap-and-trade system. Grubb, Brewer, Sato, Heilmayr, and Fazekas (2010) suggest lessons that could be applied in the design of a cap-and-trade system in the USA on the basis of the experience with the EU ETS. Ellerman (2008) and Ellerman, Convery, and de Perthuis (2010) discuss the EU Emission Trading Scheme; also see numerous items on the websites www.ceps.eu and www.climatestrategies.org.

International competitiveness and leakage issues – as well as other climate–trade issues – are addressed in Biermann and Brohm (2005), Brewer (2003; 2004a; 2007; 2010), Gros, Egenhofer, Fujiwara, Georgiev, and Guerin (2010), Grubb and Neuhoff (2006), Hufbauer, Charnovitz and Kim (2009), Van Asselt and Brewer (2010), Van Asselt, Brewer, and Mehling (2009), and Tamiotti, Teh, Kulacoglu, Olhoff, Simmons, and Abasa (2009). Also see the website www.TradeAndClimate.net.

APPENDIX 5.1

Titles and subtitles of the Waxman–Markey bill, as passed[a]

The bill was over 900 pages long when it was introduced and over 1400 pages when it was passed; Title III concerning the cap-and-trade system was 410 pages long. The following list of the titles and subtitles therefore only conveys a bare outline of its contents.

TITLE I – CLEAN ENERGY
Subtitle A – Combined Efficiency and Renewable Electricity Standard
Subtitle B – Carbon Capture and Sequestration
Subtitle C – Clean Transportation
Subtitle D – State Energy and Environment Development Accounts
Subtitle E – Smart Grid Advancement
Subtitle F – Transmission Planning
Subtitle G – Technical Corrections to Energy Laws
Subtitle H – Energy and Efficiency Centers and Research
Subtitle I – Nuclear and Advanced Technologies
Subtitle J – Miscellaneous

TITLE II – ENERGY EFFICIENCY
Subtitle A – Building Energy Efficiency Programs
Subtitle B – Lighting and Appliance Energy Efficiency Programs
Subtitle C – Transportation Efficiency
Subtitle D – Industrial Energy Efficiency Programs
Subtitle E – Improvements in Energy Savings Performance Contracting
Subtitle F – Public Institutions
Subtitle G – Miscellaneous

TITLE III – REDUCING GLOBAL WARMING POLLUTION [amendments to the Clean Air Act]
Subtitle A – Reducing Global Warming Pollution

[Clean Air Act] TITLE VII – GLOBAL WARMING POLLUTION REDUCTION PROGRAM
Part A – Global Warming Pollution Reduction Goals and Targets
Part B – Designation and Registration Of Greenhouse Gases
Part C – Program Rules
Part D – Offsets
Part E – Supplemental Emissions Reductions From Reduced Deforestation
Subtitle B – Disposition of Allowances

PART H – DISPOSITION OF ALLOWANCES

[Clean Air Act] TITLE VIII – ADDITIONAL GREENHOUSE GAS STANDARDS
Part A – Stationary Source Standards
Part C – Exemptions From Other Programs
Part E – Black Carbon
Part F – Miscellaneous
Subtitle D – Carbon Market Assurance
Subtitle E – Additional Market Assurance

TITLE IV – TRANSITIONING TO A CLEAN ENERGY ECONOMY
Subtitle A – Ensuring Real Reductions in Industrial Emissions
Subtitle B – Green Jobs and Worker Transition
Subtitle C – Consumer Assistance
Subtitle D – Exporting Clean Technology
Subtitle E – Adapting to Climate Change

[a] This was the content of the bill as it was formally approved by the House on June 26, 2009. Title and subtitle names are taken verbatim from the bill.
Source: Condensed from the original bill by the author from US Library of Congress (2009b).

APPENDIX 5.2

Comparison of four climate change / energy bills in the Senate, 2010

- **S. 1462, the American Clean Energy Leadership Act (ACELA)** of 2009, was introduced by Senator Bingaman and reported by the Senate Committee on Energy and Natural Resources on July 16, 2009 (S.Rept. 111–48). S. 1462 [was] a broad energy bill aimed at promoting the development of clean energy technologies, increasing energy efficiency, and promoting domestic energy resources. Incentives for new technology include[d] a renewable energy standard (RES) for electric utilities. The bill [did] not directly address greenhouse gas emissions: provisions for a greenhouse gas cap-and-trade system were instead included in S. 1733, the Clean Energy Jobs and American Power Act, sponsored by Senators Kerry and Boxer, and reported by the Senate Committee on Environment and Public Works on February 2, 2010.
- **S. 2877, the Carbon Limits and Energy for America's Renewal (CLEAR) Act**, was introduced by Senators Cantwell and Collins on December 11, 2009 and [was] referred to the Senate Committee on Finance. S. 2877 would [have] establish[ed] a program to control only carbon dioxide (CO_2) emissions (covering 80% of US GHG emissions), requiring fossil-fuel producers (e.g., coal mines, gas wellheads) and importers to submit "carbon shares" for the CO_2 emissions related to the fossil fuels they produce or import. The President would [have] limit[ed] . . . the quantity of carbon shares available for submission each year, and the Department of Treasury would [have] distributed all of the carbon shares through monthly auctions.
- **S. 3464, the Practical Energy and Climate Plan Act** of 2010, was introduced by Senators Lugar, Graham, and Murkowski on June 9, 2010 and [was] referred to the Senate Committee on Finance. S. 3464 [was] a broad energy bill aimed at promoting the development of clean energy technologies, increasing energy efficiency, and promoting domestic energy resources. Instead of a renewable energy standard (RES) like that contained in S. 1462, S. 3464 contain[ed] a "Diverse Energy Standard" which would [have] permit[ted] the use of a broad range of electric generation technologies including renewables, but also including nuclear energy and advanced coal generation with carbon capture and storage. Other provisions include[d] building and vehicle

efficiency standards and nuclear energy loan guarantees. The bill [did] not contain a mandatory scheme to limit greenhouse gas emissions.

- **A discussion draft of the American Power Act** (APA) was released on May 12, 2010 by Senators Kerry and Lieberman. A comprehensive energy and climate change policy proposal, the draft would [have] set GHG reduction goals similar to those in H.R. 2454 (the bill most comparable to the APA draft), which passed the House in June 2009. The APA employ[ed] a market-based cap-and-trade scheme for electric generators and industry with a separate price mechanism to cover emissions from transportation fuels. The draft proposal would [have] allocate[d] a significant amount of allowance value to energy consumers, low-income households, and the promotion of low-carbon energy technologies. In addition, the draft would [have] provide[d] incentives for the expansion of nuclear power, carbon capture and storage technology, and advanced vehicles.

Source: Reprinted from US Congressional Research Service (2010a).

APPENDIX 5.3

Organizations and firms engaged in the legislative process concerning cap-and-trade bills in Congress, 2008–2010

N.B. This is not an exhaustive list; however, it is suggestive of the diversity of the organizations and interests represented. Not all of these organizations engaged in "lobbying" in a formal, legal sense – either directly or indirectly.

Group/Name	Position/ Source	*Group*/Name	Position/ Source
Energy and electric utility associations		*Health associations*	
America's Natural Gas Alliance	O/c	American Lung Association	S/a
American Petroleum Institute	O/c	American Public Health Association	S/a
American Wind Energy Association	S/a	Association of State and Territorial Health Officials	S/a
National Hydropower Association	S/d	National Association of County and City Health Officials	S/a
Energy and electricity utilities: firms		National Environmental Health Association	S/a
American Electric Power	S/g	*Labor organizations*	
Austin Energy	S/a	AFL-CIO	S/a
Avista	S/a	Blue Green Alliance	S/a
BP Solar	S/a	Building and Construction Trades Department, AFL-CIO	S/a
Con Edison	S/a	Communications Workers of America	S/a
Constellation Energy	S/a	Laborers International Union of North America	S/a
Covanta Energy Corporation	S/d	Service Employees International Union	S/a
Duke Energy Corporation	S/a	United Auto Workers	S/a
Edison Electric Institute	S/a	United Steelworkers	S/a

(cont.)

Group/Name	Position/ Source	*Group*/Name	Position/ Source
Exelon Corporation	S/a	Utility Workers Union of America	S/a
Exxon Mobil Corporation	O/h	*Environmental organizations*	
FPL Group	S/a	National Religious Partnership for the Environment	S/a
National Grid	S/a	American Rivers	S/a
NRG Energy	S/a	Association of Fish and Wildlife Agencies	S/a
Nuclear Energy Institute	S/a	Audubon	S/a
PG&E Corporation	S/a	Audubon Committee	S/a
PNM Resources	S/a	Center for Clean Air Policy	S/a
PSE&G	S/a	Ceres	S/a
Solar Power Industries	S/a	Clean Water Action	S/a
Agricultural associations		Coalition for Emission Reduction Projects	S/f
American Corn Growers Association	S/b	Defenders of Wildlife	S/a
American Farmland Trust	S/b	Earthjustice	S/a
Growth Energy	S/b	Environment America	S/a
National Association of Wheat Growers	S/b	Environmental Defense Action Fund	S/a
National Farmers Union	S/b	Izaak Walton League of America	S/a
Renewable Fuels Association	S/b	League of Conservation Voters	S/a
Chemical associations and firms		National Association of Clean Air Agencies	S/a
Dow Chemical Company	S/a	National Parks and Conservation Association	S/a
DuPont	S/a	National Wildlife Federation	S/a
Metal and mining associations and firms		Natural Resources Defence Council	S/a
Alcoa	S/a	Pew Center on Global Climate Change	S/a
American Iron and Steel Institute	O/e	Sierra Club	S/a
National Mining Association	O/c	Sustainable Forestry Management	S/a
Peabody Energy	O/c	The Nature Conservancy	S/a
Rio Tinto	S/a	The Wilderness Society	S/a
Diversified industrials		Theodore Roosevelt Conservation Partnership	S/a
General Electric	S/a	Trout Unlimited	S/a
Siemens	S/a	Union of Concerned Scientists	S/a

Group/Name	Position/ Source	*Group*/Name	Position/ Source
Information technology firms		World Resources Institute	S/a
Applied Materials	S/a	World Wildlife Fund	S/a
Hewlett-Packard	S/a	*International development organizations*	
National Semiconductors	S/a	CARE	S/a
Sun Microsystems	S/a	Oceana	S/a
Symantec	S/a	Oxfam	S/a
Health associations		*Religious organizations*	
American Lung Association	S/a	American Jewish Committee	S/a
American Public Health Association	S/a	Baptist Center for Ethics	S/a
Association of State and Territorial Health Officials	S/a	Coalition on the Environment and Jewish Life	S/a
National Association of County and City Health Officials	S/a	Evangelical Climate Initiative	S/a
National Environmental Health Association	S/a	Evangelical Environmental Network	S/a
Financial firms		Evangelical Lutheran Church in America	S/a
Kleiner, Perkins, Caufield, Byers	S/a	National Council of Churches	S/a
Jones Lang LaSalle	S/a	Religious Action Center on Reform Judaism	S/a
Consumer goods and services		The United Methodist Church – General Board of Church and Society	S/a
Aspen Skiing Company	S/a	United States Conference of Catholic Bishops	S/a
Clif Bar	S/a	*Other organizations*	
eBay	S/a	United States Climate Action Partnership (USCAP)	S/i
Gap	S/a	League of Women Voters	S/a
Johnson & Johnson	S/a	National League of Cities	S/a
Johnson Controls	S/a	US Conference of Mayors	S/a
Levi Strauss	S/a	Vote Vets	S/a
Nike	S/a	Wider Opportunities for Women	S/a
Seventh Generation	S/a		
Starbucks	S/a		
The North Face	S/a		
Timberland	S/a		
Wal-Mart	S/f		

(cont.)

Position: O = Opposed; S = Supported

Sources: I am indebted to Maria Drabble for assistance in compiling this list from the following sources, which are indicated for the individual items in the list.

a: US House of Representatives, Select Committee on Energy Independence and Global Warming (2009a).

b: US House of Representatives, Select Committee on Energy Independence and Global Warming (2009b) [Supporting with Peterson Amendment]

c: Mulkern (2010).

d: Mulkern (2010).

e: Bravender (2010).

f: Marshall (2008).

g: Samuelsohn (2010).

h: Marshall (2010).

i: E&E News (2010).

CHAPTER 6

Facilitating energy technology innovation and diffusion

Here we have a serious problem: America is addicted to oil. . . . The best way to break this addiction is through technology.

President George W. Bush (2006)

We're faced with the following choices: We can become the leader of a new industrial revolution and lay the foundation of our future economic prosperity . . . or we can hope the price of oil will go back to $30 a barrel, deny climate change is happening and let other countries take the lead in energy innovation.

Secretary of Energy Stephen Chu (2009)

This is our generation's Sputnik moment.

President Barack Obama (2011)

This chapter ranges widely across a variety of questions about the political economy of energy technology issues: In Sections 6.1 and 6.2 basic questions about market failures and government policies are addressed: How do the market failures associated with technological innovation and diffusion differ from those associated with cap-and-trade? How do the benefits and costs of technology policy vary over time and among groups? What are the economic rationales for government policies that facilitate innovation and diffusion of climate-friendly technologies? How do government programs and policies address technology-related market failures? In Section 6.3: What kinds of energy technology subsidy programs are there, and how are they allocated among industries? Section 6.4: What are the elements of the government's Climate Technology Program, and how have its priorities changed with changes in the administration? Section 6.5 considers questions about budgeting for energy programs: How much is allocated to various programs? How have the absolute and relative amounts for renewable energy and energy efficiency programs varied over time? How is the budget decided? How do congressional "earmarks" affect the funding of energy programs? What were the energy program priorities in the American Recovery and Reinvestment Act (ARRA) that was promulgated in the midst of the financial crisis of 2008–2011?

In Section 6.6: What are the central issues about US policies concerning international technology transfers for climate change mitigation and/or adaptation? Finally, in Section 6.7 there is a brief statement about economic interests, governmental institutions, and ideologies about the proper roles of markets and governments in promoting technological progress – how all these factors interact and produce the outcomes that are observed.

The analytic approach of the chapter is both qualitative and quantitative: It is concerned with not only types of market failures and types of technologies and policies that have been adopted to address the failures, but also the sizes of government subsidies of particular industries and other government programs. Understanding budget numbers is thus essential to understanding many important and contentious key issues about the US response to climate change.

The chapter addresses the problem of "scale" which is essential to understanding the scope of the problem of climate change and solutions to it – that is the magnitudes of technological deployment and the amounts of financing of programs that are at issue. The magnitudes are sufficiently great that there are often references to the need for an energy technology revolution, as in the chapter heading quote from former Secretary of Energy Chu. There are also occasional references to the need for the government to ramp up a massive project akin to the Marshall Plan of assistance to Europe for reconstruction after World War II, or references to the program to put an American in space and then on the moon in response to the Soviet Union's launch of its Sputnik manned space craft in 1957, as in the chapter heading quote from President Obama.

All of these topics are difficult to present without occasional use of such terms as "clean technologies" or "climate-friendly technologies." However, they have of course become highly politicized terms, so that definitional questions have periodically surfaced in the Congress and the administration about what technologies should and should not be included in official notions of "clean technologies" or "climate-friendly technologies." The definitional issues are not merely semantic matters; they take on tangible financial significance in the context of legislation and administration policies, especially when there are differences among them (see Box 6.1). Of special interest and controversy has been whether to include nuclear power, natural gas, large hydro power projects, and/or coal-fired power plants with substantially reduced GHG emissions (sometimes referred to as "clean coal" technologies). I have tried to use these and other terms in neutral, descriptive ways with due regard for the context in which they are being discussed, without imposing my own cost-benefit calculations or policy preferences.

Box 6.1 What are "clean-energy technologies"?

House of Representative's definition

"The term 'clean energy technology' means a technology that – (A) produces energy from solar, wind, geothermal, biomass, tidal, wave, ocean, and other renewable energy resources (as such term is defined in section 610 of the Public Utility Regulatory Policies Act of 1978); (B) more efficiently transmits, distributes, or stores energy; (C) enhances energy efficiency for buildings and industry, including combined heat and power; (D) enables the development of a Smart Grid (as described in section 1301 of the Energy Independence and Security Act of 2007 7 (42 USC. 17381)), including integration of renewable energy resources and distributed generation, demand response, demand side management, and systems analysis; (E) produces an advanced or sustainable material with energy or energy efficiency applications; (F) enhances water security through improved water management, conservation, distribution, and end use applications; or (G) improves energy efficiency for transportation, including electric vehicles."

Source: US Library of Congress (2009b).

Senate's definition

"The term 'clean energy technology' means a technology related to the production, use, transmission, storage, control, or conservation of energy that will – (A) reduce the need for additional energy supplies by using existing energy supplies with greater efficiency or by transmitting, distributing, or transporting energy with greater effectiveness through the infrastructure of the United States; (B) diversify the sources of energy supply of the United States to strengthen energy security and to increase supplies with a favorable balance of environmental effects if the entire technology system is considered; or (C) contribute to a stabilization of atmospheric greenhouse gas concentrations through reduction, avoidance, or sequestration of energy-related emissions."

Source: US Senate (2009), 10.

Administration's definition

". . . the president's fiscal 2011 budget appears to soften what can be considered 'clean' under the Clean Technology Fund to potentially include natural gas. The language includes high-efficiency gas plants that displace coal generation and reduce global warming emissions at least 50 percent among those that can be considered for funding.... Last year, the House sent a clear mandate that clean technology money go only for 'zero carbon' and new renewable technologies. Lawmakers included a detailed list of what could be considered clean that specifically excluded coal, oil, natural gas, tar sands, oil shale and large hydro plants."

Source: Howell (2009).

6.1 Technologies: benefits and costs

Box 6.2 presents a well-known list of currently available or emerging technologies, which have been calibrated according to their potential to achieve significant reductions in greenhouse gas emissions. Individually and collectively, they could make notable reductions in global greenhouse gas emissions if they were deployed on a sufficient scale. In sum, the list conveys the important message that there are technological solutions which are technically feasible and which are or soon will be available.

While such a list is an important contribution to our understanding of the potential technological solutions to the problem, an economic analysis of their costs or relative cost-benefits is also needed. Figure 6.1 thus presents a well-known chart of the relative marginal GHG abatement costs of many technologies. It has been developed in several versions for the world as a whole and for individual countries; this one is for the United States in particular. Although it inevitably depends on a variety of assumptions and forecasts, the chart is a useful summary of basic patterns in the relative attractiveness in economic terms of a wide range of technologies. It is notable that many technological solutions involving energy efficiency actually entail economic savings for the users, as those at the left side of the chart indicate (for qualifications and contrary evidence to this assumption, see Allcott and Greenstone 2012).

Shale gas

Two technological developments in the oil and gas industry – hydraulic fracturing and horizontal drilling – have made recovery of shale gas economically feasible in many parts of the USA as well as other countries, and its increased exploitation has attracted much attention and controversy. Whether or not it is a "revolution" or a "game changer" or is "turning the energy world upside down" is a matter of semantics as well as

Box 6.2 Technological wedges: summary of list of comparably scaled technologies for climate change mitigation developed by Pacala and Socolow[a]

End-user efficiency and conservation:

(1) Increase fuel economy of automobiles

(2) Reduce automobile use by telecommuting, mass transit, urban design

(3) Reduce electricity use in homes, offices and stores

Power generation:

(4) Increase efficiency of coal-fired power plants

(5) Increase gas baseload power (reduce coal baseload power)

Carbon capture and storage (CCS):

(6) Install CCS at large, baseload coal-fired power plants

(7) Install CCS at coal-fired plants to produce hydrogen for vehicles

(8) Install CCS at coal-to-synfuels power plant

Alternative energy sources:

(9) Increase nuclear power

(10) Increase wind power

(11) Increase photovoltaic power

(12) Use wind to produce hydrogen for fuel cell cars

(13) Substitute biofuels for fossil fuel

Agriculture and forestry:

(14) Reduce deforestation, increase reforestation and afforestation, add plantations

(15) Increase tillage conservation in cropland

[a] The estimated quantities of each technology required for a "wedge" have been omitted in this summary in order to focus attention on the *types of technologies* rather than the amounts of them needed to meet a specific goal.

Sources: Compiled by the author from Socolow and Pacala (2006), 52; Pacala and Socolow (2004), 4.

facts about how its production, distribution, and consumption evolve over the next many years.

In the present context we are of course especially interested in the implications for climate change in the increasing production of shale gas in the USA and the prospect for its widespread production – and trade – in many countries in all regions of the world.

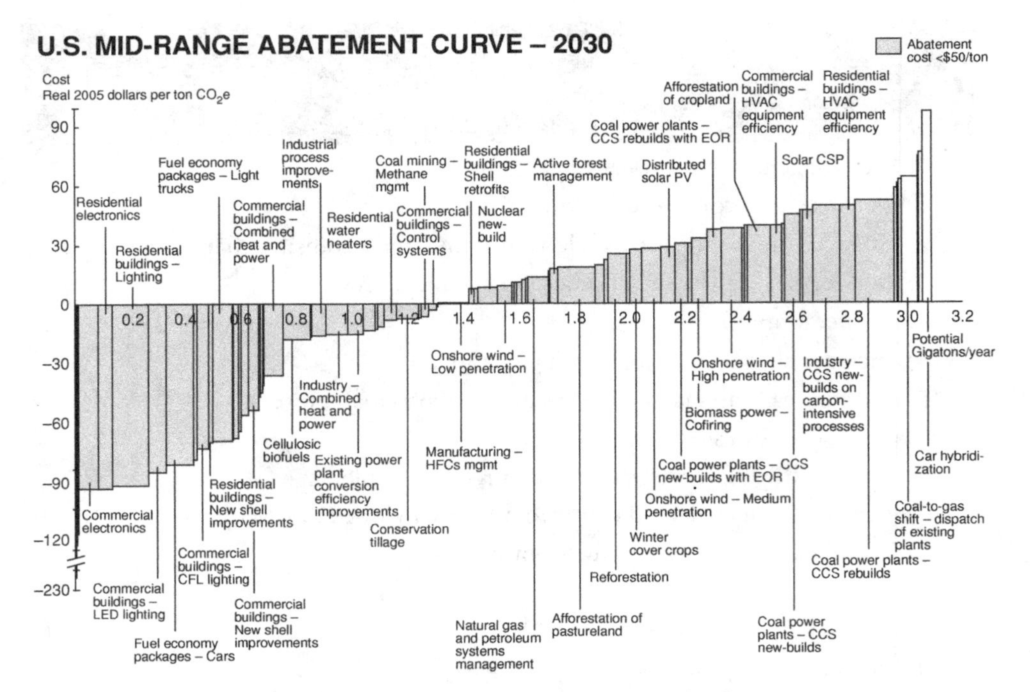

Figure 6.1 McKinsey marginal cost curve for GHG abatement technologies

Source: Exhibit from "Reducing US Greenhouse Gas Emissions: How Much at What Cost?" December 2007, McKinsey and Company, www.mckinsey.com. Reprinted with permission.

Because shale gas is cheap relative to other widely used energy sources and abundant in many parts of the world, it is likely to become an increasingly important source of energy. (For further information about shale gas, especially in relation to climate change, see Brewer (2013a; 2014); US Energy Information Administration (2011); Bipartisan Policy Center (2011); International Energy Agency (IEA 2011); Monitz et al. (2012); Schrag (2012); Wigley (2011).)

An attractive feature of shale gas from the standpoint of climate change mitigation is that, like natural gas in general, greenhouse gas emissions from shale gas *consumption* are much lower than those from coal, for instance in electric power plants (about 40–50% lower). However, there are serious issues about greenhouse gas emissions, in particular methane, during *exploration* and *extraction.* Methane has long been known to be a much more potent greenhouse gas than carbon dioxide; methane's Global Warming Potential (GWP) is more than twenty times greater than CO_2 at 100 years and more than seventy times greater at twenty years. Over the longer term, a key issue is whether cheap, abundant shale gas will undermine investment in renewable energy sources. Already in the USA there has been a significant decrease in the price of natural gas. With a consequent decline in electricity prices produced in natural-gas-fired power plants, the competitive position of wind, solar, and other renewable energy sources has been weakened; and the future shares of those technologies in the energy mix are thus also undermined. In some scenarios, therefore, while the substitution of shale gas for coal to produce electricity may yield a net reduction in greenhouse gas emissions in the short run, the increasing share of shale gas and concomitant smaller share of renewables may yield a net increase in emissions.

Government support for shale gas received a strong – but not unqualified – endorsement in a report on climate change policy of the US President's Council of Advisors on Science and Technology (2013). In particular, it mentioned the "fugitive" methane emission problem. To summarize a complicated and controversial set of issues, there is no doubt that such emissions have potent climate change effects. The questions are about whether, how much, and how such emissions will be prevented, and of course how much it will cost and who will pay. The answers to all of these questions depend on a combination of industry practices and government regulations – and public attitudes and actions.

The upsurge in shale gas production in the USA has also prompted renewed interest – and much conflict – over US natural gas international trade policies (see Box 6.3).

Geo-engineering

Beyond technologies that are currently commercially available and/or in advanced stages in the R&D process, there have been discussions of the potential deployment of new, untested technologies in the future, including large-scale, advanced geo-engineering possibilities. Discussions of these technologies have thus far been highly speculative and mostly limited to technical experts, as their effectiveness, costs, and potential side effects are only now beginning to be scrutinized. However, they are likely to receive increasing attention – including among political and business leaders as well as the public – if the sense of urgency about the threat of climate change becomes more intense and the desire to mitigate and adapt to it becomes more urgent. See Box 6.4 for illustrative geo-engineering possibilities.

The processes of innovation and diffusion of these and other technologies – current and future – are constrained by market failures; government policies to address them are therefore central to the climate change agenda.

Box 6.3 Shale gas as a US climate change – international trade issue

US government policy concerning international trade in natural gas offers an interesting example of how climate change policy is sometimes embedded in other policies. With the increases in the extraction of shale gas and decreases in prices noted in the body of the chapter, interest in the potential for US exports of natural gas has increased. In fact, the previously prevailing expectation that the USA would become an increasingly important importer of natural gas has been replaced by an expectation that it will become a significant exporter.

However, exports of natural gas from the USA require approval by the Department of Energy. This policy and its implementation have been under intense scrutiny. On the basis of the Natural Gas Act of 1938, as amended by the Energy Policy Act of 2005, the current policy allows natural gas exports to a select small group of countries and considers others on an *ad hoc* basis. A country must have a free trade agreement (FTA) with the USA in order to engage in natural gas trade with it on a reliable, ongoing basis, without the uncertainties of an ad hoc application process that is outside an FTA.

Only twenty countries have FTAs with the USA – including countries in NAFTA and the Caribbean Basin Initiative (CBI), plus a few countries with bilateral agreements such as those with Bahrain, Oman, and Korea. However, there are two countries – Costa Rica and Israel – of the twenty, whose FTAs do not include provisions concerning natural gas. Thus, as of early 2013 there were eighteen countries that qualified for more or less automatic natural gas trading rights. All other countries could only receive US exports if the US exporter obtained an export license from the Department of Energy – a process that was not certain to lead to success.

Of course, these trade policies were of interest in other countries – not only potential importers from the USA such as Japan, but also countries who were themselves actual or potential major exporters of natural gas as a result of having discovered and exploited their own abundant shale gas reserves. Among the countries that likely have relatively large shale gas reserves in addition to the USA are Argentina, Australia, China, Mexico, Russia, and South Africa.

What is at stake for climate change, therefore, is the widespread use in many countries of a relatively low-carbon alternative to coal – if the fugitive methane emissions problem is solved – and the diffusion of the associated exploration and extraction technologies, as well as the technologies to reduce methane emissions.

Source: Brewer (2013a; 2013b; 2014).

Box 6.4 Potential geo-engineering technologies

The following examples of geo-engineering possibilities for addressing climate change have been taken from a report of the US Government Accountability Office (2010), which drew extensively upon a report of the UK Royal Society (2009). Also see US National Academy of Sciences (2011).

A September 2009 study from the Royal Society – the United Kingdom's national academy of science – categorized most geo-engineering proposals into two approaches: solar radiation management (SRM), which would offset temperature increases by reflecting a small percentage of the sun's light back into space, thus reducing the amount of heat absorbed by the earth's atmosphere and surface, and carbon dioxide removal (CDR), which would address what scientists currently view as the root cause of climate change by removing carbon dioxide – a greenhouse gas – from the atmosphere.

Examples of SRM approaches in the study include the following:

- increasing the reflectivity of the earth's surface through activities such as painting building roofs white, planting more reflective crops or biomass, or covering desert surfaces with reflective material;
- increasing the reflectivity of the atmosphere by whitening clouds over the ocean or injecting reflective aerosol particles into the stratosphere to scatter sunlight; and
- space-based methods to use shielding materials to reflect or deflect incoming solar radiation.

Examples of CDR approaches in the study include the following:

- enhancing biological, physical, or chemical land-based carbon sinks to capture and store carbon in biomass or soil (carbon sequestration), or in chemically reactive minerals (land-based enhanced weathering);
- enhancing biological, physical, or chemical ocean-based carbon sinks through the introduction of nutrients to promote phytoplankton growth (ocean fertilization), physically altering ocean circulation patterns to transfer atmospheric carbon to the deep sea, or adding chemically reactive minerals to increase ocean alkalinity (ocean-based enhanced weathering); and
- technology-based methods to remove carbon dioxide from the atmosphere (air capture) and then store the carbon dioxide – for example, in geological formations (geological sequestration).

Source: US Government Accountability Office (2010), 2–3.

6.2 Market failures in technological innovation and diffusion

As with the innovation and diffusion of other types of technologies, the development of technologies to mitigate climate change is constrained by market failures; climate-related energy technologies are not unique in this regard. However, the market failures that constrain technological innovation and diffusion as solutions to the climate change problem are different from the market failures that contribute to the existence of the problem (Jaffe, Newell, and Stavins 2005). As for the role of market failures in creating the problem: "Pollution [in the form of greenhouse gases] creates a negative externality, and so the invisible hand allows too much of it." As for the role of market failures in constraining solutions: "Technology creates positive externalities, and so the invisible hand produces too little of it." Hence, the two sets of market failures are "compounding": the market failures that contribute to the existence of the problem consist of externalities because prices established in energy markets and other markets do not include the costs of emissions to people who are not parties to the market transactions; the market failures that constrain the innovation and diffusion of technologies that would mitigate the emissions pertain to externalities that would allow non-participants in market-based transactions to benefit from the adoption of the technologies.

The market failures at issue in the present chapter concerning the innovation and diffusion of climate-friendly technologies are not only different in kind, compared with those discussed in the previous chapter, they are also more numerous and more diverse. As explained further in Box 6.5, technological innovation and diffusion are constrained by a wide range of market deficiencies reflecting knowledge spillovers, learning-by-using, learning-by-doing, network externalities, investors' imperfect information, and the inadequacies of capital markets to fund socially valuable but financially risky projects, especially large-scale and long-term projects. This latter problem is sometimes referred to as the "valley of death" problem, whereby promising projects with socially important, long-term benefits that have progressed through research and development stages then "die" without progressing to the diffusion stage because they do not receive enough funding from private sources for them to be commercialized.

These market failures provide economic rationales for a wide range of government programs that are designed to overcome the barriers to technological innovation and diffusion (Brown 2001). Many of the programs involve direct government participation in research and development projects for individual technologies – often in partnerships with business; often in one of the government laboratories that specialize in energy efficiency or renewable energy technologies; often in government-funded research at

Box 6.5 **Market failures that constrain technological innovation and diffusion**

"[I]ndependent of the externality associated with pollution [such as greenhouse gases], innovation and diffusion are both characterized by externalities as well as other market failures. . . . "

Knowledge spillovers. "A successful innovator will capture some rewards, but those rewards will always be only a fraction – and sometimes a very small fraction – of the overall benefits to society of the innovation. Hence innovation creates positive externalities in the form of 'knowledge spillovers' for other firms, and spillovers of value or consumer surplus for the users of the new technology."

Adoption externalities due to dynamic increasing returns. The value accruing to one user depends on the number of other adopters. The benefit of a new technology is thus dependent on the scale of adoption. "Dynamic increasing returns can be generated by learning-by-using, learning-by-doing, or network externalities."

> *Learning-by-using* occurs when "the adopter of a new technology creates a positive externality for others in the form of the generation of information about the existence, characteristics, and success of the new technology."
>
> *Learning-by-doing* "describes how production costs tend to fall as manufacturers gain production experience. If this learning spills over to benefit other manufacturers without compensation it can represent an additional adoption externality."
>
> *Network externalities* "exist if a product becomes technologically more valuable to an individual user as other users adopt a compatible product (as with telephone and computer networks)."

Incomplete information. "Both innovation and diffusion of new technology are characterized by additional market failures related to incomplete information."

> "[T]he uncertainty associated with the returns to investment in innovation is often particularly large."
>
> "[I]nformation about the prospects for success of given technology research investments is asymmetric, in the sense that the developer of the technology is in a better position to assess its potential than [are] outsiders." Investors thus demand a premium for their exaggerated, uninformed view of the risks.

"[I]mperfect information can slow the diffusion of new technology."

"[I]nformation has important 'public goods' attributes: once created it can be used by many people at little or no additional cost."

"Incomplete information can also foster principal-agent problems, as when a builder or landlord chooses the level of investment in energy efficiency in a building, but the energy bills are paid by a later purchaser or tenant."

"These market failures with respect to adoption of new technology are part of the explanation for the apparent 'paradox' of underinvestment in energy-saving technologies that appear cost-effective but are not widely utilized."

Source: Excerpted from Jaffe, Newell, and Stavins (2005), 166–168. Used with permission.

universities or other organizations. There are also a variety of government policies, such as tax credits or other subsidies, which create incentives for firms and individuals to change their behavior in various stages of the process from innovation to development to diffusion of technologies; in simple economic terms, such policies address market failures by creating incentives for producers of goods and services and/or consumers of goods and services. For instance, there have been tax credits for producers of electricity from wind or solar power; in addition, there have been tax credits for consumers of hybrid electric automobiles.

In the US government, such policies are developed and implemented through four different, but overlapping and interacting, policymaking processes as far as climate change is concerned:

- *Energy policy legislation*, in which the Department of Energy and the energy committees of Congress play key roles, but which also involves numerous other agencies in the executive branch and committees in Congress;
- *The Climate Change Technology Program*, which is developed and implemented by an interagency group in the executive branch, subject to congressional funding;
- *The annual budget cycle*, which includes as many as fifteen executive agencies' climate change programs, as well as the appropriations committees and their subcommittees, and the budget committees in both houses of Congress – and which also sometimes involves the tax committees of both houses;
- *International agreements* concerning bilateral, regional, and multilateral arrangements involving technology innovation and diffusion – i.e. policymaking in the

intersection of the energy technology issue cluster and the international cooperation issue cluster.

The following sections of the chapter offer examples of how interests, ideologies, institutional settings, and the influence of particular actors in the process affect the policy outcomes.

6.3 Energy policies

Given the wide range of market failures that inhibit technological innovation and diffusion, and given the diverse types of technologies that are being encouraged to address climate change issues, it is not surprising that there are also numerous and varied types of policy instruments that have been adopted (Jacobsson and Johnson 2000; OECD and IEA 2003). The policy instruments are not only varied in terms of their nature, e.g. whether direct subsidies are in the form of grants or tax-based incentives for producers or consumers; they are also varied in terms of the economic sectors and industries to which they apply. In the primary sector of the economy, some apply to agriculture, others to forestry, and yet others to mining; many apply to the manufacturing sector, including motor vehicles and petroleum refining in particular, and of course, many apply to the electric power sector, including the producers and the end users of electricity. For a chronology and précis of major pieces of energy legislation and other government actions, please see Appendix 6.1.

Subsidies

Among the diverse policy instruments, industry-specific subsidies have been particularly controversial in recent years. The subsidies not only have potentially important consequences for the levels of greenhouse gas emissions, they also have important consequences for individual firms and for entire industries (IEA, OECD, and World Bank 2010). In some instances, industries' and firms' competitive positions depend directly on the amounts of government subsidies. For instance, US government programs that provide liability insurance subsidies for the nuclear power industry are obviously especially beneficial for the electric power industry, but more so for electricity-producing firms that have nuclear plants and less so for those firms that do not. Or, for another example in the same industry, firms that manufacture components for nuclear reactors as well as electric generators are affected differently, as compared with firms that manufacture generators but not reactor components.

Other industry-specific issues have recurred from time to time, especially in regard to fossil fuel subsidies for the oil and gas industry and the levels and continuity of subsidies for the wind power, solar power, and biofuels industries. A nettlesome analytic challenge with significant political economy consequences is that it is notoriously difficult to be precise about the costs of subsidies to the government because they occur in many forms – direct cash payments to producers or consumers in some cases, favorable tax treatment for producers and consumers, and of course research and development programs. By one count (US Office of Management and Budget 2011a) in the FY2012 budget proposal there were as many as twelve different forms of subsidies for fossil fuels (coal, oil, and gas) with an estimated potential savings of $46 billion to the government over ten years from phasing them out. Another study (US Energy Information Administration 2008) estimated annual subsidies for electricity production were about $3.0 billion for coal, $2.3 billion for nuclear, $0.7 billion for wind, and $0.2 billion for solar. There have been increases for some incentives and decreases for others since then, and in any case there are other subsidy programs not included in these figures. During that same year, the study found that tax expenditure subsidies were $4.8 billion for fossil fuels and $4.0 billion for renewable energy and energy efficiency.[1] However, as funding levels for the American Recovery and Reinvestment Act of 2009 diminished over time, the total levels of subsidies for renewable energy and energy efficiency programs began to decline.

The funding levels for industry-specific subsidies are driven by political as well as economic and technological considerations. Cost-effectiveness and technological expertise are often less important than the economic size and location of an industry and its access to government decision-makers. Some industries are obviously more influential than others, and within any given industry some firms are more influential than others. Large, entrenched, and geographically extensive industries – such as the oil and gas and the automotive industries – have been in strong positions to receive government subsidies of many types, including the government bailout of auto firms. "Newer" industries, however, such as solar, wind, and geothermal, found it difficult in their early histories to compete for energy industry subsidies. Over time, though, as solar and wind have grown, they have been in a stronger position. Changes in administrations and increasing public support for renewable energy sources and energy efficiency programs combined to create a fundamentally new political environment for subsidizing those sectors of the economy.

1 In the administration's request to Congress for the FY2011 budget, there was a proposal for a reduction of the long-standing subsidies for the oil and gas industry; but Congress significantly cut back the proposed reductions. A similar sequence played out in subsequent budget cycles, e.g. FY2014.

Other factors also come into play in determining the magnitudes and forms of subsidies. The nuclear power industry is sui generis in this regard: its origins in a secretive and well-funded World War II weapons program – plus the unusually controversial and large-scale safety, environmental, and security issues associated with civilian nuclear power – have all led to pressures for both unusual regulatory and subsidy programs. For instance, the federal government has a large insurance program that subsidizes the industry by taking on much of the risk of the human and economic costs of catastrophic accidents. Another example of an industry in an especially strong position to receive large government subsidies is the biofuels industry – both ethanol as a substitute for gasoline and biodiesel fuels. Because of the long tradition of agricultural subsidies and the influential positions in Congress of senators and representatives from agricultural areas, the biofuels industry enjoyed large subsidies for many years. Over time, nevertheless, the subsidies for the nuclear and biofuels industries have been eroded by increasing doubts about their fundamental viability over the long term.

In addition to issues about which industries receive subsidies and the levels of those subsidies, there are also issues about the cost-effectiveness of types of subsidies, especially the relative cost-effectiveness of grants versus tax credits. The issue became particularly salient in the context of the ARRA, which included substantial subsidies for a variety of energy programs and projects (see below). A study by the Bipartisan Policy Center (2011), for instance, found that grants were about twice as cost-effective as tax credits. But the issues about the forms that subsidies do or should take is more complex than this. For example, for low-tech products such as home insulation, the challenge is to devise subsidies that will appeal to home-owners; for the issue is the widespread uptake of well-established technologies. Subsidies thus take the form of financial assistance to consumers. Moreover, the potential for relatively rapid and nationally widespread distribution of home insulation subsidy funds – in addition to the labor-intensive nature of the installation process – all of these features made the home insulation subsidy program an obviously attractive recipient for the 2009 American Recovery and Reinvestment Act funding during the economic crisis. In contrast, where the technology is in its early stages and involves large-scale facilities that will be in service for decades, the challenge is different; and the subsidies are different. The subsidies are in research and development projects and pilot projects that can be used to test the operational feasibility of engineering design choices.

Yet another issue about government subsidies for particular industries is their variability and uncertainty over time (Narayanamurti, Anadon, and Sagar 2009). The issue

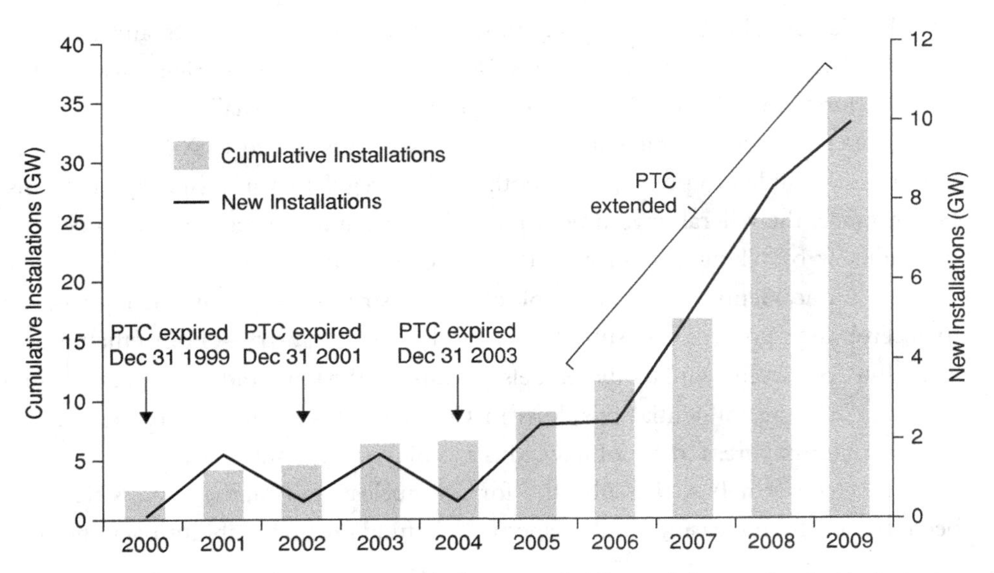

Figure 6.2 Changes in US government wind power subsidies and changes in wind industry installations

Source: Bipartisan Policy Center (2011). Used with permission.

has been central to the growth of the wind power, solar power, and biofuels industries. For example, the relationship between changes in government wind power subsidies and the number of installations of wind turbines is evident in Figure 6.2.

Variations in industry interests

Just as with cap-and-trade and other regulatory issues, the economic interests of firms and industries are of course highly varied in terms of the effects of government technology programs. Many of the patterns of conflict and cooperation are similar for cap-and-trade and technological issues, with fossil fuel industries and renewable energy and energy efficiency (RE&EE) industries on opposite sides. But there are also some economically, politically, and environmentally significant differences in the political economy patterns of the two issue clusters as well; for instance, the role of the nuclear power industry is different across the two issue clusters.

Nuclear power

As a low-carbon source of electric power, the nuclear industry has an interest in the adoption of cap-and-trade regulations and carbon taxes, which thus aligns it naturally with RE&EE industries and environmental NGOs. However, safety concerns and government subsidies often dominate consideration of the future of the industry. In fact, environmental NGOs have been split over whether to support or oppose increased reliance on nuclear power, as a matter of government policy. Further, because of the magnitudes of the subsidies at issue – whether actual or proposed – the amounts of money for nuclear industry subsidies in the budget process have often dwarfed the amounts for RE&EE programs. In the administration's FY2011 budget proposal, for instance, the amount requested for loan guarantees to support the construction of nuclear power plants was $54.5 billion over ten years, which represented an increase of $36 billion. However, following the crisis at the Fukushima nuclear power plant in Japan in March 2011, the future of the nuclear power industry in many countries, including the United States, has been highly uncertain.

The political economy of subsidies in both the nuclear power industry and the fossil fuel industries differs from other industries in two respects. (a) Both industries have enjoyed relatively large subsidies for a long time, and thus the issues are about whether to increase or decrease the levels of the subsidies from a relatively high trend line with many entrenched interests. They are both like agriculture in that respect. (b) Subsidy issues for both industries became bargaining chips in the Obama administration's search for votes in the Senate for cap-and-trade legislation. For instance, in his January 2010 State of the Union speech, President Obama endorsed increased federal government support for expansion of the nuclear power industry through increased funding for loan guarantees, and he also endorsed exploitation of offshore oil and natural gas reserves. These were carefully targeted "carrots" that were offered to gain increased Republican support in the Senate for the pending cap-and-trade legislation, and they were two of only a very few items in the President's speech that garnered widespread Republican as well as Democratic applause. A year later, in his January 2011 State of the Union speech, he again supported expansion of the nuclear power industry; but he simultaneously targeted subsidies for the oil industry for elimination and increases in energy efficiency and renewable energy programs in their stead.

6.4 The Climate Change Technology Program

Although the Clinton administration decided not to push for ratification of the Kyoto Protocol or the establishment of a national cap-and-trade program in view of congressional opposition to such measures (see Chapter 5), it did develop a Climate Change Technology Initiative and advocated increases in funding of a variety of renewable energy and energy efficiency programs (US Congressional Research Service 1999a; 1999b; 1999c; 2001). The administration was able to advance the programs incrementally. The programs, however, were in a politically perilous situation when the George W. Bush administration entered office in January 2001.

The prospects for them, in fact, became more propitious in June 2001, when the George W. Bush administration announced support for a National Climate Change Technology Initiative. After another name change a year later to the Climate Change Technology Program (CCTP), the program had five explicit goals (US Climate Change Technology Program 2003, 3):

- reducing emissions from energy end-use and infrastructure;
- reducing emissions from energy supply;
- capturing and sequestering carbon dioxide;
- reducing emissions of other GHGs; and
- enhancing capabilities to measure and monitor GHG emissions.

The administration listed a large number of projects and other activities organized in terms of these five goals (US Climate Change Technology Program 2005; US Department of Energy 2003). There was also an elaborate management structure involving numerous multi-agency groups, with the Secretaries of Energy and Commerce as nominal co-leaders of the overall program. The program was further elaborated in the form of a detailed statement of a Strategic Plan (US Department of Energy 2006), with a sixth goal added, namely "bolstering basic science contributions to technological development."

In terms of analytic clarity and institutional structure, the program was thus highly developed, and its long-term goals were ambitious. In more tangible and immediate terms, however, it encountered problems on two fronts – congressional opposition to funding for some of its activities, and difficulties implementing high-profile projects. As for funding, although Congress included "authorizing" provisions for funding the program in its Energy Policy Act of 2005, the actual funding levels (in the form of "budget authority") subsequently approved in annual budgets by Congress were below the administration's requests. In addition, the FutureGen project for reducing greenhouse gas emissions from coal-fired power plants and the Generation IV project for nuclear power both encountered a combination of financial and technical problems and were derailed at least temporarily.

After it entered office in January 2009, the Obama administration reframed the issues, and changed the program priorities. Its "Strategy for American Innovation" (US National Economic Council 2009) included among its several parts and goals the following concerning energy technology:

> *Catalyze Breakthroughs for National Priorities.* There are certain sectors of exceptional national importance where the market is unlikely to produce the desirable outcomes on its own. These include developing alternative energy sources . . . and manufacturing advanced vehicles. In these industries where markets may fail on their own, government can be part of the solution.
>
> *Unleash a clean energy revolution.* Historic investments in smart-grid, energy-efficiency, and renewable technologies like wind, solar, and biofuels will help unleash a wave of ingenuity and progress that creates jobs, grows our economy, and ends our dependence on oil.

When President Obama submitted his fiscal year 2012 budget (US Office of Management and Budget 2011a), renewable energy and energy efficiency programs were discussed in a section titled "Competing and Winning in the World Economy" and in a subsequent document "Strategy for American Innovation: Securing Our Economic Growth and Prosperity" (US National Economic Council, Council of Economic Advisers, and Office of Science and Technology Policy 2011). This statement of an economic growth strategy was also explicitly a statement of energy technology policy and implicitly a statement of an important element of climate change policy. Its extent and implications for climate change are evident in the excerpts from it in Box 6.6.

The scope of such a strategy and its implications for government programs is further evident in the wide variety of agencies involved (see Box 6.7).

6.5 Budgets

Issues about the scale of the effort that is needed to address climate change inevitably lead to budget issues, which are the focus of this section. Readers who are not familiar with the budgeting process may want to refer to Appendix 6.2, which introduces the terminology and institutional features of the process, particularly in Congress.

In addition to addressing issues about the scale of the government's effort in relationship to the need, budgets reveal the priorities that administrations and the Congress give to programs (see Gallagher and Anadon 2012, for an extensive and updated data set). Figure 6.3 – which is in constant dollars and therefore controls for inflation – reveals several features of the overall spending levels and the priorities for particular technologies within the total.[2] As for the total, in more than three decades it has not

2 All the budget numbers in the chapter are presented in fiscal year terms – i.e. financial years, which start on October 1 and end on September 30 in the calendar year with the same number as the fiscal year. Because of this and the fact that new presidents enter office in January – which is

Box 6.6 Excerpts from the "Strategy for American Innovation"

In areas of well-defined national importance, public investments can help catalyze advances, leveraging key breakthroughs and US leadership. The 21st century brings several critical areas – including energy . . . – where the demand for breakthroughs is clear. The Administration's *Strategy for American Innovation* will harness public mechanisms to help meet our common goals, sparking commercial innovations and American ingenuity as we seek to meet the grand challenges of the next century and add impressive new chapters to the history of American progress.

Unleash a clean energy revolution

For our national security, economy, and environment, it is crucial to develop clean energy technologies. President Obama is committed to US leadership in the energy economy of the future. The President's strategy will meet our energy goals and put the US at the cutting edge of the renewable energy, advanced battery, alternative fuel, and advanced vehicle industries.

Double the nation's supply of renewable energy by the end of 2012

The Administration is committed to doubling the supply of renewable energy by the end of 2012. Federal tax credits and financing support, including the Section 1603 and Section 48C programs, have leveraged the manufacture and deployment of gigawatts of new renewable energy investments in innovative solar, wind, and geothermal energy technologies. Aided by these incentives, electricity generation from renewables (excluding conventional hydropower) is projected to surpass twice its 2008 level, meeting the Administration's goal.

Spur innovation through new energy standards

The President has set a national goal of generating 80 percent of the nation's electricity from clean sources by 2035. The proposed Clean Energy Standard will mobilize hundreds of billions of dollars in private investment, spur the deployment of clean energy technologies, and create market demand for new innovations. The Administration is also working to meet the renewable fuel mandate set by Congress, which requires the use of 36 billion gallons of renewable fuel by 2022. The EPA finalized a rule to implement the Renewable Fuel Standard on February 3, 2010,

already one-third of the way into a fiscal year – it is difficult for them to make major changes in spending levels and priorities for the first fiscal year whose number corresponds to the first calendar year of their presidency. For more about the intricacies and terminology of the budget process, see Appendix 6.2.

and the *Growing America's Fuels* strategy focuses on a number of the innovations that will help us achieve that goal.

Create Energy Innovation Hubs
Bringing together scientists and innovative thinkers from different disciplines to form highly-integrated research teams can create research breakthroughs on tough problems. The Administration established three Energy Innovation Hubs in FY 2010 to tackle challenges in nuclear energy modeling, energy efficiency in buildings, and the generation of fuel from sunlight. The Administration's FY 2012 Budget calls for doubling the number of Energy Innovation Hubs, from three to six, to tackle additional energy challenges.

Invest in clean energy solutions
The Department of Energy's Advanced Research Projects Agency-Energy (ARPA-E) has awarded nearly $400 million to more than 100 research projects that seek fundamental breakthroughs in energy technologies. The President's FY 2012 Budget proposes to expand support for ARPA-E.

Promote energy efficient industries
The Administration is spurring private sector innovation through new fuel efficiency and greenhouse gas emissions standards, with new efforts to develop standards over the 2017–2025 model years for light vehicles and new standards over medium- and heavy-duty vehicles. As the single largest energy consumer in the US economy, government procurement provides an additional, important mechanism to catalyze demand for innovative energy technologies. In October 2009, President Obama signed an Executive Order that calls on agencies to cut the federal government's fleet petroleum use by 30 percent by 2020.

Invest in Advanced Vehicle Technology
The President's FY 2012 Budget proposes to make the United States the world's leader in manufacturing and deploying next-generation vehicle technologies, expanding funding for vehicle technologies by almost 90 percent to nearly $590 million and enhancing existing tax incentives. Recovery Act and prior year investments are already making progress on advanced technology vehicles through research initiatives like an ARPA-E grant to develop a battery that will go 300 miles on a single charge. The FY 2012 Budget will significantly broaden R&D investments in

technologies like batteries and electric drives – including an over 30 percent increase in support for vehicle technology R&D and a new Energy Innovation Hub devoted to improving batteries and energy storage for vehicles and beyond. In addition, the President is proposing to transform the existing $7,500 tax credit for electric vehicles into a rebate that will be available to all consumers immediately at the point of sale.

Source: US National Economic Council, Council of Economic Advisers, and Office of Science and Technology Policy (2011). Also see National Economic Council (2009).

reached the peak level of the late 1970s (established during the Carter administration). The total was reduced by about 70 percent during the Reagan administration, and then fluctuated around that lower level during the George H.W. Bush, Clinton and first George W. Bush administrations. By the end of the second George W. Bush administration, it had begun a sharp upward trend, followed by the spike of the ARRA increases (which were actually spread over many years, and not all concentrated in the single fiscal year). Following the ARRA spike, during the Obama administration the pre-ARRA upward trend has continued. However, total authorizations in constant dollar terms in FY2013 (not including that year's share of ARRA) were less than half those of 1978.

As for the priorities for specific technologies, there have been shifts roughly corresponding to the trends in the totals. In the late 1970s nuclear and fossil programs were favored, with renewables beginning to experience increases; then as total expenditures declined during the mid-1980s, renewables' share as well as absolute amount diminished significantly. By about FY2000 both renewable and efficiency programs had begun to increase in both relative and absolute terms.

ARRA

The ARRA program reflected dramatically different priorities from the prior years' normal budgeting cycles: renewable programs spiked up and so did fossil fuel programs, though much of the latter was for carbon capture and sequestration (CCS) and thus not reflective of traditional fossil fuel programs. Of the $3.399 billion for all fossil programs in the ARRA, $1.569 billion (or about 46 percent) was for CCS. The ARRA programs allotted almost no additional funds for nuclear R&D. The funding amounts for these

Box 6.7 Agencies and their activities in the Climate Change Technology Program

Department of Agriculture
Carbon Fluxes in Soils, Forests and Other Vegetation, Carbon Sequestration, Nutrient Management, Cropping Systems, Forest and Forest Products Management, Livestock, and Waste Management, Biomass Energy and Bio-based Products Development

Department of Commerce *including:*
National Institute of Standards and Technology
International Trade Administration
National Oceanic and Atmospheric Administration
Instrumentation, Standards, Ocean Sequestration, Decision Support Tools

Department of Defense
Aircraft, Engines, Fuels, Trucks, Equipment, Power, Fuel Cells, Lasers, Energy Management, Basic Research

Department of Energy
Energy Efficiency, Renewable Energy, Nuclear Fission and Fusion, Fossil Fuels and Power, Carbon Sequestration, Basic Energy Sciences, Hydrogen, Electric Grid and Infrastructure

Department of Health and Human Services *including:*
National Institutes of Health
Environmental Sciences, Biotechnology, Genome Sequencing, Health Effects

Department of the Interior
Land, Forest, and Prairie Management, Mining, Sequestration, Geothermal, Terrestrial Sequestration Technology Development

Department of State *including:*
US Agency for International Development
International Science and Technology Cooperation, Oceans, Environment International Assistance, Technology Deployment, Land Use, Human Impacts

Department of Transportation
Aviation, Highways, Rail, Freight, Maritime, Urban Mass Transit, Transportation Systems, Efficiency and Safety

Environmental Protection Agency
Mitigation of CO_2 and Non-CO_2 GHG Emissions through Voluntary Partnership Programs, including Energy STAR, Climate Leaders, Green Power, Combined Heat and Power, State and Local Clean Energy, Methane and High-GWP Gases, and Transportation; GHG Emissions Inventory

National Aeronautics and Space Administration
Earth Observations, Measuring, Monitoring, Aviation Equipment, Operations and Infrastructure Efficiency

National Science Foundation
Geosciences, Oceans, Nanoscale Science and Engineering, Computational Sciences

Source: US Climate Change Technology Program (2010).

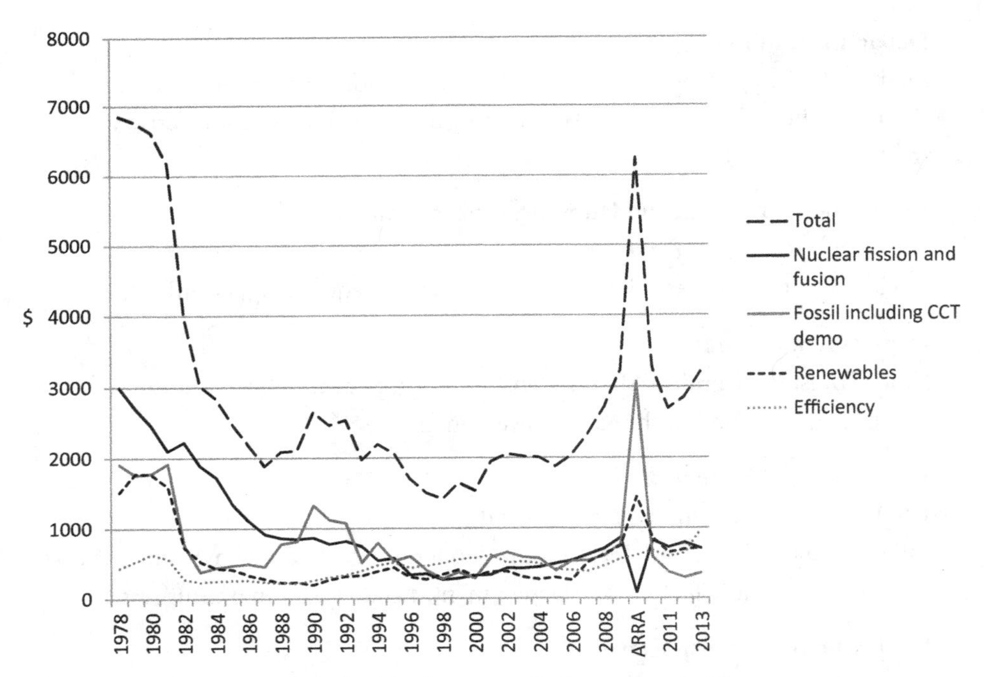

Figure 6.3 Long-term trends in funding for energy R&D programs (millions – constant FY2005 dollars)

Sources: Computed by the author from data in Gallagher and Anadon (2012).

Table 6.1 How much clean energy is there in the American Recovery and Reinvestment Act (ARRA) – and how does it compare with other countries' economic stimulus plans?

A. Amounts in US ARRA

Category	US$ billions (percent of total)
Total: ARRA	787.0
Subtotal: energy efficiency and low-carbon power	94.0 (11.9)
Building energy efficiency	27.4 (3.5)
Renewable energy (wind, solar, thermal . . .)	22.5 (2.9)
Water/waste [e.g. energy efficiency and methane recovery]	15.6 (2.0)
Electric grid	11.0 (1.4)
Rail	9.6 (1.2)
Carbon capture and storage	4.0 (0.5)

Source: Compiled by the author from HSBC (2009); also see Pew Center on Global Climate Change (2009c).

B. Comparisons of US and other countries' programs

Country	Total recovery plan (US$ billion)	Low-carbon and energy efficiency subtotal (US$ billion)	Low-carbon and energy efficiency subtotal (percent of total)	Number of years of program – beginning in 2009
United States	787	94	12	10
China	586	201	34	2
South Korea	38	31	81	4
European Union	537	53	10	2–3
Canada	32	3	8	5
Mexico	8	1	10	1

Source: Compiled by the author from HSBC (2009); also see Pew Center on Global Climate Change (2009c).
Although the total for the USA was the highest among the countries, it was spread over a relatively long ten years, as compared with the one to five years of the other countries. More to the point of the relative importance of spending on climate-friendly programs, South Korea's four-fifths and China's one-third reflected the relative high priority those countries gave to such programs. All the other countries were much lower, with the United States at 12 percent and the remaining countries below that.

programs became salient and controversial in the context of the American Recovery and Reinvestment Act of 2009 during the economic crisis, and again in the context of delayed congressional action on the FY2011 budget following the change of the majority party in the House of Representatives. As for the ARRA, as Figure 6.3 indicates, there was a historically significant increase in energy program spending in FY2010. Panel A in Table 6.1 indicates the full scope of ARRA funding for energy efficiency and other climate-related programs. Of the $787 billion total in the ARRA, $94 billion (about 12 percent) was designated for energy efficiency and low-carbon programs – of which

$22 billion was for renewable energy programs such as wind, solar, and geothermal. An important issue about these amounts (especially from a macro-economic stimulus perspective) was the timing of the disbursements – especially whether they would be soon enough to facilitate an economic recovery. The relatively large size of the funding for "building energy efficiency" programs was partly responsive to this concern, as the installation of insulation and energy efficiency equipment, including retrofitting existing buildings, can be done relatively quickly. In any case, the relatively long time period of ten years compared with other countries' economic recovery programs is notable in Panel B of Table 6.1. It is also notable there that although Canada, Mexico, and the European Union devoted similar proportions of their stimulus packages to climate-friendly programs, compared with the USA, the programs of South Korea and China were much more climate-friendly, with the former at 81 percent and the latter at 34 percent, compared with the USA at 12 percent.

As the economic crisis subsided, ARRA spending approached its end, and the Obama administration began its second term in 2013, there was much interest in the government's energy program. The President gave it prominence in his inauguration speech, in the annual State of the Union Speech and in specialized speeches in early 2013. Thus, the tangible manifestations of the administration's energy policy directions and priorities in the proposed FY2014 budget was subjected to unusually intense scrutiny by a wide variety of stakeholders and commentators. The interest was further heightened by the impasse between the President and Congress on many other budget matters – an impasse that had begun several years earlier. In Table 6.1 the data on the FY2014 budget, particularly in comparison with the FY2013 budget, took on special political significance in the annual budgeting process. By the beginning of the second Obama administration, in the normal annual budgeting cycle, renewable R&D was being funded at about the same level as nuclear programs while energy efficiency programs at close to $1 billion a year were higher than both. In the FY2014 request, the Obama administration asked for $2.3 billion for the Office of Energy Efficiency and Renewable Energy, compared with $0.8 billion for the Office of Nuclear Energy (US Office of Management and Budget 2013).

Earmarks

Earmarking projects in the congressional budgeting process has a long and controversial history. As applied to energy efficiency and renewable energy projects, it has been a subject of much interest in recent years, as the amounts of money at stake in the budget for those projects have increased. There is a natural tension between the cost-effectiveness and bureaucratic-professional-accountability issues in executive agencies,

Box 6.8 Congressional earmarks for energy-efficiency and renewable-energy projects in the budget process

During its deliberations on the fiscal year 2010 budget for the Department of Energy programs in Energy Efficiency and Renewable Energy (EERE), the Conference Committee of the House and Senate Appropriations Committee approved more than 250 earmarked projects totaling $292,135,000. Though accounting for only about 14 percent of the $2.2 billion approved for the EERE budget that year and of course an even smaller fraction of the $27 billion approved for the entire Energy Department, the practice attracts much attention and remains popular among members of Congress because of the high-profile nature of some of the projects in the members' home states. For example, the Energy Efficiency and Renewable Energy projects from the 2010 budget with $5 million or more per project budget included:

- Alternative and Unconventional Energy Research and Development – Utah
- Tropical Feed Stocks and Biomass Conversion – Hawaii
- Hawaii Energy Sustainability Program – Hawaii
- Hawaii Renewable Energy Development Program – Hawaii
- Offshore Wind Initiative – Maine
- Center for Biomass Utilization – North Dakota
- Center for Nanoscale Technology – North Dakota

The projects are thus widely distributed around the country, but with a disproportionate number in a few states, such as the home states of party leaders and committee chairs and members.

Source: Compiled by the author from US Congress (2010), 11a–11f (in the manually marked-up version).

on the one hand, and the location-specific political interests of individual members of Congress, on the other. The tensions periodically become public, for instance when first President Bush and then candidates McCain and Obama all called on Congress to restrain the practice – to little avail. See Box 6.8 for examples of earmarking.

6.6 International technology issues

Many energy technology issues involve international aspects and thus overlap with international issues as discussed further in the next chapter (see especially US Department of State 2007). There are therefore the same underlying issues about how governments address market failures, as in the present chapter, but in addition how they address free-rider problems, as in the next chapter. Furthermore, the overlaps include both technology innovation and technology diffusion issues, as in the present chapter title. Out of this rather complex array of issues, the analysis here focuses on a few issues of special significance in US responses to climate change.

Because many energy technology innovation issues involve large capital outlays and have uncertainties associated with very long-term investments, there is much interest in the potential for international collaboration to spread out investment costs and risks among countries. There is also an interest in avoiding the redundancies and wastes of multiple projects with similar objectives being undertaken in several countries. Such concerns have led the US government to enter into a wide variety of bilateral and regional arrangements as well as technology-specific arrangements. Among them was the Asia-Pacific Partnership on Clean Development and Climate (the APP), which had more than a dozen flagship projects, which were notable for the variety of industries and technologies included. It is also noteworthy that emissions of several greenhouse gases in addition to carbon dioxide were of concern – a reminder of the extent to which the issues of climate change are idiosyncratic to particular industries. The entire APP, however, was terminated in April 2011, with some of its activities transferred to other programs.

Another set of issues concerns barriers to international transfers of technologies through private international trade, investment, and licensing transactions – barriers that restrict international transfers that could contribute to cost-effective efforts to address climate change mitigation and/or adaption efforts. Several studies have considered this problem and related features of the multilateral regime centered in the World Trade Organization (Brewer 2008a; 2008b; 2008c; 2009b; 2012). US policies have been central to conflicts with Brazil, the EU, and Japan, while China's policies have been at issue in relations with the USA and other countries (see Bridges at www.ictsd.org; also see www.TradeAndClimate.net).

As the trade-related issues have been increasingly defined in terms of foreign market access and export promotion, more US government agencies' programs have been drawn into the wide spectrum of climate change policies. For instance, the programs of two US government agencies – the ExportImport Bank (ExIm) and the Overseas Private Investment Corporation (OPIC) – which subsidize, respectively, exports and outward foreign direct investments by US-based firms – have come under more scrutiny to determine their roles in promoting international transfers of climate-friendly and/or climate-hostile technologies. Both agencies use a combination of below-market-rate loans, loan guarantees, and insurance against "political risks" such as wars; both are therefore subsidy programs of a highly specialized type. The question about them in the present context is whether they subsidize business transactions in goods and services that mitigate greenhouse gas emissions or increase them.

In an effort to shift those and other agencies' priorities more toward climate-friendly activities and away from climate-hostile activities, the Obama administration announced a Renewable Energy and Energy Efficiency (RE&EE) Export Initiative that includes eight agencies, with the intention to "provide new financing options for RE&EE exporters; enhance market access for US RE&EE technologies in foreign markets; increase the number of trade promotion events for US RE&EE companies and significantly [improve] the effectiveness of and efficiency of US government export promotion programs" (US White House 2010). This export initiative is guided by an inter-agency committee with co-chairs from the Departments of Commerce and Energy.

6.7 Implications: interests, institutions, and ideologies

Because of the economic significance of energy industries in the national economy – and particularly the predominance of some industries in local, state, and regional economies – it is inevitable that any government programs that either encourage or discourage production or consumption patterns in them will be highly politicized. For there are not only *absolute* winners and losers in the distributive effects of such government policies, there are also *relative* winners and losers: some winners gain more than other winners; some losers lose more than other losers. Furthermore, the distributive effects of government policies differ among firms within industries as well as among industries, and the relative gains/losses affect firms' competitive positions vis-à-vis their rivals (see Salorio 1991; 1992).

The institutional context of government policymaking is different for the technology issues of this chapter, compared with the cap-and-trade issues of the previous chapter: the energy, agriculture, and transportation departments are particularly important in policymaking in the issues of the present chapter, while the Environmental Protection Agency and other regulatory agencies are more involved in the issues of the previous chapter. As for congressional committees, although the same legislative committees such as those concerned with energy, agriculture, and transportation are important to both clusters of issues, the appropriations committees and their subcommittees are especially important for the issues of this chapter.

The issues of the two chapters are also different in terms of ideology. The ideological issues of the previous chapter are largely about government regulation, whether in the form of command and control policies or the caps and other rules in cap-and-trade systems, with additional issues about the role of markets in the trade component of cap-and-trade. The ideological issues about technology policy in the present chapter are more about the nature, magnitudes, and roles of government subsidies. Tax policy, though, is involved in both issues, yet with different emphases: a central issue about taxes in the previous chapter is their advantages and disadvantages relative to cap-and-trade as a way to reduce GHG emissions; a central issue in the present chapter is the advantages and disadvantages of taxes relative to subsidies as a way to encourage innovation and diffusion of climate-friendly energy technologies.[3]

These patterns of clusters having similar features as well as differences are reflected in the overlapping areas of the clusters. Thus, the issues of the previous and present chapters not only overlap with each other, they also overlap with the international cluster of the next chapter, with their distinctive features created by a highly pluralistic, decentralized system of nation states.

Suggestions for further reading and research

Richter (2010) offers an excellent analysis of energy issues in relation to climate change, though not with a US emphasis. Resources for the Future (www.rff.org) regularly produces informative and focused studies on key energy policy issues. The refereed journal, *Energy Policy*, has articles on many aspects of energy policy issues in the USA and other countries. Periodic studies of energy issues by the Congressional Research Service (www.crs.gov) and the Congressional Budget Office (www.cbo.gov) are consistently informative on current topics of special interest in the legislative process.

The energy policymaking process, particularly in the Congress, can be followed in *E&E Daily* (www.eenews.com) and the weekly reviews of the Environmental and Energy Study Institute (www.eesi.org). Also see *Bloomberg New Energy Finance* (www.bnef.com) and the synopses of developments and reports at www.epoverviews.com.

3 Though conceptually distinct and exhibiting different patterns of political economy, taxes as an alternative to emissions cap-and-trade and taxes as an alternative to energy subsidies are nevertheless causally related. For taxes and cap-and-trade have in common the creation of incentives for producers and consumers to adopt different technologies. This important causal link has sometimes been overlooked in the US discussions of climate change policy.

Extensive budget data on Department of Energy programs are available in Gallagher and Anadon (2012).

The Department of Energy's Office of Renewable Energy and Energy Efficiency has an unusually informative and user-friendly web site at www.eere.energy.gov. The US Climate Change Technology Program (www.uscctp.gov) has been recently revamped after several years of neglect. The annual budget documents of the US Office of Management and Budget and the reports of the Appropriations Committees of the House and Senate are essential for access to official presidential requests and congressional actions. The Union of Concerned Scientists (www.ucsus.org) presents careful analyses of energy budget issues as well as reports on other energy policy issues.

Mount (1999) offers an insightful analysis of US government energy policies and market failures – and related disagreements between economists and environmentalists about the government's capacity to address successfully the challenges of climate change. Rowlands (2009) discusses US renewable energy industry issues in their North American context.

Two industry associations provide extensive coverage of policy developments concerning renewable energy and energy efficiency – the American Council on Renewable Energy (ACORE) at www.acore.org and the American Council for Energy Efficiency at www.aceee.org.

APPENDIX 6.1

Chronology of major US energy policy legislation and other actions

1970s

The quadrupling of world oil prices in 1973–1974 prompted a series of actions. *Project Independence* was initiated by President Nixon in late 1973. It outlined a series of plans and goals for Americans to achieve energy independence within a decade. In December 1973, President Nixon set up the Federal Energy Office as a crisis-management agency. In 1974, Congress established the Federal Energy Administration, which absorbed and enlarged the energy office.

The Energy Policy and Conservation Act of 1975 was passed following proposals by President Ford. It mandated vehicle fuel economy standards, extended oil price controls to 1979, and created a strategic petroleum reserve. The Department of Energy was also mandated to undertake programs and studies to advance research on alternative fuels. The Department of Energy Organization Act of 1977 created a cabinet-level agency following the suggestion of President Jimmy Carter. The National Energy Conservation Policy Act of 1978 directed the United States Department of Energy (DOE) to set Minimum Energy Performance Standards (MEPS) to replace those set by the Energy Policy and Conservation Act (EPCA) in 1975. The amendment to the EPCA changed the energy standards from voluntary to mandatory, and these new federal standards pre-empted those established by state authorities. The DOE was also charged with establishing procedures for the submission, approval, implementation, and monitoring of residential energy conservation plans by state utility regulatory authorities.

1980s

The Alternative Fuels Act of 1988 amended a portion of the Energy Policy and Conservation Act to pursue the use of alternative fuels. The Act encouraged the development, production, and demonstration of alternative motor fuels and alternative-fuel vehicles. The Act specified "alternative fuel" as any fuel not derived from petroleum, including ethanol, methanol, natural gas, liquefied petroleum gas, hydrogen, and electricity.

1990s

The Energy Policy Act of 1992 set goals, created mandates, and amended utility laws to increase clean energy use and improve overall energy efficiency in the United States. The Act consists of twenty-seven titles detailing various measures designed to lessen the nation's dependence on imported energy, provide incentives for clean and renewable energy, and promote energy conservation in buildings. The act directed the federal government to increase energy conservation in federal buildings when feasible, and to integrate the use of alternative-fuel vehicles in federal and state fleets. Title XXII authorized tax incentives and marketing strategies for renewable energy technologies in an effort to encourage commercial sales and production (Kenney 2008).

2000–2008

The Energy Policy Act of 2005 was the first major national energy legislation passed by Congress in over a decade. It included consumer tax credits for purchasers of hybrid automobiles and other fuel-efficient motor vehicles, and for production of energy using renewable sources such as wind, solar, biomass, geothermal, and ocean tides. There were also provisions for R&D programs and other subsidies for nuclear power and coal-fired power plants.

The Energy Independence and Security Act of 2007 focused on energy efficiency and renewable energy issues. It established a new Corporate Average Fuel Economy (CAFE) standard of 35 miles per gallon for cars and light trucks by the 2020 model year. It set a Renewable Fuels Standard (RFS) beginning with 9 billion gallons in 2008 and increasing to 36 billion gallons by 2022. It included new standards for lighting and appliances.

The Food, Conservation, and Energy Act of 2008 provided for loan guarantees and other support programs for biofuels.

2009–2012

The American Recovery and Reinvestment Act of 2009 established tax credits, loan guarantees, RD&D, and other programs to support renewable energy and energy efficiency projects in the building, transportation, manufacturing, electricity production and distribution, and other sectors.

Sources: Kenney (2008); Union of Concerned Scientists (2005); US Department of Energy (2008; 2009); US Senate Committee on Energy and Natural Resources (2010).

APPENDIX 6.2

Introduction to the budget process and terminology

Presidential submission and the fiscal year calendar. The President is required by law to submit a budget to the Congress by the first Monday in February of each calendar year. The formal proposal is the result of an inter-agency coordination/negotiation process that begins about a year earlier. The Director of the Office of Management and Budget (OMB) in the Executive Office of the President is responsible for overseeing the preparation of the proposals that are submitted by the President to the Congress. The proposed budget pertains to the fiscal year that begins on October 1 of the year in which it is presented to Congress and ends on September 31 of the following year. The fiscal year is given the number of the calendar year in which it ends. E.g. the budget proposed by President Obama in February 2014 for FY2015 pertained to the period from October 1, 2014 to September 31, 2015.

Budget presentation format. Standard budget presentations typically include three numbers for any given program or other line item:

- *Actual* amount (i.e. "outlays") for the *previous* fiscal year;
- *Enacted* amount for the *current* fiscal year, i.e. the amount enacted by Congress and agreed by the President, which is in the process of being spent by the executive agencies; and
- *Requested* amount, i.e. the amount being *proposed* by the President for the coming fiscal year, subject to congressional revision.

Functional budgeting. The core climate policy programs, as well as related ancillary programs, can be partly defined in terms of their place in the scheme of functions – and "super-functions," and "sub-functions" – that are identified in the current standard budget format. Some of the functional categories are redefined from time to time, as new issues, agencies, and programs emerge, as programs and agencies are reorganized, and as programs and organizational units morph into different forms.

Executive agencies. Because functional and programmatic responsibilities related to climate change are shared by more than one agency, there are "cross walks" in the form of matrices that identify the distributions of program budgets among agencies. At least fifteen executive agencies participate in programs that are explicitly identified as climate change programs in the budget. In addition, several units of the Executive Office of the President are directly involved in budgeting and other policymaking processes. Finally, of course, yet other agencies have responsibilities that bear on climate change programs at least marginally, either on a continuing basis or from time to time.

Congressional processes. There are three types of congressional committees and decisions that are routinely involved in the budgeting process – i.e. authorizing, appropriations, and budget. *Authorizing* committees periodically write and revise legislation that establishes programs, and the legislation often includes provisions concerning amounts of money to be spent. For instance, a program of the Agriculture Department to subsidize farmers for preserving forested land to serve as carbon sinks passes through the agriculture authorizing committees; a program of the Energy Department subsidizing technology to reduce carbon dioxide emissions from coal-fired power plants passes through the energy committees. Proposals to change emissions standards in motor vehicles are considered by the transportation committees. In addition to legislation that authorizes programs, there must also be appropriation bills to provide the funds.[4]

Appropriations committees annually decide on "budget authority" levels for individual programs and agencies, often with much specificity, and these funding levels are embodied in thirteen appropriation bills. These appropriation bills are key legislative vehicles, and they are considered by appropriations subcommittees in the House and a similar array in the Senate. The bills are also considered by the full appropriations committees of both houses. The chairs of these appropriations subcommittees and the committees – as well as the ranking minority party members and other members, and the individual members' and the committees' staff – are highly influential in their particular policy areas.

Yet a third process and set of committees are involved. *Budget* committees annually set overall targets for the entire budget, including expenditures and revenues, and they set targets for major functional categories (such as environment and natural resources). The purpose of the budget resolutions is to impose fiscal restraint by an explicit focus on the macro-economic issue of the relationship between expenditures and revenues. The budget resolution not only establishes a total for expenditures, it also establishes

4 The appropriation amounts must not be greater than – but they can be less than – any applicable authorizing legislation that has been passed. If the appropriated amounts are much less than the authorized amounts, however, they can be the occasion of conflicts between the authorizing and appropriating committees, because the former see significant under-funding as subverting the intention of the legislation that authorized the creation and operation of a program.

subtotals for functions and agencies. Sometimes the budget resolution is used as a legislative vehicle for trying to determine specific sub-functional or programmatic funding levels. In any case, the annual budget resolution is only for internal congressional use, and its amounts are not legally binding on the authorizing or appropriation bills.

Tax committees – i.e. the House Ways and Means Committee, Senate Finance Committee, and Joint Committee on Taxation – are also involved in climate-change-related policy issues whenever tax rates or tax credits are at issue, for instance, proposals to increase taxes on carbon emissions or to reduce taxes through credits for purchases of certain types of vehicles or home appliances.

Conference committees. Differences between the House and Senate versions of a particular item (including authorizing legislation, appropriation bills, and budget resolutions) are reconciled by conference committees consisting of members of both the House and the Senate, and their "reports" (i.e. agreed amounts) must be approved by a majority in each house. Conference committees have their own political dynamics, which are only partly a function of the preferences of the two houses' committees; cross-function and cross-agency bargaining sometimes occurs within the conference committee as committee members consider funding levels for a broad range of programs, functions, and executive agencies, particularly in budget resolution conference committees, where some specific items are of special local interest in the states or districts of individual committee members.

Congressional earmarks. Individual members of Congress sometimes introduce earmarks, which specify that particular projects are to be funded at particular locations. Of course, the project is often located in the member's own state or congressional district and/or has other direct economic benefits for the members' constituents. In effect, an earmark is a legislative procedure that enables a member to add a line item to a budget.

Presidential and executive agency actions. The authorization and appropriation bills, as agreed by both the House and the Senate, must also be agreed to by the President. Formally, a presidential veto can be overridden by Congress only if each house votes to do so by a two-thirds majority. The President does not have a line-item veto. Informally, of course, there is a continual negotiating process between the president and other officials of the executive branch, on the one hand, and congressional committee chairs, party leaders, and other members in both houses – a process that can last for a year or even more for any given fiscal year budget.

There are possibilities for additional spending as programs are implemented during the fiscal year, as a result of "supplemental" appropriations (which have to be passed by Congress and agreed by the president). There are also possibilities for executive branch reductions or delays in previously agreed amounts (within the constraints of laws regulating congressional oversight of "rescissions" and "deferrals").

Budget authority, obligations, and outlays. The three terms – budget authority, obligations, and outlays – reflect differences between legislation by Congress and implementation by executive agencies, and also variability in the exact timing of expenditures. The appropriation bills passed by Congress and signed into law by the president establish *budget authority* for executive agencies to spend money. Not all of the funds that are appropriated necessarily have to be spent in the same fiscal year, though they do have to be spent within a time limit – such as three years – specified by the appropriation bill and/or within the limits of authorizing legislation pertaining to the program. Each year, executive agencies undertake *obligations* to spend money, perhaps for instance in the form of agreements with corporations, universities, or other (private sector or public sector) organizations for R&D projects. Again, these can be over one or more years, but within the limits of the budget authority. After the end of a fiscal year – and only then – can the actual *outlays* of funds during that year be determined, and they are the result of the cumulative amounts of obligations undertaken on the basis of budget authority for that year and previous years.

Continuing Resolutions have become increasingly common as Congress has failed to pass a budget by the beginning of the fiscal year on September 1. Continuing Resolutions are stop-gap measures according to which Congress creates the legal authority for executive agencies to spend money even in the absence of the full array of fourteen appropriation bills, with their specific provisions for agencies' programs. Continuing Resolutions became especially pertinent to climate change issues in 2011 when they were used by new members of the House of Representatives as vehicles for them to register their opposition to numerous administration policies, including its climate change policies (see Chapter 5).

Sources: This description draws upon Congressional Quarterly (2003); Schick (2007); US Congressional Budget Office (2004); US Congressional Research Service (2003a; 2003b); US House of Representatives, Committee on the Budget (2001). Also see Wildavsky and Caiden (2003).

CHAPTER 7

Strengthening international cooperation

Climate change threatens us all; therefore, we must bridge old divides and build new partnerships to meet this great challenge of our time.

US President Barack Obama (2009)

There are two fundamental political economy facts that must be taken into account in efforts to address climate change through international cooperation:

- Climate change mitigation through reductions in GHG emissions is a global public good. The benefits of mitigation are thus available to all countries, regardless of whether they participate in the mitigation efforts.
- Because the international nation-state system is highly decentralized and generally lacking in formalized enforcement mechanisms, it is a challenge to gain participation and compliance in international climate change agreements.

It is therefore tempting for countries to be free riders through non-participation and/or non-compliance, and these core conditions pose formidable obstacles to the achievement of environmentally effective, economically efficient, and politically viable international agreements.

Yet, progress in the development of an international climate regime has nevertheless been made incrementally over more than two decades. In order to understand the evolution of the system and the constraints on it, it is necessary to consider some key issues and the context in which they are addressed. As we have seen in the previous two chapters, international issues overlap with regulating emissions (Chapter 5) and advancing technologies (Chapter 6); but in addition, there is a wide range of other international issues – including institution-building, financing adaptation and mitigation projects in developing countries, and monitoring, reporting, and verifying emissions and mitigation actions.

Many of these issues have traditionally been addressed as "developed-developing country" – or "north-south" – issues, though the differences in interests and divisions on votes are in fact much more complicated. Yet, there are certainly some subgroups of developing countries with distinctive interests; for instance, the need for adaptation and the costs of adaptation, in particular, are especially salient in developing countries with extensive low-lying coastal areas. Some small island countries such as Mauritius even face the prospect of being inundated and disappearing as a result of rising sea levels, while Bangladesh, with large low-lying coastal areas with significant concentrations of populations, faces the prospect of perhaps as many as a hundred million of its citizens needing to relocate farther inland.

Such adaptation issues are often addressed in the context of ethical issues concerning the collective historical responsibilities of the United States and other wealthy industrialized countries that are responsible for most of the accumulated greenhouse gas emissions and thus the elevated concentration levels in the atmosphere that are the core of the climate change problem. As of 2010, the US share of total *cumulative* world GHG emissions during 1950–2010 was 27 percent and the EU-27 countries' share was 22 percent; if the total emissions since the advent of the industrial revolution in the eighteenth century were included, the European and US shares would of course be larger. Among "developing" countries, China's cumulative 1950–2010 share was 10 percent and India's was 3 percent, while Brazil, Indonesia, and South Africa each contributed about 1 percent.

In terms of *annual* emissions, however, China's levels now exceed those of the USA. In fact five developing countries are among the top ten current emitters: Brazil, India, Indonesia, and Mexico, as well as China (computed by the author from World Resources Institute 2010a). Over time, as the annual emissions levels of some developing countries, especially China and India, have increased significantly, the issues about emissions and responsibilities for them have shifted from a simple "developed-developing" dichotomy. As a result, all countries that have relatively high levels of current emissions – whether the "developed" countries of the "north" or the "developing" countries of the "south" – are under increasing pressure from countries that are especially vulnerable to sea level rises and droughts. As we shall see below, the "developed-developing" dichotomy has not been explicit in some recent international agreements.

The distinction between *cumulative* emissions and *annual* emissions, nevertheless, remains crucial to an understanding of the issues, as does the distinction between *emissions levels* and *concentration levels* (see Table 1.3 in Chapter 1 for additional details).

Throughout the present chapter, again, we focus on issues from a political economy perspective. Among the questions considered are those in Section 7.1: How has the multilateral institutional system evolved since the Rio conference in 1992? What positions has the USA taken at key points, particularly at the annual Conferences of the Parties (COPs) of the UN Framework Convention on Climate Change (UNFCCC)? In Section 7.2: What have US policies been concerning financial assistance to developing countries for mitigation and adaptation programs? In Section 7.3: How can monitoring, reporting, and verifying procedures contribute to the development and effectiveness of the multilateral institutions? In Section 7.4, which concerns basic institutional design issues for the international climate change regime: What economic incentives could be used to overcome the free-rider temptations that tend to undermine participation and compliance in international agreements? The concluding section (7.5) focuses on how US international policies are constrained by the domestic political economy factors discussed in previous chapters – a topic which is also taken up in the concluding Chapter 8.

7.1 Multilateral institutional evolution: The Conference of the Parties (COP) of the UN Framework Convention on Climate Change (FCCC)

International conferences devoted specifically to climate change have been occurring since 1979. Here a few of particular interest are highlighted; see Appendix 7.1 for a lengthy list of multilateral meetings held under the aegis of the United Nations.

On the basis of agreements reached at the Rio de Janiero conference in 1992, President George H.W. Bush signed and then submitted the Framework Convention on Climate Change to the US Senate, which ratified it by a unanimous vote of 95 to 0. The FCCC remains the principal multilateral legal framework for international negotiations and programs; its objective is avoiding dangerous anthropogenic interference with the climate system. It recognizes "common but differentiated" responsibilities of all parties; it thus includes commitments of "developed" country Annex I parties – e.g. the United States – to establish national action plans with measures that aim to reduce GHG emissions. Those countries also have obligations to provide technical and financial assistance to non-Annex I "developing" countries. Non-Annex I parties have general obligations, including for GHG mitigation, adaptation planning, and reporting, but they have no specific quantified emissions commitments. Some developing countries have since notified voluntarily-promulgated emissions targets to the UNFCCC, on the basis of the Copenhagen Accord discussed below.

Five years after President George H.W. Bush signed the FCCC, President Clinton signed the Kyoto Protocol in 1997. The same year, however, the Senate voted 97–0 in favor of the Byrd–Hagel resolution, which was a "sense of the Senate" resolution, and therefore a resolution with no legally binding effect. But it had important political consequences, as noted below. In its entirety it said (US Senate 1997):

> Resolved, That it is the sense of the Senate that – (1) the United States should not be a signatory to any protocol to, or other agreement regarding, the United Nations Framework Convention on Climate Change of 1992, at negotiations in Kyoto in December 1997, or thereafter, which would – (A) mandate new commitments to limit or reduce greenhouse gas emissions for the Annex I Parties, unless the protocol or other agreement also mandates new specific scheduled commitments to limit or reduce greenhouse gas emissions for Developing Country Parties within the same compliance period, or (B) would result in serious harm to the economy of the United States; and (2) any such protocol or other agreement which would require the advice and consent of the Senate to ratification should be accompanied by a detailed explanation of any legislation or regulatory actions that may be required to implement the protocol or other agreement and should also be accompanied by an analysis of the detailed financial costs and other impacts on the economy of the United States which would be incurred by the implementation of the protocol or other agreement.

Nevertheless, the Clinton administration signed the Kyoto Protocol, but did not subsequently submit it to the Senate for ratification. These three events – the administration signing it, not submitting it to the Senate for ratification, and the Senate vote on the Byrd–Hagel resolution – became politically significant symbols of the state of climate change issues in the USA not only in domestic politics but also in international climate change negotiations. The events were widely interpreted as indicating the USA could not take much action to address climate change issues because of the domestic political impasse symbolized by the Kyoto Protocol. Climate change became not only an executive-legislative institutional issue but also a partisan issue.

The unanimous vote in the Senate was widely misinterpreted because at least some of the Democratic votes were *not* a vote against doing something about climate change; in fact, they were concerned that the FCCC and Kyoto Protocol did not go far enough to address the problem. But the *perception* of a unanimous Senate sentiment that was against doing something at least internationally became an important ingredient of the domestic dialog. Furthermore, the attention, the associated institutional and partisan conflicts, and the generalized pessimism about the prospects for action on climate change engendered by these circumstances distracted attention from other possible actions and delayed them for several years.

In any case, early in 2001 the George W. Bush administration made clear that it would also not submit the Protocol to the Senate for ratification and furthermore made explicit its opposition to the Protocol itself. Yet, the possibility, even expectation, of eventual

US participation – perhaps after the end of the George W. Bush administration – was nevertheless tacitly still on the international agenda for several years, at least as far as other countries were concerned. The heads of the governments of the UK, Germany, and France, as well as representatives of the EU Commission and Parliament, persisted throughout the George W. Bush administration in actively trying to encourage what they considered would have been more constructive US engagement in the international negotiating process.

The Protocol, which entered into force in 2005 without US participation, set a goal of reducing the total of industrialized countries' emissions of six GHGs – in carbon dioxide equivalent terms – to 5 percent below 1990 levels during the first commitment period of 2008–2012 (with an average for the period of −7 percent for the United States, −8 percent for Europe, and −6 percent for Japan, with −8 percent to +10 percent for other countries). It also provided for "flexibility mechanisms," including trading of emissions credits from emission-reducing projects, through the Joint Implementation and Clean Development Mechanism.

In 2007 at the annual COP, which was in Bali, Indonesia, there were important procedural and substantive agreements that restructured the international agenda and negotiating processes in order to make progress on negotiating a follow-on agreement to the Kyoto Protocol after its scheduled end in 2012. That process did not include the USA because of its non-participation in the Protocol; however, a second, parallel negotiating process was established to include the USA in wider negotiations on the future of the international climate regime. In particular, the Bali Action Plan established the Ad Hoc Working Group on Long-term Cooperative Action (AWG-LCA), including the USA, with a mandate for parties to the FCCC to negotiate toward new GHG mitigation actions and commitments in the post-2012 period. The plan identified four main elements for negotiation and inclusion in future agreements, namely mitigation, adaptation, technology, and finance.

The Copenhagen Accord, COP-15

In 2009, the COP-15 meeting in Copenhagen, Denmark, was a landmark event in the evolution of the international climate regime. Instead of a new multilateral agreement by the nearly 200 governments in attendance, there was an agreement among a small group of national government leaders, including President Obama, who met outside the formal COP process. The process and the resulting agreement were shocks to the international negotiating process.

As a prelude to the meeting at which the Copenhagen Accord was agreed among Brazil, China, India, South Africa, and the USA, there was a tense meeting chaired by Denmark, host of the COP, and attended by Obama as well as representatives of several other countries. The following unofficial account (Rapp, Schwägerl and Traufette 2010; © *Der Spiegel* 2014) of that meeting is based on a recording which offers a glimpse into a pivotal event in the evolution of international climate change policymaking. The discussion among the government leaders at the meeting provides an encapsulated representation of central features of the international context of US government policy-making as well as statements of the positions taken by the US President. Because of the importance of the meeting and because the transcript offers a highly unusual opportunity to learn the details of a closed meeting of high-level officials, it is quoted at length:[1]

> *Thursday, December 17, 2009:* Environment ministers and [other officials] had presented their bosses with a 200-page bundle of documents, because they had been unable to agree on emissions levels, reduction measures and control measures. When the heads of state and government arrived on Thursday, they were shocked by the chaos their subordinates had left for them after 10 days of negotiations.... Together, [the leaders] asked the Danish host [Prime Minister Rasmussen] to reduce the maze of documents to a few, key pages. They still contained bold statements, such as the goal of a 50 percent reduction in global CO_2 emissions by 2050 (compared with a 1990 benchmark). That kind of a commitment would have required that the United States, China and India also agree to cut their greenhouse gas emissions in half.
>
> *Friday, December 18, 2009:* Danish Prime Minister Rasmussen opened the meeting....
>
> [T]he Chinese negotiator looked at the document from the previous evening and said: '... given the importance of the paper, we do not want to be rushed.... We need some more time.'
>
> There were still two important placeholders, X and Y, in the draft agreement. They marked the spots where the percentage targets for reductions in greenhouse gas emissions, for the industrialized nations and emerging countries respectively, were to be entered. 'We cannot go over and say nice things but x and y wait please one year or so,' [German Chancellor] Merkel said. The German chancellor was determined to secure a commitment from China and India to participate in the climate protection efforts. But China and India were unwilling to make that commitment. Behind the backs of the

1 Although the recording of the meeting was not officially authorized and is thus not the basis of an authorized transcript, it does offer an unusual opportunity to "observe" indirectly the interactions of world leaders as they tried to reach agreement on next steps in the evolution of the international climate regime. See Rapp, Schwägerl, and Traufette (2010) about the origins of the transcript.

Europeans, they had apparently reached their own agreement with Brazil and South Africa. 'We have all along been saying "Don't prejudge options!"', said a representative of the Indian delegation, prompting Merkel to burst out: 'Then you don't want legally binding!' This, in turn, prompted the Indian negotiator to say angrily: 'Why do you prejudge options? All along you have said don't prejudge options and now you are prejudging options. This is not fair!' Chinese negotiator He Yafei stood by this remark.

British Prime Minister Brown . . . tried to mediate. 'I think it's important to recognize what we are trying to do here,' he said. 'We are trying to cut emissions by 2020 and by 2050. That is the only way we can justify being here. It is the only way we can justify the public money that is being spent to do so. It is the only way we can justify the search for a treaty.' Norwegian Prime Minister Jens Stoltenberg pointed out that it was the Indians who had proposed the inclusion of concrete emissions reductions for the industrialized nations in the treaty. But India had made an about-face within hours and was no longer interested in [its] own proposal.

The Chinese diplomat [He Yafei] refused to give in to the Europeans' demands, saying: 'Thank you for all these suggestions. We have said very clearly that we must not accept the 50 percent reductions. We cannot accept it.'

The West, [French President] Sarkozy said, had pledged to reduce greenhouse gas emissions by 80 percent. 'And in return, China, which will soon be the biggest economic power in the world, says to the world: Commitments apply to you, but not to us.' Sarkozy . . . then said: 'This is utterly unacceptable!' And then the French president stoked the diplomatic conflict even further when he said: 'This is about the essentials, and one has to react to this hypocrisy!' A hush came over the room.

Finally, . . . US President Barack Obama [spoke]. Like the Europeans, the US president was also intent on securing a commitment to protect the climate from the new economic superpowers, China and India. 'I think it is important to note that there are important equities that have to be considered,' he said, with a distinctive note in his voice that suggested the foresight of a statesman. Obama reminded his fellow leaders that the industrialized nations are also dependent on the will of their citizens to contribute to saving the climate. 'From the perspective of the developed countries, in order for us to be able to mobilize the political will within each of our countries to not only engage in substantial mitigation efforts ourselves, which are very difficult, but to also then channel some of the resources from our countries into developing countries, is a very heavy lift,' Obama said. Then, speaking directly to China, he added: 'If there is no sense of mutuality in this process, it is going to be difficult for us to ever move forward in a significant way.' . . . But then Obama [contradicted] the Europeans . . . , saying that it would be best to shelve the concrete reduction targets for the time being. 'We will try to give some opportunities for its resolution outside of this multilateral setting. . . . And I am saying that, confident that, I think China still is as desirous of an agreement, as we are. Obama even [implicitly but clearly] downplayed the importance of the climate conference, saying '... [W]e are not staying until tomorrow. I'm just letting you know. Because all of us obviously have extraordinarily important other business to attend to.'

[The Chinese representative] He Yafei decided to give the group a lesson in history: 'People tend to forget where [the climate problem] is from. In the past 200 years of industrialization developed countries contributed more than 80 percent of emissions. Whoever created this problem is responsible for the catastrophe we are facing.'... 'I have a procedural question,' He Yafei said. 'I kindly ask for a suspension of a few minutes for consultation. We need some time of consultation.' What he meant was that he wanted to make a phone call to his prime minister.

'How long?' [Chancellor] Merkel asked. The chairman, Rasmussen, made the decision. 'We meet again (at) half past four. Forty minutes.'

... But the meeting did not reconvene. The key decisions were made elsewhere – without the Europeans. The Indians had reserved a room one floor down, where Prime Minister Singh met with his counterparts, Brazilian President Lula da Silva and South Africa President Jacob Zuma. Wen Jiabao was also there. Shortly before 7 p.m., US President Obama burst into the cozy little meeting of rising economic powers. At that meeting, everything that was important to the Europeans was removed from the draft agreement, particularly the concrete emissions reduction targets.

The result of the meeting among the USA, Brazil, China, India, and South Africa was the Copenhagen Accord, which was initially made public by President Obama during a press conference in Copenhagen. It was negotiated by a small, self-selected group of countries. The Accord did not reflect long-standing European preferences for specific emissions targets and timetables, which had been the principal concerns for more than a decade of European leadership in international climate change fora. It was not a result of a formally constituted COP process. Rather, it was merely a statement signed by the heads of government or state of five countries. But it served the US president's domestic political purposes: it was an international agreement that could be touted as a tangible result of US international "leadership" – without making any commitments that could cause him trouble with Congress or the public. It was not a treaty and thus not subject to Senate ratification; nor was it likely to be a salient issue in the then-forthcoming congressional elections eleven months later.

As for its content, the Accord was marked by a combination of brevity and imprecision. The brevity was a result of the absence of agreement on a large array of issues. The imprecision was the result of the absence of agreement on key amounts of money and levels of emissions. See Box 7.1 for a summary of the Accord.

Box 7.1 Summary of provisions of the Copenhagen Accord

1. The signing governments recognize "the scientific view that the increase in global temperature should be below 2 degrees Celsius" [compared with pre-industrial levels].
2. "Agree that deep cuts in global emissions are required."
3. Agree that "enhanced action and international cooperation on adaptation is urgently required" and "that developed countries shall provide adequate, predictable and sustainable financial resources, technology and capacity-building to support the implementation of adaptation action in developing countries."
4. "Annex I Parties [to the UNFCCC] commit to implement individually or jointly the quantified economy-wide emissions targets for 2020 in Appendix I" [which was blank at the time of the agreement but subsequently filled in with numerical targets indicated below]. "Delivery of reductions and financing by developed countries will be measured, reported and verified . . . ."
5. "Non-Annex I Parties [to the UNFCCC] will implement mitigation actions, including those . . . in Appendix II" [which was blank at the time of the agreement but subsequently filled in with numerical targets indicated below]. "Mitigation actions taken by Non-Annex I Parties will be subject to their domestic measurement, reporting and verification. . . . [Mitigation] actions [undertaken with international support] will be subject to international measurement, reporting and verification . . . ."
6. ". . . agree on the need to provide positive incentives to [actions to reduce deforestation and forest degradation] through the immediate establishment of a mechanism including REDD-plus, to enable the mobilization of financial resources from developed countries."
7. "Developing countries . . . should be provided incentives to continue to develop on a low emission pathway."
8. "Scaled up, new and additional, predictable and adequate funding . . . shall be provided to developing countries . . . [by developed countries for] . . . mitigation, including [for] REDD-plus, adaptation, technology development and transfer and capacity-building . . . ." The developed countries agreed to provide an amount "approaching" $30 billion during 2010–2012, and $100 billion per year beginning in 2020 (the latter to come from a "wide variety of" public and private sources).

9. A High Level Panel was established to study the "potential sources" for the $100 billion a year noted above in 8 – a group that subsequently prepared such a report.
10. Agreed to establish a Green Climate Fund to support the activities in 6 and 8 above.
11. Agreed to establish a Technology Mechanism "to accelerate technology development and transfer" for support of mitigation and adaptation.
12. Review in 2015 of the implementation of the Accord, including the possibility of reducing the temperature increase to 1.5 degrees Celsius.

Source: Adapted by the author from UNFCCC (2009). The numbers of the items here correspond to the numbered provisions in the Accord.

Despite the irregular circumstances of its origins, the Copenhagen Accord became an important building block in the further construction of an international climate regime. By a year later, 114 countries had indicated their "agreement" with the Accord, and an additional twenty-six had reported their intention formally to indicate their agreement, for a total of 140. The twenty-seven countries of the EU, plus fifteen other Annex I countries, including the United States, had submitted their supplementary statements for Appendix I (see the US statement in Box 7.2). As for non-Annex I countries, forty-four had submitted their Appendix II supplementary statements by early 2011 (see China's statement in Box 7.3). The emissions pledges associated with the Copenhagen Agreement and subsequent agreements have been analyzed in detail in simulations that have found the cumulative impacts of such pledges, even if they were actually met over time, to be insufficient for keeping the average global warming to 2 degrees centigrade. (See Appendix 7.2 for a description of a widely noted simulation system that estimates the gap between the stated emission goals and projections based on actual and projected policies.)

In terms of the evolution of US climate diplomacy, the negotiations in Copenhagen represented a landmark event in three respects: First, the USA rejected the European emphasis on the centrality and urgency of establishing binding internationally agreed emission targets and timetables.[2] Second, the USA recognized the importance of

2 These actions by the Obama administration at the Copenhagen meeting in 2009 marked the second time US administrations directly thwarted long-standing European goals for the international climate agenda. The first was in 2001 when the George W. Bush administration announced that the Kyoto Protocol was "dead" without previously discussing their views or notifying the Europeans in advance of the announcement.

Box 7.2 US commitments in the Copenhagen Accord

The US "quantified economy-wide emissions targets for 2020" compared with emissions levels in 2005 are "[i]n the range of 17 percent, in conformity with anticipated US energy and climate legislation, recognizing that the final target will be reported to the Secretariat in light of enacted legislation."[1]

[1] "The pathway set forth in pending legislation would entail a 30 percent reduction in 2025 and a 42 percent reduction in 2030, in line with the goal to reduce emissions 83 percent by 2050."[a]

[a] N.B. The legislation that was envisioned was not enacted (see Chapter 5 for details).

Source: UN Framework Convention on Climate Change (2011a). "%" symbols in the original documents were changed to "percent" for stylistic consistency.

Box 7.3 Chinese commitments in the Copenhagen Accord

China will endeavor to lower its carbon dioxide emissions per unit of GDP by 40–45 percent by 2020 compared to the 2005 level, increase the share of non-fossil fuels in primary energy consumption to around 15 percent by 2020 and increase forest coverage by 40 million hectares and forest stock volume by 1.3 billion cubic meters by 2020 from the 2005 levels.

Source: UN Framework Convention on Climate Change (2011b).

reaching an agreement with large developing countries, and it intensified a process of direct engagement with them – especially China, India, Brazil, and South Africa. Third, the Copenhagen meetings publicly signaled a significant shift in the US government's approach to climate diplomacy concerning international venues. It was a shift away from a focus on large, formal multilateral venues of the annual COP meetings in which nearly 200 countries participate and where consensus among all of them is sought, and a shift toward greater reliance on bilateral, regional, sectoral, and other venues.[3] In particular, the USA has increasingly favored the relatively small venues of the G-20

3 Climate change policymaking in the USA thus took on a similar approach to the long-standing US approach to international trade–investment–technology transfer issues, namely to formalize a large number of bilateral, regional, and sectoral agreements that overlap in many respects with the multilateral system.

countries and the Major Economies Forum (MEF) of seventeen countries.[4] The result for the USA and the rest of the world has been that the international climate negotiating process and the evolving institutional arrangements are much more fragmented than previously. Nevertheless, the multilateral negotiating process based on the FCCC has continued, and in 2010 it yielded the Cancun Agreements.

Cancun Agreements, COP-16

The Cancun Agreements of 2010 reflected a resuscitation of the multilateral negotiating process that had been circumvented by the USA a year earlier in Copenhagen, and it also marked a re-engagement of the USA in that multilateral process, after the congressional elections of that year. The Cancun Agreements were thus widely regarded as a symbolic triumph of multilateral diplomacy. However, they were also recognized as incorporating some substantive achievements, particularly in dampening somewhat the "north–south impasse" by recognizing "developing" countries' voluntary GHG emission targets. They furthermore included recognition of the "developed" countries' pledges of short-term (so-called "fast-track") financial assistance of $30 billion over three years and a longer-term pledge of $100 billion per year by 2020 (see Box 7.4 for the details).

Yet, the Cancun agreements also deferred decisions about the future of the Kyoto Protocol after the expiration of its first commitment period in 2012, and they only provided sketchy outlines of arrangements for financing programs and technology development and transfer programs, without indications of the amounts of funds to be committed to either one. Despite their limitations, the Cancun Agreements were widely seen as accomplishing at least incremental progress, in part because they reflected a re-engagement of the USA in the formal multilateral negotiating process. Such a re-engagement was not only symbolically important at the time; it also offered hope for more tangible and constructive US engagement in future COPs. (For further commentary on the Cancun Agreements, see assessments by Diringer (2010); Stavins (2010b); and World Resources Institute (2010d).)

Durban Platform, COP-17

With US acquiescence but without the USA playing an active leadership role, COP-17 in Durban, South Africa, in 2011 agreed to launch a new negotiating process within the

4 A derivative group – the Clean Energy Ministerial (CEM) – has become an important and unusual arrangement that includes not only the major energy-consuming countries but also leading countries in clean-energy technologies (Denmark, Finland, Norway, and Sweden).

Box 7.4 Excerpts from the Cancun Accords

The Cancun Agreements are embodied in two documents. One is a lengthy report on the "Outcome of the Work of the Ad Hoc Working Group on Long-Term Cooperative Action (AWG-LCA)," which contains the key decisions and other items, and the second is a brief report on the "Outcome of the Work of the Ad Hoc Working Group on Further Commitments for Annex I Parties Under the Kyoto Protocol." The latter merely kept open the possibility that there would be an extension or sequel following the expiration of the first commitment period of the Protocol at the end of 2012. The following excerpts are from the AWG-LCA document. The numbers refer to its paragraph numbers. In all references, it is the Conference of the Parties of the FCCC that has recorded its agreements; thus the Conference of the Parties:

"1. *Affirms* that climate change is one of the greatest challenges of our time and that all Parties share a vision for long-term cooperative action... [that] addresses mitigation, adaptation, finance, technology development and transfer, and capacity-building in a balanced, integrated and comprehensive manner to enhance and achieve the full, effective and sustained implementation of the Convention, now, up to and beyond 2012;

"2. *Further affirms* that:

"(a) Scaled-up overall mitigation efforts that allow for the achievement of desired stabilization levels are necessary, with developed country Parties showing leadership by undertaking ambitious emission reductions and in providing technology, capacity-building and financial resources to developing country Parties....

"(b) Adaptation must be addressed with the same priority as mitigation and requires appropriate institutional arrangements to enhance adaptation action and support....

[Mitigation]

"4. *Further recognizes* that deep cuts in global greenhouse gas emissions are required according to science, and as documented in the Fourth Assessment Report of the Inter-governmental Panel on Climate Change, with a view to reducing global greenhouse gas emissions so as to hold the increase in global average temperature below 2°C above pre-industrial levels, and that Parties should take urgent action to meet this long-term goal, consistent with science and on the basis of equity...

"36. *Takes note* of quantified economy-wide emission reduction targets to be implemented by Parties included in Annex I to the Convention as communicated by them . . .

"49. *Takes* note of nationally appropriate mitigation actions to be implemented by non-Annex I Parties as communicated by them . . .

[Mitigation Monitoring, Reporting and Verification]

"61. *Also decides* that internationally supported mitigation actions will be measured, reported and verified domestically and will be subject to international measurement, reporting and verification . . .

"62. *Further decides* that domestically supported mitigation actions will be measured, reported and verified domestically . . .

[Deforestation]

"70. *Encourages* developing country Parties to contribute to mitigation actions in the forest sector by undertaking the following activities, as deemed appropriate by each Party and in accordance with their respective capabilities and national circumstances:

"(a) Reducing emissions from deforestation;
"(b) Reducing emissions from forest degradation;
"(c) Conservation of forest carbon stocks;
"(d) Sustainable management of forest;
"(e) Enhancement of forest carbon stocks . . .

[Markets to promote mitigation]

"80. *Decides* to consider the establishment, at its seventeenth session, of one or more market-based mechanisms to enhance the cost-effectiveness of, and to promote, mitigation actions . . .

[Finance]

"95. *Takes note* of the collective commitment by developed countries to provide new and additional resources, including forestry and investments through international institutions, approaching USD 30 billion for the period 2010–2012, with a balanced allocation between adaptation and mitigation . . .

"98. *Recognizes* that developed country Parties commit, in the context of meaningful mitigation actions and transparency on implementation, to a goal of mobilizing jointly USD 100 billion per year by 2020 to address the needs of developing countries . . .

"102. *Decides* to establish a Green Climate Fund, to be designated as an operating entity of the financial mechanism of the Convention . . . , with arrangements to be concluded between the Conference of the Parties and the Green Climate Fund to ensure that it is accountable to and functions under the guidance of the Conference of the Parties, to support projects, programmes, Policies and other activities in developing country Parties . . .

"107. *Invites* the World Bank to serve as the interim trustee of the Green Climate Fund, subject to a review three years after operationalization of the fund . . .

[Technology development and transfer]

"117. *Decides* to establish a Technology Mechanism . . . , which will consist of the following components:

"(a) A Technology Executive Committee . . .

"(b) A Climate Technology Centre and Network . . . "

Source: Excerpted from UN Framework Convention on Climate Change (2010).

context of the Ad Hoc Working Group on the Durban Platform for Enhanced Action. Its mandate is "to develop a protocol, another legal instrument or a legal outcome under the [UNFCCC] applicable to all Parties," with agreement by 2015 and entry into force by 2020 (UNFCCC 2011c). It reflects a willingness at the time of only agreeing to agree on an unspecified something within four years.

Yet, it also marked a new paradigm for a future agreement in as much as it did not explicitly incorporate the long-established distinction between "developing" and "developed" countries, with "common but differentiated responsibilities." It also called for negotiations on a broad agenda, including "mitigation, adaptation, finance, technology development and transfer, transparency of action, and support and capacity-building." This represented an expansion of the work on the four pillars agreed at the Bali COP, and thereby created expectations of further tangible progress in the future.

Doha agreements, COP-18

The results of the COP-18 meeting in Doha included finalizing arrangements for the second commitment period of the Kyoto Protocol (2013–2020) – which was an important accomplishment – and consolidating the two separate negotiating "tracks" that had been in existence since 2005 in order to accommodate the USA and other countries outside the Kyoto Protocol negotiating process (UNFCCC 2012). In addition, a variety of administrative matters concerning finance, deforestation, and technology transfer programs were addressed, again as further progress on the Bali four pillars. One other potentially important development was an agreement to develop institutional arrangements to address the "loss and damage" faced by many countries, especially small and vulnerable islands; these are both serious international legal issues and international economic assistance issues. More generally, however, the Doha meeting did not produce any significant changes in the previously established incremental path of mostly marginal adjustments in existing institutional arrangements.

7.2 Financing adaptation and mitigation in developing countries

An increasingly salient item on the agenda of multilateral conferences – at Doha and otherwise – has been financing issues. In order to understand the US position on these issues, it is necessary to understand both the nature of the pledges made by the USA in the context of international climate negotiations and also the progression of the requested appropriations through the annual congressional budgeting process (see especially Fransen, Stasio, and Nakhooda 2012).

As to the publicly pledged amounts, US administrations have made three such pledges.[5] The first was a pledge of the George W. Bush administration, along with the governments of Japan and the UK, to contribute a total of $5 billion over three years (2008–2010) to a new Clean Technology Fund (CTF) at the World Bank. The exact share of each of the three governments was not explicitly made public; however, one-third would have been $1.67 billion or an average of $555 million each year from each country. If the intention had been to make the shares proportional to the countries' GDPs, then of course the US share would have been much larger. In any case, the official US pledge was ultimately $1.492 (World Resources Institute 2010b; 2010c). The actual US contributions were at most $509 million during the US fiscal year 2010 (October 2009 to September 2010) and $351 million in fiscal year 2011. The administration's request to Congress for FY2011 was $400 million, but that was reduced (see Table 7.1).

5 This discussion of "pledges" uses the word as it is used in common discourse, not as a term of international law.

Table 7.1 US multilateral budget for multilateral foreign assistance programs (for climate change mitigation and/or adaptation) (US dollars, millions)

	FY2010 enacted	FY2011 request	House Appropriation Committee	Senate Appropriation Committee
Global Environment Facility	86.5	175	143	149
Clean Technology Fund	300	400	300	370
Special Climate Fund	75	235	150	126
Total	461.5	810	593	745
Percentage change from administration's request	NA	NA	−28.6 percent	−10.6 percent
Percentage change from previous year	NA	NA	+28.5 percent	+56.8 percent

Source: Compiled by the author from US Office of Management and Budget (2010a); Natural Resources Defense Council (2010); Geithner (2010); World Resources Institute (2012).

A second US pledge, which was made with other Annex I countries during the 2009 conference in Copenhagen, was to participate with such countries in a total contribution of $30 billion over the three-year period 2010–2012. This is the pledge that was explicitly noted a year later in the Cancun Agreements as the "Fast Start" pledge. Again, however, the shares to be contributed by the United States and the other countries were not initially specified. The actual funds transmitted by the USA in the form of appropriations through the annual budget process were $2.0 billion in FY2010 and $3.1 billion in FY2011.[6] However, these amounts were not entirely – and arguably not mostly – "new and additional."

In addition to these shortfalls in actual US contributions compared with the pledges, there has been doubt about whether these amounts represented "new and additional" funds beyond those in previous years in the international assistance program. Also, the USA includes its contributions to the World Bank Clean Technology Fund as part of its "Fast Start" pledge.

A third pledge made by the "developed" countries – this one for the longer term – was also initially made at the Copenhagen conference and reiterated in the Cancun

6 Five countries of the European Union indicated they would contribute $3.36 billion (equivalent to approximately €2.33 billion) per year, and Japan indicated it would contribute $5 billion per year. The total for all countries for three years was about $23 billion, i.e. $7 billion less than the previously announced amount (World Resources Institute 2010b; 2010c).

Agreements, namely $100 billion per year by 2020. In addition to uncertainty about the US share, an important element of ambiguity in this pledge is that it implicitly includes private sources of financing as well as governmental sources. Since the government cannot make private entities actually provide such funds, it is not clear exactly what the implications for governments' responsibilities are to "commit . . . to a goal of mobilizing jointly $100 billion per year by 2020 to address the needs of developing countries."[7]

There is a pattern, then, of US administrations making pledges of certain amounts of money by a certain time and then not meeting the terms of pledges. The domestic US institutional reality underlying the pattern is that administrations make the pledges in international venues, but they then must gain congressional approval of the funding through the annual budget appropriation process. In the context of the annual budgeting process, there is a long tradition in international program budgeting for presidents to request more than Congress is willing to appropriate – tendencies that have long been especially pronounced in funding international economic assistance programs. Since most of the funding for international climate change programs is embedded in the international economic assistance budgets of several agencies, there has been a strong pattern of congressionally enacted levels of funding being substantially below the administration's requested amounts. The pattern is evident in Table 7.1.

7.3 Monitoring, reporting, and verifying emissions and actions

The acronym "MRV" (for monitoring, reporting, and verifying) has entered the FCCC negotiating agenda lexicon in several contexts. Developed countries, especially the USA, are concerned that any financial support they provide for developing countries' mitigation or adaptation efforts is tied to the recipient countries' progress in reducing their emissions levels; the developed countries therefore want to know – and be able to report credibly to domestic constituents and stakeholders – the levels of emissions and reductions in the recipient countries on a country-by-country basis. For their part, developing countries want to see evidence that developed country commitments about financing have in fact been met. Though these two expressed desires for MRV are connected through a *quid pro quo* in international negotiating

7 A UN High-level Advisory Group on Climate Change Financing (2010) undertook an extensive analysis of the topic and concluded that it is "a challenging but feasible" goal.

processes, they clearly require quite different technical capabilities and administrative procedures.

The need for monitoring, reporting, and verifying countries' GHG emissions has been embodied in international agreements – for instance in paragraphs 4 and 5 of the Copenhagen Accord. At his press conference at the Copenhagen meetings, President Obama noted that "... [W]e can actually monitor a lot of what takes place [in greenhouse gas emissions] through satellite imagery and so forth. So I think we are going to have a pretty good sense of what countries are doing" (US White House 2009). Just before and after the Copenhagen meetings, the governments of the UK and France indicated their interest in GHG monitoring technologies, not only for monitoring emissions within the EU but also in other parts of the world (*Telegraph* 2010). At the EU level, the European Commission on behalf of the EU and the European Space Agency are in the process of enhancing their capabilities in this regard. The Japanese also already have in place some capabilities of this type, and the government of India announced its intention to have its first such satellite in operation by 2012 and its second in 2013 to monitor GHG emissions "across the country and globe" (*Times of India* 2010). Verification procedures are thus likely to be issues in international negotiating arenas and administrative processes for many years. They are already attracting interest in domestic deliberations on US national climate change policymaking.

There is much information in the public domain about the technical capabilities for monitoring emissions and removals by sinks in agriculture, forestry, and other land uses – including the UN program REDD, whose purpose is Reducing Emissions from Deforestation and Forest Degradation in Developing Countries (see especially US National Research Council 2010). Though less is publicly known about existing or prospective capabilities for satellite-based monitoring of other kinds of site-specific sources of greenhouse gas emissions, it is clear that questions about international cooperation, sharing, and transfers of these technologies will be on the international climate change negotiating agenda. There is interest – particularly in Europe – in the possibility of creating a new international agency that would be responsible for such monitoring. This could be in addition to whatever system the EU itself develops for its own regional purposes. Finally, regardless of any issues about the prospective international institutionalization of satellite-based monitoring methods, there are likely to be dual-use issues – namely, the use of such monitoring capabilities not only for greenhouse gas emissions monitoring, but also for military uses; and dual-use issues apply to US systems and to the systems of other countries.

7.4 Core international institutional design issues

During the more than two decades of efforts to develop an international climate regime, of course there has been much discussion among policymakers, scholars, and stakeholders about the core features such a regime should have. The roles of economic incentives to encourage participation in international regimes (i.e. discourage free riding) and encourage compliance have been among the most contentious institutional design issues. This is so in part because the two largest emitters of greenhouse gases – the USA among "developed countries" and China among the "developing countries" – are commonly regarded as free riders. Issues about how to reduce free riding have thus received much attention from scholars as well as policymakers. The analyses of Barrett (1994; 1997; 2001; 2003; 2007; 2010) and Victor (2001; 2004; 2011) have been especially prominent on these issues; also see Buchner, Carraro, Cersosimo, and Marchiori (2002) as well as Carraro (2007), Frankel (2007), McKibbin and Wilcoxon (2007), Pizer (2007), and Schelling (2007) in Aldy and Stavins (2007b).

In order to change free-riding incentives, there are possibilities for creating linkages to other issues and regimes concerning, for instance, international trade and international technology transfer. As for linkages to international trade, in Chapter 5 we noted that the US Congress has considered the possible use of offsetting border measures on imports from countries whose measures to address climate change mitigation would be considered insufficient from a US perspective. The specific institutional context was congressional consideration of cap-and-trade legislation in 2009–2010, and the possible use of offsetting border measures was raised especially in relation to US–China trade (Van Asselt and Brewer 2010; Van Asselt, Brewer, and Mehling 2009).

Fundamentally similar – but mirror-image – issues have arisen in other countries vis-à-vis the USA. This has been particularly true in Europe, where the EU Emissions Trading System has been in operation for many years and where there are sensitivities about both the environmental problem of potential international leakages of GHG emissions if/as production facilities shift internationally and the associated economic problem of international competitiveness. There have been recurrent calls in Europe for the imposition of offsetting border measures on imports from the USA in order to reduce the environmental and economic impacts of the absence of serious US measures to mitigate its GHG emissions (Gros, Egenhofer, Fujiwara, Georgiev, and Guerin 2010; also Brewer 2007; 2010).

Beyond issues about the use of offsetting border measures, however, there are a large number of other issues about the interactions of the international climate regime

Box 7.5 Issues in the nexus of climate change and international trade[a]

- Offsetting border measures that address international competitiveness and free-rider concerns (see Chapter 5);
- Tariffs and non-tariff barriers to trade, investment, and technology transfer of climate-friendly goods and services (see Chapter 6);
- The adequacy of energy efficiency and renewable energy product labeling and standards in international trade;
- Government programs that subsidize exports, foreign direct investments, and technology transfers, especially to emerging economies;
- Government subsidies for renewable energy and energy efficiency goods and services;
- Domestic purchasing restrictions on government procurement of renewable energy goods and services that could otherwise be imported;
- GHG emissions in the international aviation and maritime shipping industries, and the agendas of the International Civil Aviation Organization (ICAO) and the International Maritime Organization (IMO).

[a] Note that this is intended to be an illustrative list, not a comprehensive list.
Sources: Adapted from Brewer (2008b; 2010); also see www.TradeAndClimate.net.

and the international trade regime, and thus the design of the international climate regime. For instance, the US government launched complaints against China through the World Trade Organization (WTO) Dispute Settlement process because of China's subsidies for the manufacture of wind power and solar power equipment (USTR 2013; WTO 2013). Further, of the total array of fifty-something WTO agreements, all or nearly all are relevant to climate change issues and related energy issues, including international transfers of climate-friendly technologies (Brewer 2003; 2004a; 2004b; 2009b; 2010; 2012). These include agreements such as the Technical Barriers to Trade Agreement, the General Agreement on Trade in Services, the Trade Related Investment Measures Agreement, the Subsidies and Countervailing Duties Agreement, and the Anti-Dumping Agreement, as well as the Dispute Settlement Understanding (see Box 7.5).

In addition to these WTO agreements, there are industry-specific international agreements that are within the UN system but outside the WTO – in particular the airline industry, whose international trade rules are under the auspices of the International

Civil Aviation Organization (ICAO), and the maritime shipping industry, whose international trade rules are under the auspices of the International Maritime Organization (IMO). The rising levels of GHG emissions in both industries are increasingly salient, at least among climate change policymakers and analysts. There are also regional trade organizations such as NAFTA, whose trade agreements have implications for climate change mitigation and adaptation. Finally, there are many hundreds of bilateral trade and investment agreements with potential implications for climate change mitigation or adaptation. In all of these domains, a key institutional design issue for the international climate regime is how that regime should interact with other international regimes, including the international trade–investment–technology transfer regime.

Beyond the details of such institutional design issues about particular aspects of the interactions of the international climate regime and the international trade regime, there are yet more basic issues about the relevant criteria for developing a multilateral climate regime. A list of such criteria, including "the importance of incentives for participation and compliance," as proposed by Aldy, Barrett, and Stavins (2003; also see Aldy and Stavins 2007b) is indicative of the magnitude of the challenge to climate change diplomacy (see Box 7.6).

7.5 Implications: domestic politics and international free riders

As a result of the US record in international climate change mitigation and adaptation efforts, the USA is widely perceived outside the USA as being a laggard – and occasionally as an obstructionist – especially in multilateral venues. Even though a reputed international leader on many other issues, the USA has consistently been considered to be behind most other "developed countries" – and many "developing countries" as well – in its willingness to address climate change seriously. It is thus widely seen as lacking moral authority on climate change issues. Indeed, it is widely seen as a "free rider," and its role as a "leader" in climate diplomacy has thus been viewed as controversial at best.

However, as we noted repeatedly in considerable detail in Chapters 5 and 6, US administrations are constrained in their ability to respond to international concerns by the domestic institutional context of policymaking. We have also noted in the present chapter, for instance, that the two-thirds positive vote required in the Senate for treaty ratification, and the need for the House of Representatives as well as the Senate to approve annual funding of international economic assistance programs restrict administrations' responsiveness to international pressures. Administrations, it

Box 7.6 Criteria for a multilateral climate change agreement

(1) Environmental outcome – including the time path of emissions or concentrations of greenhouse gases.
(2) Dynamic efficiency – maximizing the aggregate present value of net benefits of taking actions to mitigate climate change impacts.
(3) Dynamic cost-effectiveness – the least costly way to achieve a given environmental outcome.
(4) Distributional equity – the distribution of both benefits and costs across populations within a generation and across generations, and can account for responsibility for climate change, ability to pay to reduce climate change risks, and other notions of equity.
(5) Flexibility in the presence of new information – a flexible policy infrastructure built on a sequential decision-making approach that incorporates new information.
(6) Participation and compliance – incentives for participation and compliance are important, since a climate policy architecture that cannot promote participation and compliance will not satisfactorily address the climate change problem.

Source: Excerpted by the author from Aldy and Stavins (2007a), 10–11; also see Aldy, Barrett, and Stavins (2003). Reprinted with permission.

should be noted, also have a habit of pleading with their international counterparts that congressional constraints prevent the administration from making concessions in international negotiations that they do not want to make anyway. It is a well-known US negotiating tactic.

The constraints and the opportunities for leadership – domestic as well as international – are the subject of the next chapter.

Suggestions for further reading and research

A series of books by Aldy and Stavins (2007b; 2009; 2010) include contributions by many of the most recognized experts on a wide range of topics, and they are essential reading on international institutional design and negotiation issues, as is Victor (2011). Young (1982; 1986; 1989; 2002) has undertaken extensive analyses of the conceptual and factual aspects of international environmental regimes; also see Breitmeier, Young, and Zürn (2006); and Oye and Maxwell (1995). For analyses of the international climate regime, in particular, see Helm and Sprinz (2000), and the contributions to Aldy and Stavins (2007b) by Barrett (2007), Frankel (2007), McKibbin and Wilcoxen (2007), Michaelowa (2007), Pizer (2007), Schelling (2007) and Victor (2007). Antholis and Talbott (2010) provide details on US climate change diplomacy.

The deficiencies of the Kyoto Protocol from a game theoretic and political economy perspective have been emphasized in numerous publications by Barrett (1994; 1997; 2001; 2003; 2007; 2010); also see Buchner, Carraro, Cersosimo, and Marchiori (2002) and Victor (2001; 2004; 2007). Cornes and Sandler (1996) provide an extensive analysis of "club goods."

Monitoring, reporting, and verification issues are addressed in Fransen (2009); the capabilities of satellite-based remote sensing of GHG emissions as well as other technologies, are discussed in US National Research Council (2010).

There is now a sizable body of studies of the interactions of the international climate regime and the international trade regime – including Brewer (2003; 2004a; 2004b; 2007; 2009b; 2010; also Brewer and Lunden 2006). For issues concerning international transfers of climate-friendly technologies, see Brewer (2008a; 2008b; 2008c; 2009b). Issues in the nexus of international trade–investment–technology transfers with climate change are the focus of the web site at www.TradeAndClimate.net.

APPENDIX 7.1

Chronicle of international climate change negotiations and agreements

1979 – The first World Climate Change Conference estimated that a doubling of carbon dioxide concentrations over pre-industrial levels would eventually lead to a 1.4–4.5-degree centigrade increase in global mean temperature.

1985 – A scientific conference in Villach, Austria, reviewed decades of observations and research, and called for policy analysis and actions to slow the rate of GHG-induced climate change.

1987 – In the Montreal Protocol, fifty-seven governments agreed to phase out production of substances that deplete stratospheric ozone. Many of these substances, such as CFCs, are also powerful and long-lasting greenhouse gases implicated in climate change.

1988 – Experts to the Toronto Conference on the Changing Atmosphere called for a reduction of global CO_2 emissions by 20 percent from 1988 levels by the year 2005.

1988 – Governments established the Intergovernmental Panel on Climate Change (IPCC) under the joint auspices of the UN World Meteorological Organization and the UN Environment Program, to assess climate change research for governmental decision-making.

1990 – The United Nations General Assembly established the Intergovernmental Negotiating Committee for a Framework Convention on Climate Change.

1992 – The United Nations Framework Convention on Climate Change (UNFCCC) was opened for signature at the United Nations Conference on Environment and Development (UNCED) in Rio de Janeiro, Brazil. The treaty cites common but differentiated responsibilities and respective capabilities of all parties, with an objective of avoiding dangerous anthropogenic interference with the climate system. It includes commitments of developed country Annex I parties to establish national action plans with measures that aim to reduce GHG emissions to 1990 levels by the year 2000. It includes obligations for parties listed in Annex I (including the United States) to provide technical and financial assistance, report GHG emissions, and additional obligations. The Global Environment Facility (GEF) was named the interim financial mechanism of the UNFCCC. Non-Annex I parties have general obligations, including for GHG mitigation, adaptation planning, and reporting.

1992 – The United States became the first industrialized nation to ratify the UNFCCC.

1994 – The UNFCCC entered into force, following ratification by fifty countries. [As of December 2010, there were 194 signatories – 193 countries plus the European Union.]

1995 – The United States became the first industrialized nation to ratify the UNFCCC.

1995 – In Berlin, Germany, the first meeting of the Conference of the Parties (COP-1) reviewed the adequacy of commitments under UNFCCC Articles 4.2(a) and (b) and concluded they were inadequate. It therefore adopted the Berlin Mandate, initiating negotiations for the post-2000 period to strengthen the GHG commitments of Annex I parties, but no new commitments for non-Annex I parties. The COP also agreed to a Pilot Phase for Joint Implementation, and to establish two entities: the Subsidiary Body on Implementation (SBI) and the Subsidiary Body on Scientific and Technological Advice (SBSTA).

1997 – The Kyoto Protocol to the UNFCCC was adopted, signed by more than 150 countries. It set a goal of reducing industrialized countries' GHG emissions to 5 percent below 1990 levels during the first commitment period of 2008–2012, and listed assigned amounts of allowable GHG emissions by parties in Annex B. It provided for flexibility mechanisms, including trading of assigned amounts, Joint Implementation, and the Clean Development Mechanism. It outlined a compliance mechanism, and required reporting by parties. Many implementing rules remained to be negotiated, covering operations of the flexibility mechanisms, how to account for land-based carbon sequestration, the nature of the compliance regime, etc. The Protocol would enter into force when fifty-five countries, including at least 55 percent of 1990 GHG emissions, had submitted papers of ratification [which occurred in 2005 without US participation].

1998 – The COP agreed to the Buenos Aires Plan of Action, with a deadline of 2000 to finalize rules to implement the Kyoto Protocol. The United States continued to press developing countries to take on voluntary commitments to reduce GHG emissions.

2000 – In the Hague, Netherlands, the sixth COP discussions collapsed, suspended without agreement on rules to implement the flexibility mechanisms in the Kyoto Protocol. Parties agreed to resume talks at COP-6 *bis* in July 2001.

2001 – At COP-6 *bis*, the United States participated for the first time as an observer, not a party to the Kyoto Protocol discussions. Decisions were made on use of the flexibility mechanisms (emissions trading, joint implementation, and the Clean Development Mechanism), carbon sinks, emission penalties for non-compliance, and to establish three new financial mechanisms: the Special Climate Change Fund, the Least Developed Country Fund, and the Adaptation Fund.

2001 – COP-7 adopted the Marrakesh Accords, establishing most rules and guidelines for the Kyoto Protocol to operate, especially for the three flexibility mechanisms: the Clean Development Mechanism, Joint Implementation, and Allowance Trading. To support adaptation in developing countries, the agreements included: (1) replenishment of GEF to address needs of developing countries due to adverse effects of climate change or of response measures; (2) establishment of Special Climate Change Fund (SCCF) to support adaptation and technology transfer; (3) establishment of a Least Developed Country Fund (LDC Fund), with guidance on its operation; and (4) establishment of an Adaptation Fund under the Kyoto Protocol. The parties also established an LDC work program and the LDC Expert Group (LEG), funding for National Adaptation Programs of Action and additional implementation support. The United States participated for the first time as an Observer in deliberations related to the Kyoto Protocol.

2002 – COP-8 issued a modest Delhi Declaration on Climate Change and Sustainable Development.

2003 – COP-9 reached several breakthrough decisions on credits for carbon absorption by forest sinks, as well as the Special Climate Change Fund and the Least Developed Countries Fund.

2004 – COP-10 increased focus on adaptation and approved the Buenos Aires Program of Work on Adaptation and Response Measures. Brazil and China submitted their first National Communications to the UNFCCC.

2005 – The Kyoto Protocol entered into force after Russia's ratification met the requirement for ratification by parties representing at least a 55 percent super-majority of CO_2 emissions (the requirement for at least fifty-five parties to the UNFCCC having already been met).

2005 – The United States announced the Asia-Pacific Partnership on Clean Development and Climate (APP), to cooperate on reducing the GHG intensity of their economies through voluntary technology exchanges. The APP included the United States, Australia, Canada, China, India, Japan, and South Korea, and included participation by the private sector. [It was subsequently disbanded and some of its activities transferred to other institutional venues.]

2005 – In Montreal, Canada, the first Conference of the Parties serving as the Meeting of the Parties to the Kyoto Protocol (CMP) met. After the US delegation walked out of the meeting, the COP agreed to two parallel tracks to consider actions in the post-2012 period, the Ad Hoc Working Group on Further Commitments for Annex I Parties under the Kyoto Protocol (AWG-KP), and another dialog to be established under the UNFCCC.

2006 – In Nairobi, Kenya, COP-12 and CMP-2 reached agreements concerning the Adaptation Fund, the Nairobi Work Programme on Adaptation, and the Nairobi Framework on Capacity Building for the CDM.

2007 – USA initiated the Major Economies Meetings (MEM) to negotiate a new post-2012 framework among a small group of countries, to develop a long-term global goal and "to complement ongoing UN activity."

2007 – In Vienna, Parties to the Kyoto Protocol agreed to consider a range of GHG reduction targets of 25 percent to 40 percent below 1990 levels for industrialized countries by 2020, though this range is resisted by Canada, Japan, and Russia.

2007 – At the first Major Economies Meeting (MEM), hosted by the United States, US pledged $2 billion over three years for a Clean Technology Fund (CTF) under the World Bank, expecting to raise $10 billion among donors to support concessional financing for energy projects in developing countries.

2007 – COP-13 agreed to the Bali Action Plan and established the Ad Hoc Working Group on Long-term Cooperative Action (AWG-LCA) with a mandate for parties to the UNFCCC to negotiate toward new GHG mitigation actions and commitments in the post-2012 period and to reach agreement by the end of 2009 (at the COP-14 meeting in Copenhagen, Denmark). The Bali Action Plan called for "a shared vision for long-term cooperative action" and identified four main elements: mitigation, adaptation, technology, and finance. Additional decisions placed management of the Adaptation Fund under the World Bank, and initiated demonstrations and commitments to reduce deforestation.

2008 – In Accra, Ghana, an exchange of views under the AWG-LCA continued on alternative approaches to "shared vision," mitigation, adaptation, technology, and finance. Differentiation among non-Annex I parties continued to be contentious, with China and the G-77 maintaining solidarity. Some developing countries argued that the AWG-LCA and AWG-AP were not mandated to consider amendments to the UNFCCC or Kyoto Protocol, only implementation of them. Some delegations supported worldwide sectoral approaches, which some developing countries argued would be inappropriate for them. Developing countries frequently called for new mechanisms for each issue, and opposed "conditionality" on financial and technology transfers (such as protection of intellectual property rights). The AWG-KP agreed on a comprehensive "basket approach" to including multiple GHGs in the second commitment period, and noted new groups of gases and new gases (e.g. NF_3) identified by the IPCC AR4. It noted that the Montreal Protocol phased out production of CFCs and HCFCs, but not their emissions. There was agreement that analysis would proceed on various "spillover" effects of mitigation actions.

2008 – The government of Japan proposed that all parties adopt a "shared vision" of achieving at least 50 percent reduction of global GHG emissions by 2050. Global GHG emissions should peak in the next ten to twenty years. It proposed criteria for entering additional countries into Annex I (i.e. to become countries with commitments), to create comparability of efforts for GHG targets among Annex I parties, according to sectoral emissions, efficiencies, and reduction costs, and for new GHG commitments among three groups of developing countries.

2008 – In Poznan, Poland, a high-level segment of COP-14 expressed political statements on a "shared vision for long-term cooperative action," and agreed to intensify negotiations. Parties agreed that a full negotiating text should be available by June 2009. Parties also resolved issues regarding the Adaptation Fund, though developing countries did not achieve commitments for additional adaptation monies.

2008 – The Government of Mexico, among the first non-Annex I parties to offer a GHG reduction commitment, announced a goal to halve GHG emissions from 2002 levels by 2050. Brazil pledged to cut deforestation by at least 50 percent by 2017.

2009 – COP-15 in Copenhagen, Denmark, was partially circumvented by the Copenhagen Accord, according to which a small subset of countries, not including the European Union, agreed to a series of mostly general items, some of which were subsequently made more specific. (See Section 7.1 and Boxes 7.1, 7.2, and 7.3 of this chapter for further information.)

2010 – COP-16 in Cancun, Mexico. (See Section 7.1 and Box 7.4 of this chapter for further information.)

2011 – COP-17 in Durban, South Africa. (Also see Section 7.1 of this chapter for further information.)

2012 – COP-18 in Doha, Qatar. (Also see Section 7.1 of this chapter for further information.)

Source: Excerpted (with additional editing) from US Congressional Research Service (2010b), for the years through 2008; sources for 2009–2012 are indicated in Section 7.1 of this chapter.

APPENDIX 7.2

Simulation model used to measure the gap between emissions targets and pledges

This appendix introduces the C-ROADS simulation system that has been used to estimate the effects of the Copenhagen Accord and the Cancun Accords on emissions in relation to the stated goal of no more than a 2-degree celsius temperature rise. C-ROADS is an acronym standing for "Climate Rapid Overview and Decision Support Simulator." "It allows for the rapid summation of national greenhouse gas reduction pledges in order to show the long-term impact on the climate" (ClimateInteractive 2010). It was developed by a group from ClimateInteractive, Ventana Systems, and MIT. Its overall analytic structure is depicted below in Figure A7.2.

The Climate Scoreboard showed that expected emissions if the Copenhagen Accord pledges were to be realized were more consistent with warming of close to 4 degrees celsius rather than the 2 degrees celsius goal articulated in the Accord. Without the Copenhagen pledges the world would have expected even more warming, close to

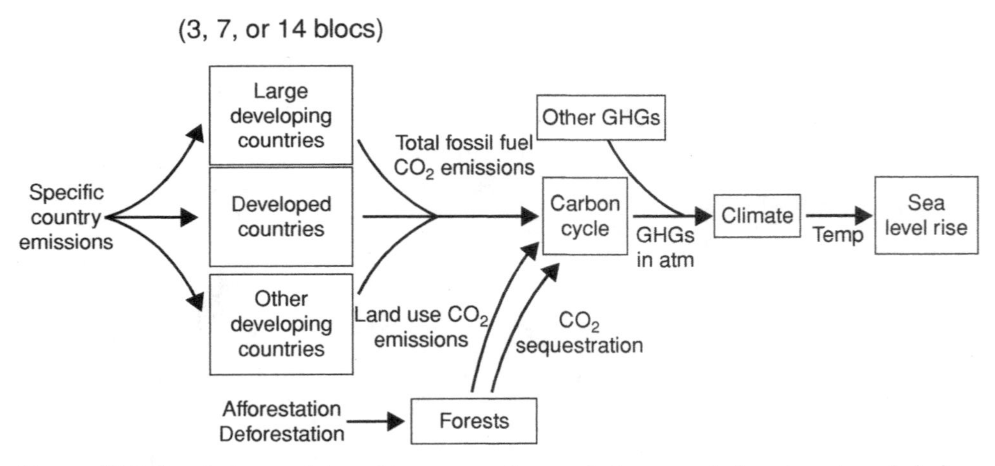

Figure A7.2 Simulation model used to measure the gap between emissions targets and pledges

Sources: ClimateInteractive (2010). Used with permission.

4.9 degrees celsius. [The Copenhagen Accord] represented some progress, but left much more work to be done.

After [the Cancun meeting], at least with regard to countries' commitments to reduce emissions, the situation remain[ed] unchanged. The Cancun Agreements codif[ied] the targets of the Copenhagen Accord rather than pushing beyond them. No country made more ambitious pledges in Cancun than what had already been on the table in Copenhagen.

Source: ClimateInteractive (2010). Used with permission.

APPENDIX 7.3

Copenhagen Accord, Annex I countries

The US commitment was an emission reduction in 2020, compared with 2005, as follows: "In the range of 17 percent, in conformity with anticipated US energy and climate legislation, recognizing that the final target will be reported to the Secretariat in light of enacted legislation."[1] [There was no "enacted legislation" as of early 2011 and little prospect of any for several years.]

[1] "The pathway set forth in pending legislation would entail a 30 percent reduction in 2025 and a 42 percent reduction in 2030, in line with the goal to reduce emissions 83 percent by 2050." [The "pending legislation" was not subsequently passed.]

European Union: 20 percent compared with 1990; or 30 percent. "As part of a global and comprehensive agreement for the period beyond 2012, the EU reiterates its conditional offer to move to a 30 percent reduction by 2020 compared to 1990 levels, provided that other developed countries commit themselves to comparable emission reductions and that developing countries contribute adequately according to their responsibilities and respective capabilities."

Japan: 25 percent reduction compared with 1990 "which is premised on the establishment of a fair and effective international framework in which all major economies participate and on agreement by those economies on ambitious targets."

The other Annex I countries – except Canada and Australia – used 1990 as the base year and made commitments varying from 5 to 30 percent reductions by 2020.

Source: UN Framework Convention on Climate Change (UNFCCC) (2010).

APPENDIX 7.4

Copenhagen Accord, non-Annex I countries

"China will endeavor to lower its carbon dioxide emissions per unit of GDP by 40–45 percent by 2020 compared to the 2005 level, increase the share of non-fossil fuels in primary energy consumption to around 15 percent by 2020 and increase forest coverage by 40 million hectares and forest stock volume by 1.3 billion cubic meters by 2020 from the 2005 levels."

"India will endeavour to reduce the emissions[1] intensity of its GDP by 20–25 percent by 2020 in comparison to the 2005 level."

[1] "Emissions from the agricultural sector will not form part of the assessment of emissions intensity."

Brazil will take actions leading to:

"Reduction in Amazon deforestation (range of estimated reduction: 564 million tons of CO_2 eq in 2020);

- "Reduction in 'Cerrado' deforestation (range of estimated reduction: 104 million tons of CO_2 eq in 2020);
- "Restoration of grazing land (range of estimated reduction: 83 to 104 million tons of CO_2 eq in 2020);
- "Integrated crop-livestock system (range of estimated reduction: 18 to 22 million tons of CO_2 eq in 2020);
- "No-till farming (range of estimated reduction: 16 to 20 million tons of CO_2 eq in 2020);
- "Biological N_2 fixation (range of estimated reduction: 16 to 20 million tons of CO_2 eq in 2020);
- "Energy efficiency (range of estimated reduction: 12 to 15 million tons of CO_2 eq in 2020);
- "Increase the use of biofuels (range of estimated reduction: 48 to 60 million tons of CO_2 eq in 2020);
- "Increase in energy supply by hydroelectric power plants (range of estimated reduction: 79 to 99 million tons of CO_2 eq in 2020);

- "Alternative energy sources (range of estimated reduction: 26 to 33 million tons of CO_2 eq in 2020);
- "Iron & steel (replace coal from deforestation with coal from planted forests) (range of estimated reduction: 8 to 10 million tons of CO_2 eq in 2020);

"It is anticipated that these actions will lead to an expected reduction of 36.1 percent to 38.9 percent regarding the projected emissions of Brazil by 2020."

Source: UN Framework Convention on Climate Change (2010). "%" symbols in the original documents were changed to "percent" for stylistic consistency.

PART IV

THE FUTURE

CHAPTER 8

Options for the future: realities, visions, and pathways

[W]e ask [the United States] for your leadership. But if for some reason you're not willing to lead, leave it to the rest of us. Please get out of the way.

Kevin Conrad, Representative of Papua New Guinea (2007)

I urge this Congress to pursue a bipartisan, market-based solution to climate change, like the one John McCain and Joe Lieberman worked on together a few years ago. But if Congress won't act soon to protect future generations, I will. I will direct my Cabinet to come up with executive actions we can take, now and in the future, to reduce pollution, prepare our communities for the consequences of climate change, and speed the transition to more sustainable sources of energy.

President Barack Obama (2013)

The first quote above marked a nadir of US diplomatic "leadership" on any issue in several decades. In fact, the delegate's plea to the USA either to lead or step aside so others could lead at the FCCC Conference of the Parties was followed by applause and cheers by hundreds of delegates for his having made the point so publicly in an official venue and in the presence of the leaders of the US delegation.[1] The event provides an apt transition from the previous chapter's focus on international cooperation to this chapter's focus on leadership, both internationally and domestically, and options for the future.

The second quote above from President Obama's State of the Union speech in February 2013, at the beginning of his second term, marked a turning point in the domestic politics of US government responses to climate change. For it made explicit and public the shift that had already been occurring, as the administration took more steps to address the climate change problem in the aftermath of Congress's failure to pass cap-and-trade legislation in 2010. Also in early 2013, the President's Council of Advisors on Science and Technology (2013) presented a wide range of suggestions for

1 Such overt public demonstrations of displeasure by audiences in official diplomatic events are extremely rare; this was one of only a very few times in half a century when the USA was subjected to such a widely supported public rebuke in a formal public setting.

how to address the climate change problem (see Appendix 8.1). Within Congress, a congressional Bicameral Task Force on Climate Change was organized by leaders in the Senate and the House, and a "climate change clearinghouse" with a weekly "open forum" meeting was organized in the Senate. There was a flurry of interest (albeit rather briefly), among a few members of Congress as well as many policy specialists, in the possibility of instituting a carbon tax as an alternative to a cap-and-trade system (Stavins 2012). There was thus renewed interest in alternative approaches to climate change in the national government's policymaking processes.

This concluding chapter accordingly presents a wide range of ideas about how climate change issues can be approached. In order to advance consideration of alternatives, the chapter presents excerpts of policy preferences produced by a wide variety of politicians, scholars, and others with diverse ideological and partisan preferences. Appendix 8.1 contains a statement by the President's Council of Advisors on Science and Technology, as noted above. Appendix 8.2 compares the positions on climate change in the 2012 election party platforms and candidates' answers to questions from a coalition of science organizations. Appendix 8.3 presents a series of policy recommendations from an organization of Republicans for Environmental Protection (now renamed ConservAmerica). Additional, diverse perspectives are offered in other appendices.

In addressing these issues, the chapter draws upon previous chapters, particularly for highlighting political and economic realities that constrain the pursuit of many options. The chapter is thus grounded in the factual analyses of the previous chapters; at the same time, much of the discussion in the present chapter is about the normative implications of the factual analyses. The theme is that US leaders must find ways to overcome not only market failures and international system free-rider problems, but also constraints in the US domestic political and economic systems. The emphasis is on national government policies; the chapter is organized around the three topics in the chapter subtitle – namely realities, visions, and pathways.[2]

- *Realities* about the problem and solutions to it (Section 8.1): What are the key projections into the future of greenhouse gas emissions, concentration levels, and temperatures? What technological developments can affect those projections? What are the US political economy realities that constrain the options? How have the politics and economics of US policymaking evolved since climate change entered the US government agenda in the 1980s?

2 Portions of this chapter have been inspired by a workshop on "climate leadership" held at MIT in 2010 and organized by Drew Jones, Sara Sheley, and Travis Franck of ClimateInteractive.

- *Visions* of the future to guide action (Section 8.2): How have climate change issues been framed? What have political leaders done to try to alter people's visions of the future?
- *Pathways* to move from current and prospective realities to the conditions represented by visions of alternative futures (Section 8.3): What are the issues that need to be addressed with practical options? What are examples of proposals for addressing them?

8.1 Realities: the changing agenda and political economy

Realistic and responsible options for the future need to take into account the information and projections of climate science, as well as the political and economic features of the USA that are the focus of the book. Although both the climate realities and the political economy realities are daunting, there are many promising options that could be undertaken.

Emissions and concentration levels

The climate change agenda – whether for governments, businesses, or individuals – is inevitably driven to a great extent by the facts and analyses of climate science. Among the literally millions of observations and thousands of reports and publications, there are trends and projections in a few indicators that are especially important – the levels of emissions of greenhouse gases[3] that cause global warming, the levels of the accumulated concentrations of them, and the global mean surface temperature.

Figure 8.1 depicts projected levels of carbon dioxide concentration and emission levels in Panels A and B and the global mean surface temperatures in Panel C. Because the projections are based on the research of many individuals and organizations using different models of climate change, as summarized by the UN IPCC (2013), and because they are projections into the future, there is of course a range of results. Nevertheless, several patterns are evident:

- The current level of *concentration* of about 400 ppm will likely increase significantly to a range of 500–1000, with a central estimate of about 700 ppm (Panel A). Any of

3 Because black carbon (commonly known as "soot") occurs as particles, not a gas, it is usually not included in the standard data sets of greenhouse gases. However, as noted in Chapter 1, it has been identified as a major source of global warming (second only to carbon dioxide in its total impact). Its atmospheric lifetime is typically a few days and its impact is much more localized than GHG gases; it is emitted by many of the same sources as greenhouse gases.

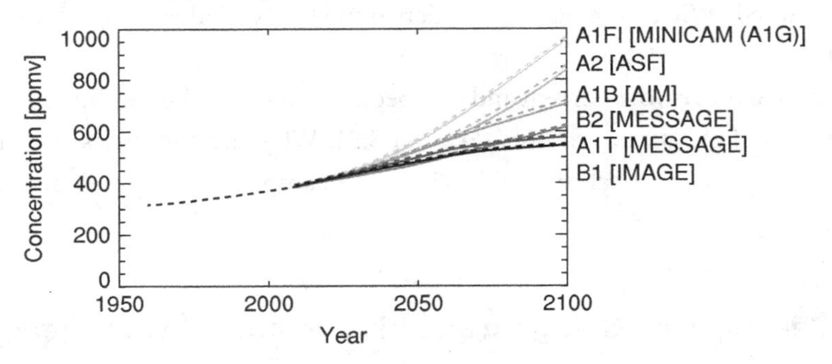

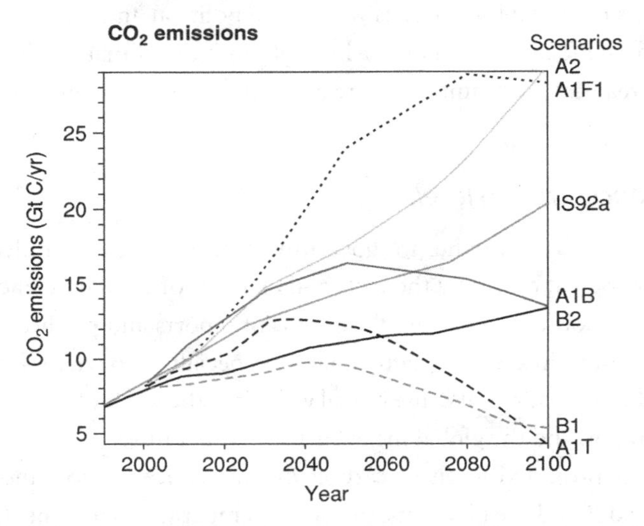

Figure 8.1 Projections of global carbon dioxide concentration and emissions levels and temperatures

Panel A. Global atmospheric concentration levels of carbon dioxide

Source: UN IPCC (2007).

Panel B. Global emission levels

Source: UN IPCC (2007).

Panel C. Projections of global mean surface temperature

Source: UN IPCC (2007).

these amounts will have dramatic impacts on a range of socio-economic conditions in all parts of the world.

- The concentration levels will obviously depend on the *emissions* levels (Panel B). Over the next two decades, world emissions are highly likely to continue to increase,

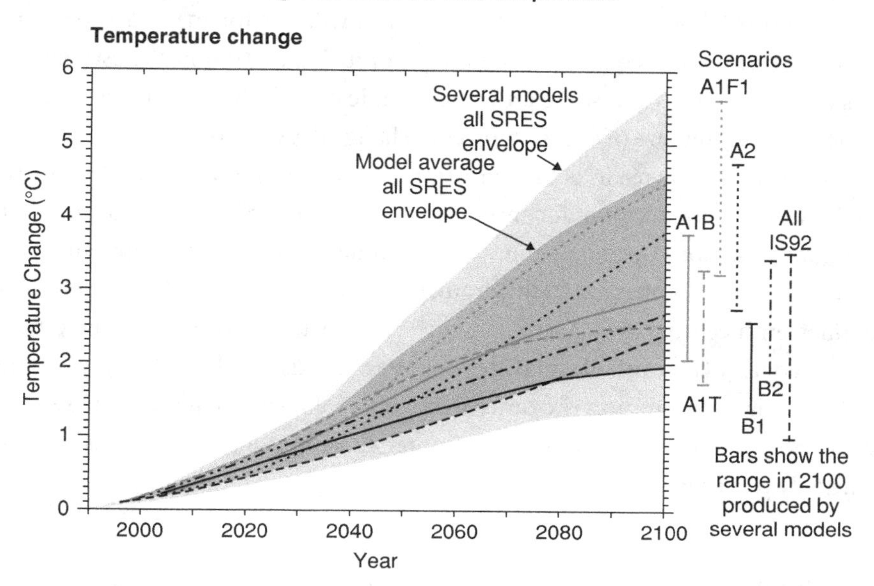

Figure 8.1 *(cont.)*

in as much as all of the projections show increases until about 2040. The key question of course is "How much?" The estimates in Panel B indicate a range of roughly 10 to 20 giga tonnes per year of carbon – or about 10 percent to 100 percent more than the 2012 level. Beyond 2040 the annual emissions levels *could* decline if there were a variety of changes in technologies, government policies, business practices, and individual habits. Note that the *concentration* levels would nevertheless continue to rise. Even if emissions levels decline – but remain positive – the concentration levels continue to increase because of the average atmospheric lifetime of the emitted carbon dioxide molecules. There are therefore inevitably differences in the contours of the graphic profiles over time in the levels of emissions and concentrations.

- Changes in the global mean surface temperature are of course also key numbers and the ones commonly used in popular discussions and in the press. In Panel C, which indicates the change in degrees Celsius, all of the estimates indicate a continuing increase over the entire period. There is thus a parallelism in the temperature and carbon dioxide *concentration* charts. This is because the temperature change is a direct function of the atmospheric concentration levels and only indirectly a function of the annual emissions (which are still contributing to climate change as long as they are positive, albeit at declining levels in some scenarios).

In sum, there is a convergence in the projections that the atmospheric concentration levels of carbon dioxide and hence temperatures will rise for many decades – in part because emissions will continue (though with much variation in the estimates of the levels of the emissions and some projections indicating declines in the levels).

These projections are of course subject to change depending on technological developments, just as changes in actual emissions, concentration levels, and temperatures have been in the past. It was, for instance, the technological changes of the Industrial Revolution – steam power, locomotives, and others, with the concomitant increased use of coal as a fuel – that led to the dramatic increases in carbon dioxide emissions in the nineteenth century. More recently, in the twentieth century, the development and widespread use of automobiles and airplanes have contributed to carbon emissions and climate change. Perhaps, on the other hand, technological changes of the twenty-first century will lower carbon dioxide and other greenhouse gas emissions below their commonly projected levels.

Technology

Three quite different types of technological change are pertinent – namely those that are recent, in progress, and imaginable.

Recent changes. Significant technological changes in the past couple of decades in "unconventional" natural gas extraction (see Chapter 6) have altered the estimates of recoverable natural gas in the USA and many other countries, and recent production levels and the projections of future production have increased dramatically. Future GHG emission levels will depend in part on the extent to which these projections become realities – and on the resolution of related issues such as methane gas releases during the production process.

Changes in progress. Carbon capture and sequestration (CCS) is already a technically feasible technology, which could be effective in reducing the carbon dioxide emissions from electric power plants that burn fossil fuels, especially coal (Chapter 6). The economic costs of CCS are still prohibitively high for commercial applications. However, there are large-scale R&D programs ongoing in several countries, including the USA, and the costs are likely to come down significantly over the next decade or so.

Imaginable changes. Large-scale geo-engineering projects that could deflect solar radiation or absorb carbon dioxide emissions are being considered (see Appendix 8.4, item 2). However, they are still in the preliminary stages of formulating their basic operational modalities and not yet in an advanced design stage (Chapter 6).

Each of these illustrative technological developments presents a different set of realities. Unconventional gas and oil at commercially viable prices are already a reality; the questions are about their volumes and prices in the future. Will shale gas prices increase so much that its displacement of coal will be less than expected? Will restraints on exports of natural gas limit its impact in countries that would otherwise import large quantities of it? Carbon capture and sequestration poses questions about its long-term safety and environmental impacts. Various types of geo-engineering may or may not be technically feasible – but in any case pose many economic, political, and ethical questions. For all of these and many other climate-relevant technological developments, there are a host of political economy questions: How much will they cost? Who will pay for them? What should be the roles of governments and markets in determining their availability?

The purpose here is not to try to resolve these questions, but rather simply to register a reminder that there are many technological developments – past, present, and future – that can change the projections of emissions, concentration levels, and temperatures and that the futures of these and other technologies depend on a wide range of political and economic factors.

Political economy

The political economy of climate change issues in the USA has been transformed over time as the agenda has expanded from its origins among a small number of scientists in Europe and the United States in the nineteenth and twentieth centuries, through its emergence in the USA as a narrowly defined issue of scientific and environmental public policy with increasing public awareness of it in the 1980s, to a partisan and ideological issue in the last decade of the twentieth century and into the first and second decades of the twenty-first century. Now, however, climate change is entering a period of pragmatic policymaking in diverse venues in the USA, where the political economy issues are more widespread and less esoteric.

Although it has not yet been fully "mainstreamed," climate change is moving in that direction. It has evolved into a multidimensional range of issues with enormous economic, political, technological, social, and security consequences. The issues now pervade nearly all sectors of the US economy and the economies of other countries – energy, mining, agriculture, manufacturing, fishing and forestry, as well as finance, construction, and other services. The mitigation and adaptation issues associated with climate change are now core issues for governmental agencies at all levels in all parts of the world, in US relations with nearly all countries, and in numerous international institutions. They are issues that will not go away, but rather will become more and more urgent.

As the agenda has expanded, so too have the specifics of the political economy of the issues become more diverse and dynamic. Consider the following highlights from the preceding chapters. (For further reference and for documentation, the most relevant chapter numbers are indicated in parentheses.)

The roles of government policies to address market failures are at the center of many specific issues related to climate change: Should the government establish a method for internalizing the costs of greenhouse emissions? Should the government subsidize R&D, production, or consumption of renewable energy and energy efficiency technologies? These and many other issues involve key *ideological issues about the mix of governments and markets* in societies (see Chapters 1 and 6, as well as Appendix 8.5 of this chapter).

The importance of *the distribution of the economic benefits and costs* of policy alternatives – among groups and over time – was emphasized in the first several chapters. The patterns of industry winners and losers – and the associated geographic patterns – have been among the most important drivers of US responses to climate change. In particular, the economic significance of the oil and gas, coal, and automotive industries in many states in the South and Midwest has been a basic feature of the domestic political economy of policymaking (Chapter 1). These and other fossil fuel industries and their national lobbying organizations have been major deterrents to the advancement of cap-and-trade and other greenhouse gas mitigation measures (Chapters 2 and 5). Political economy thus involves a combination of *political geography and economic geography.*

Yet, neither industries nor regions are economically monolithic: some firms in some industries have become leaders on climate change issues even as their rivals have remained laggards (Chapter 2), and some states have become leaders even as their neighbors have been laggards (Chapter 4). In many of these leading firms and states, there have been individual business and political leaders who have seen differently their firms' interests and their states' interests and the future business environment and the future quality of life of their citizens. It is therefore often *perceptions* of the distributions of benefits and costs of alternative policies that are important; *expectations* about future distributions are important, along with current realities.

The same can be said about public opinion (Chapter 3). Key realities about public opinion, however, seem not to be widely understood; several dimensions of public opinion are different from common assumptions about them. Indeed public opinion surveys have found that public perceptions of public opinion are often wrong. The extent to which such erroneous assumptions about public opinion are also held by members of Congress and their staff is an open question, but there is some evidence that many of them misperceive public levels of concern and support for action, even among their own constituents (Chapter 3). Among the important realities are that consistently a plurality – and in some years a majority – of the public have believed that climate change is already happening, while only about one in seven believes it will never happen. National majorities – indeed majorities in nearly every state – want the national government to do more about climate change. Majorities support greater US participation in international climate change agreements. Majorities support government subsidies for renewable energy research, development, production, and consumption.

Within the broad national consensus that the problem is serious and should be addressed more aggressively by the government, there are many disagreements. Of special significance are the *ideological* and partisan splits: whereas "liberals," Democrats, and independents tend to be more concerned and support many mitigation measures by large majorities, "conservatives" and Republicans tend to be less concerned and less supportive (there are important exceptions to such generalizations). The political economy of climate change is thus in part an element in a broad set of ideological issues about the forms and extent of *government regulations* (Chapter 5) and the size and beneficiaries of *government expenditures* (Chapter 6).

Many of the political economy issues about climate change, furthermore, have become enmeshed in high-stakes court cases, especially in the federal court system (Chapter 4). Federal District Courts and Appeals Courts, as well as the Supreme Court, have made decisions about what federal government agencies such as the Environmental Protection Agency can and cannot do, and what states can and cannot do. Thus, the political economy of climate change now involves issues of *constitutional law*, such as whether and how constitutional provisions concerning *interstate commerce* apply to state laws that regulate greenhouse gas emissions.

Other prerogatives and responsibilities of state and local governments in a federal political system are also now sometimes at issue (Chapter 4). For instance, in what ways and to what extent can state and local governments enter into agreements with counterparts in foreign countries? What are the limitations imposed by constitutional provisions concerning *foreign commerce*? The political economy of climate change has thus become enmeshed in *international and transnational* political and legal issues.

In Chapters 5 and 6, political economy issues are evident in the functioning of governmental institutions. The extent to which Congress over-represents parts of the country that have relatively energy-intensive and greenhouse-gas-intensive economies, and under-represents parts of the country that are less so, poses significant political economy issues about *governmental institutional structure and procedures.* The influence of industry associations representing the interests of some economic sectors versus the influence of public preferences poses yet more issues about the Congress as a *representative institution* (Chapter 5).

The Congress's role in the *annual budgeting process* is yet another institutional political economy topic (Chapter 6). Indeed, the essence of the budgeting process is precisely to allocate economic benefits and costs among groups – as is the periodic revision of *tax* policies. Within the executive branch, not only decisions about budget and tax issues, but also about *technology research and development program priorities,* are of course also political and economic issues (Chapter 6).

Internationally, responsibility for historical and ongoing greenhouse gas emissions and compensation for the costs of adapting to them are major political economy issues (Chapter 7), as well as ethical issues.

In short, previous chapters of the book have identified many types of political economy questions associated with the causes, consequences, mitigation, and adaption issues of climate change. The political economy topics comprise many of the realities that need to be taken into account in policymaking; political economy issues are also central to many visions of the future.

8.2 Visions: framing the issues

Political leadership often involves framing issues. In the case of climate change, this has been a major challenge, and yet at the same time there have been many opportunities for re-framing the issues. From a political economy perspective, one of the most important topics has been perceptions about the distributions of the benefits and costs of various alternatives, including doing nothing.

A do-nothing approach tends to be based on a vision of the future that denies the scientific evidence or common observations of climate changes; or in some instances, a do-nothing approach is based on a vision of the future according to which the problem is so enormous, inevitably catastrophic, and far into the future, that action now would be useless. Yet another vision sometimes underpinning a do-nothing approach is that the economic costs of effective measures to address the problem would be too great and/or that the government regulatory and other programs would be too detrimental to the economy. Such visions and their related policy positions are based on some combination of disregard for science and/or ideological opposition to government in general; they are also outside the mainstream of informed consideration of pragmatic ways to deal with the climate problem and its implications. Fortunately, there are relevant alternative visions to blithe denial, at one extreme, and hopeless despair at the other.

To be sure, though, there are significant political economy trade-off issues that need to be considered. At the most general level, there has been a long-standing assumption that there is an inherent trade-off between environmental protection and economic growth – an assumption that has been explicit for instance in standard public opinion survey questions that ask respondents for their priorities between the two (a simplistic and false dichotomy). One of the challenges of framing climate change issues has thus been to present a plausible alternative vision of the future in which a variety of measures are taken to address the problem but either do not seriously undermine economic growth, or in some ways actually facilitate economic growth. Revenue-neutral taxes are one possibility. Cap-and-trade programs that return government revenues to consumers and/or use the revenues to support low-carbon technologies are another (see Appendix 8.6).

There are nevertheless losers – most clearly in the short term – from many measures that would succeed in reducing carbon dioxide and other greenhouse gas emissions; but the losers can be compensated and helped to adjust. Government support of carbon capture and sequestration, for instance, is in part a coal industry subsidy program that is intended to facilitate its adjustment to a low-carbon economy and ameliorate its hostility to addressing climate change. In addition, winners can be highlighted, such as the number of jobs created in the solar power and wind power industries, or the number of lives saved by retiring coal-fired power plants. Although the identities of the winners and losers – and the magnitudes of their gains and losses – can be estimated in econometric modeling exercises, the responsibility for presenting the outlines of plausible alternative futures in which there are significant winners is also a task for politicians and other leaders.

There is also a challenge in realistically acknowledging that there will be costs associated with addressing the climate problem – but at the same time offering evidence that those costs will be less than the costs of not addressing it (Stern 2007). Because of technical economic issues such as the appropriate discount rate to be used to estimate the "present value" of future costs and because of the psychological issues of discounting the future in general, making the case that it is cheaper to act sooner rather than later is a challenge.

There is an ethical issue, however, that is commonly used in lieu of – or in addition to – the comparative economic cost argument, namely that present generations have an ethical responsibility to future generations. The vision of one generation satisfying its own desires with little or no regard for the consequences for future generations is – for some – a compelling negative vision to induce action on climate change. Or alternatively, the positive vision of a generation that takes its ethical responsibilities seriously and makes sacrifices to protect future generations is also one that resonates strongly with many people. Although these ethical issues are not the usual province of a political economy approach, political and economic analyses can help to clarify important political and economic facts involved in the ethical choices.

More generally, however, it should be noted that there are of course limits to a political economy approach to the analysis of climate change issues; this is particularly so in regard to visions of the future. Though there are important political and economic elements in such visions, there are also obviously other significant factors. For instance, there are cultural constraints on free riding that have been observed in many other domains (Ostrom 1990). Such constraints should not be ignored, though they may be more significant domestically than internationally. Addressing many climate change issues effectively, therefore, requires other approaches in addition to a political economy approach; for climate change poses a wide range of social-psychological issues as well as ethical issues. There are questions of imagination, for instance: Why can some people visualize alternative futures more easily, clearly, and realistically than other people? There are questions about cognitive processes: Why do some people resist or misinterpret or misrepresent the findings of science more than others? There are important issues of individual and collective responsibility as well: Why are some people more willing than others to accept responsibility, and to act on the basis of that sense of responsibility? As government, business, and the public take on the problem of climate change over the long term, answers to these questions and other questions involving changes in attitudes and behavior will no doubt become more salient.

8.3 Pathways: challenges and opportunities

Whether based on an understanding of the nature and relevance of visions of the future, the facts and analyses of climate science, or the available and potential technologies, or the facts of political economy, pathways into the future will inevitably be taken, even if they only involve "business as usual" projections. What might they be? What are some general guidelines and specific suggestions that are on the agenda?

Alternative pathways, whether conceived broadly or in terms of specific possibilities, need to take into account the following criteria; these criteria, it should be noted, are overlapping and interacting, and should be subject to much more refined specification and serious debate among climate change specialists, publics, other stakeholders, and philosophers.

- *Effectiveness* in addressing the problem – i.e. reducing the emissions and concentration levels of greenhouse gases and/or otherwise mitigating climate change and adapting to its impacts.
- *Economic efficiency* – i.e. achieving objectives in a cost-effective way.
- *Equity* – i.e. undertaking measures that are acceptable in terms of widely accepted notions of what is fair.
- *Political feasibility* – i.e. adopting measures that can gain enough support in relevant political processes so that they can be adopted and sustained.
- *Administrative feasibility* – i.e. including measures that can become operationally sustainable if organizational mechanisms are needed for their success.

With these criteria in mind, US leaders need to facilitate progress along three pathways representing fundamental changes in national and international approaches to climate change:

(1) *Changes in the government–market mix in order to reduce greenhouse gas emissions.* Such changes can be achieved through government interventions to reduce the economic externalities in emissions, and through reductions in energy consumption via producers' and consumers' actions. The changes include the adoption of a national cap-and-trade system and/or carbon taxes – in coordination with related measures in other countries.

(2) *A techno-economic revolution in energy systems around the world.* This is a revolution that has begun; but it has not yet penetrated economies enough to prevent climate change catastrophes. It is occurring in many countries, and its progression will be determined by a combination of government policies and market conditions as well as technological developments. The revolution requires changes in both US domestic and international policies. In particular, even greater increases in support should be given to energy efficiency measures – or "energy productivity" measures, as they are now often called in the USA.

(3) *Reform of international institutional arrangements.* Current international institutional arrangements are inadequate to the tasks they need to undertake; this is true of trade and aid institutions as well as climate change institutions (see Appendix 8.7). These institutions have evolved substantially over the decades, but they need to change more and faster. The issues concerning them go far beyond administrative issues such as coordination among programs and agencies – important as those

may be to administrative efficiency and accountability; they especially include fundamental issues about the political economy of incentives and free riding.

The three pathways are interdependent. Pricing carbon through cap-and-trade and/or taxes creates economic incentives for technological innovation and diffusion. Opportunities for international cooperation on technology projects create incentives for international cooperation on other mitigation or adaptation measures. These broadly defined pathways are only three of many options, however, and they are compatible with many more specific options.

Another pathway, which is already being taken, is a legalistic one. It has been particularly noteworthy in the national (i.e. "federal") court system, albeit with reservations about the legal system's capacity to deal effectively with issues of such a wide socio-economic scope (see Box 8.1).

Box 8.1 The legal path

Increasing resort to the courts may be nearly inevitable in a society where litigation is so commonplace, but it has its liabilities. Lawyers who are themselves participants in the trend towards legalization of climate change issues have noted the significance of the issues – and at least in one notable instance even doubted the courts' capacity to deal with them. This was evident in the oral arguments in a Supreme Court case, *American Electric Power Company et al. v. Connecticut et al.*, where the specific legal issue was whether the courts should intervene to establish greenhouse gas emissions levels in "public nuisance" cases involving electric power firms.

The Acting Solicitor General of the US Justice Department, representing the administration, asserted: "In the 222 years that this Court has been sitting, it has never heard a case with so many potential perpetrators and so many potential victims, and that quantitative difference with the past is eclipsed only by the qualitative differences presented today. . . . This court has never had a case involving this scale and scope."

While this was perhaps making the point a bit hyperbolically, the Chief Justice expressed a similar theme – and then further noted his concern about asking the courts to make such momentous decisions: ". . . [T]he whole problem of dealing with global warming is that there are costs and benefits on both sides, and you have to determine how much you want to readjust the world economy to address global warming, and I think that's a pretty big burden . . . to impose on a [US] district court judge" (US Supreme Court 2011).

For further consideration, there is surely no dearth of alternative approaches and lists of specific proposals. The seven chapter appendices, which have already been noted above in relation to individual points, present a selection of ideas that have been drawn from diverse sources – such as government reports, academic studies, position papers, and open letters by policy advocacy organizations. Because of their potential value, individually and collectively, as thought-provoking analyses that can be helpful in developing many pragmatic pathways, they are noted again here. They include:

- A report by the President's Council of Advisors on Science and Technology (Appendix 8.1), which proposes "six key components" that should be "central to [US] climate change strategy and policy."
- Side-by-side comparisons of statements from the party platforms and candidates in the 2012 presidential election (Appendix 8.2), which identify their basic approaches to the issues.
- A list of specific policy proposals from a group, Republicans for Environmental Protection (Appendix 8.3), which includes quotes from conservative presidents and presidential candidates.
- An open letter from a group of leading scholars from universities and think tanks in the USA and other countries (Appendix 8.4), which includes "guidance for decision makers" in the form of "ten concrete points" covering a wide range of key issues.
- An analysis of government–market mixes for addressing environmental problems (Appendix 8.5), which includes a "spectrum" of options that can be applied to climate change options.
- A discussion of design issues for cap-and-trade systems (Appendix 8.6), which identifies some of the many variables that should be taken into account in developing such systems.
- An analysis of international institutional arrangements for addressing climate change (Appendix 8.7), which focuses on a range of institutional venues for attempts to make greater progress.

In toto, these materials express a wide range of views about what is – or should be – on the climate change agenda; what approaches to the agenda are particularly promising; how specific issues can be defined; and what should or should not be done about individual issues. The materials are responsive to many parts of the expanding and multi-faceted climate change agenda. These particular selections are hardly comprehensive, however; indeed there are many other similarly thought-provoking sources noted in the chapter's suggestions for further reading and research.

8.4 Concluding remarks: incipient transformational trends

Overall, the responses to climate change thus far by the US government, business, and public have been inadequate to the challenge. Yet, there is evidence that the USA is in the early phases of the required transformations. Indeed, some cities, states, firms, and industries – and parts of the national government – have begun potentially transformational processes. Therefore, just as it is possible to note instances of individual leadership in the midst of otherwise laggard parts of government and business, it is also possible to conclude on a hopeful note by observing that there is already evidence of other incipient transformational trends. They include:

- increasing understanding of the problem and potential solutions by many political and business leaders as well as many segments of the public;
- changes in production processes and positions on public policy issues by firms in diverse industries;
- changes in consumer attitudes and behaviors, including in regard to energy efficiency;
- adoption of diverse types of measures to reduce greenhouse gas emissions at the sub-national level by states and cities;
- innovation and diffusion of energy efficiency and renewable energy technologies through a combination of government policies and private-sector investments;
- expansion of technical expertise as well as understanding of economic and other issues through the activities of many organizations in the private and public sectors and in non-governmental environmental organizations, think tanks, and universities.

The scale of such changes to date, however, is insufficient to solve the problem. What is needed now is more widespread and more rapid transformations in how Americans think about the problem of climate change and in actions they take to solve to it – transformations in all regions, industries, institutions, professions, leadership groups, and society generally. Whether the transformations that materialize will be soon enough and extensive enough to prevent irreversible catastrophic climate change remains to be seen.

In the meantime, I leave the concluding observations and exhortations to others: a bipartisan group of nationally prominent leaders; an unusually highly experienced and prominent former cabinet member; and Einstein. The first is an internationally oriented excerpt from an open letter sponsored by the organization, Partnership for a Secure America, in February 2013 and signed by thirty-eight former members of Congress, cabinet members, generals, admirals, and others with extensive experience in international security issues:

> The effects of climate change in the world's most vulnerable regions present a serious threat to American national security interests. As a matter of risk management, the United States must work with international partners, public and private, to address this impending crisis. Potential consequences are undeniable, and the cost of inaction, paid for in lives and valuable US resources, will be staggering. Washington must lead on this issue now.... We, the undersigned Republicans, Democrats and Independents, implore US policymakers to support American security and global stability by addressing the risks of climate change in vulnerable nations. Their plight is our fight; their problems are our problems.
>
> (Partnership for a Secure America 2013)[4]

The following is a domestically oriented excerpt from a speech given on Capitol Hill in April 2013 by George Shultz, who held positions in the 1970s and 1980s in the Nixon and Reagan administrations as Secretary of State, Secretary of Defense, Secretary of the Treasury, Secretary of Labor, and Director of the Office of Management and Budget (and who was an economics professor and business school dean before that). His immediate audience for the speech was predominantly members of Congress and their staff, when he said:

> [O]n the global warming issue... I respect science,... but people are saying they don't like the science and so on. So I'm saying well, never mind the science. Just use your eyes. A new ocean is being created [in the Arctic region]... And if you look at the chart on the way in which the sea ice has been disappearing, the most stunning thing in it is how in recent years suddenly there has been a shift. And the discontinuities are the thing you have to watch out for. Because something may come and hit you faster than you believe [it will] because of the operation of the discontinuity.
>
> (Shultz 2013)

And finally, the wisdom from Einstein (no date) noted at the beginning of Chapter 1: "*We can't solve problems by using the same kind of thinking we used when we created them.*"

Suggestions for further reading and research

Assessments of US climate change policy options are available in Aldy and Stavins (2007b), Council on Foreign Relations (2008), Kopp and Pizer (2007), and Victor

4 The letter signers included seventeen former members of the House and Senate (from both political parties), cabinet-level officials from two Democratic administrations (Carter and Clinton) and five Republican administrations (Nixon, Ford, Reagan, George H.W. Bush, and George W. Bush), and nine retired generals and admirals.

(2004; 2011). For bipartisan analyses of climate change and related energy issues, see the Bipartisan Policy Center (2010) at www.bipartisanpolicy.org. Republicans for Environmental Protection at www.rep.org provides examples of Republicans who actively support measures to address climate change, and it publishes analyses of energy and other issues related to climate change. Among Democratically oriented organizations, the Center for American Progress at www.cap.org, which is close to the Obama administration, also conducts studies. Antholis and Talbott (2010) discuss the policies of the two Bush administrations and the Clinton administration, and the role of the Obama administration in the Copenhagen agreement of 2009. Pooley (2010) provides a detailed factual account of US national government policymaking on cap-and-trade proposals and other issues over the period from the late 1980s through 2009.

Articles in the refereed journal *Climate Policy* are essential reading for many specific topics, as are reports from the Harvard University Belfer Center for Science and International Affairs at www.belfer.ksg.harvard.edu. A comprehensive analysis of the legal issues posed by climate change in the USA is provided by Gerrard (2007), with updates at www.abanet.org and case tracking at www.climatecasechart.com.

US failures to act on climate change are assessed by Bryner (2000) and by Schroeder and Glicksman (2010). Congressional policymaking on climate change has been evaluated by Rabe (2010a; 2010b; 2010c; 2010d) and by Smith, Vogel, Cruce, Seidel, and Holsinger (2010). Climate policy predictions are presented in Selin and VanDeveer (2007).

APPENDIX 8.1

Recommendations of the President's Council of Advisors on Science and Technology[a]

March 2013

Dear Mr. President:

When you met with your Council of Advisors on Science and Technology (PCAST) at the end of November, you noted that your Administration was in the process of developing a strategy for addressing climate change during your second term and you asked for our input.

In this letter, we suggest six key components for consideration that we deem central to your climate change strategy and policy: . . .

(1) Focus on national preparedness for climate change.

- Create a National Commission on Climate Preparedness charged with recommending an overall framework and blueprint for ongoing data collection, planning, and action.
- Designate Departments to serve as leads to oversee the annual creation of climate preparedness plans at home and abroad.
- Develop an infrastructure renewal plan that integrates climate preparedness and other benefits to the Nation's economy.
- Improve coordination and support for research efforts on climate change preparedness.

(2) Continue efforts to "decarbonize" the economy, with an initial focus on the electricity sector.

To reinforce the key decarbonization pathways, the Administration could:

- Support continuing expansion of shale-gas production, ensuring that environmental impacts of production and transport do not curtail the potential of this approach.
- Continue implementation of Clean Air Act requirements on criteria pollutants (such as SO_2 and NO_x) and hazardous air pollutants (such as mercury) to include

creating new performance standards for CO_2 emissions from existing stationary sources, which would follow the performance standards for new plants released in March 2012.

- Accelerate efforts to reduce the regulatory obstacles to deployment of CCS, and continue political support for the large CCS projects currently underway.

(3) Level the playing field for clean-energy and energy-efficiency technologies by removing regulatory obstacles, addressing market failures, adjusting tax policies, and providing time-limited subsidies for clean energy when appropriate.
Some opportunities worthy of consideration are:

- Level the playing field on access to capital through special tax benefits.
- Broaden the tax credit for wind to include all forms of renewable energy, replacing the annual renewal with a longer time horizon of 5 to 10 years.
- Eliminate market failures that prevent the adoption of technologies for energy efficiency.

(4) Sustain research on next-generation clean-energy technologies, and remove obstacles for their eventual deployment.

- We recommend that you sustain and, if possible, augment the investment in research and development in energy innovation, focusing on the critical technologies that have the potential to dramatically lower our greenhouse gas emissions in the long run.
- Nuclear power requires special attention, as the Federal Government's role is different than for all other technologies.

(6) Conduct an initial Quadrennial Energy Review (QER).

[[a] The President's Council of Advisors on Science and Technology consists of the following:
Co-Chairs: John P. Holdren, Assistant to the President for Science and Technology, and Director of the White House Office of Science and Technology Policy; and Eric Lander, President and Founding Director of the Broad Institute of Harvard and MIT.

The members include the heads of the following . . . or their representatives: Departments of State, Treasury, Agriculture, Commerce, Labor, Energy, Transportation, Homeland Security; Office of United States Trade Representative, Export-Import Bank, Small Business Administration, United States Trade and Development Agency, Overseas Private Investment Corporation, Council of Economic Advisers, Office of Management and Budget, National Economic Council, National Security Staff. Also the following may be included, "in their discretion, the heads of the following organizations or their designees": National Governors Association, United States Conference of Mayors, five members of the United States Senate, designated by the President of the Senate, and five members of the United States House of Representatives, designated by the Speaker of the House; and "not to exceed 28 citizens appointed by the President. They shall include representatives of business and industry, agriculture, and labor."]

Source: US President's Council of Advisors on Science and Technology (2013).

APPENDIX 8.2

Climate change issues in the 2012 presidential election: party platforms and candidates' answers to questions

Republican party platform[a]

"[W]e oppose any and all cap and trade legislation."

"We also call on Congress to take quick action to prohibit the EPA from moving forward with new greenhouse gas regulations that will harm the nation's economy and threaten millions of jobs over the next quarter century."

Democratic party platform[b]

"We know that global climate change is one of the biggest threats of this generation – an economic, environmental, and national security catastrophe in the making. We affirm the science of climate change, commit to significantly reducing the pollution that causes climate change, and know we have to meet this challenge by driving smart policies that lead to greater growth in clean energy generation and result in a range of economic and social benefits."

Candidate Romney's answer to "Science Debate" question[c]

"I am not a scientist myself, but my best assessment of the data is that the world is getting warmer, that human activity contributes to that warming, and that policymakers should therefore consider the risk of negative consequences. However, there remains a lack of scientific consensus on the issue – on the extent of the warming, the extent of the human contribution, and the severity of the risk – and I believe we

Candidate Obama's answer to "Science Debate" question[c]

"Climate change is the one of the biggest issues of this generation, and we have to meet this challenge by driving smart policies that lead to greater growth in clean energy generation and result in a range of economic and social benefits. Since taking office I have established historic standards limiting greenhouse gas emissions from our vehicles for the first time in history. My administration has made unprecedented investments in must support continued debate and investigation within the scientific community.... So I oppose steps like a carbon tax or a cap-and-trade system that would handicap the American economy and drive manufacturing jobs away, all without actually addressing the underlying problem. Economic growth and technological innovation, not economy-suppressing regulation, is the key to environmental protection in the long run. So I believe we should pursue what I call a "No Regrets" policy – steps that will lead to lower emissions, but that will benefit America regardless of whether the risks of global warming materialize and regardless of whether other nations take effective action. For instance, I support robust government funding for research on efficient, low-emissions technologies that will maintain American leadership in emerging industries. And I believe the federal government must significantly streamline the regulatory framework for the deployment of new energy technologies, including a new wave of investment in nuclear power."

clean energy, proposed the first-ever carbon pollution limits for new fossil-fuel-fired power plants and reduced carbon emissions within the Federal Government. Since I took office, the US is importing an average of 3 million fewer barrels of oil every day, and our dependence on foreign oil is at a 20-year low. We are also showing international leadership on climate change, reaching historic agreements to set emission limits in unison with all major developed and developing nations. There is still more to be done to address this global problem. I will continue efforts to reduce our dependence on oil and lower our greenhouse gas emissions while creating an economy built to last."

Sources:

[a] Republican Party Convention (2012).

[b] Democratic Party Convention (2012).

[c] "Science Debate" was an ad hoc effort that was co-sponsored by a large number of professional organizations, including the American Association for the Advancement of Science, the National Academies, the American Institute of Physics, and others, to solicit the candidates' positions on a wide range of issues. One was climate change.

The question posed to the candidates for their written responses was: "What is your position on cap-and-trade, carbon taxes, and other policies proposed to address global climate change – and what steps can we take to improve our ability to tackle challenges like climate change that cross national boundaries?"

Also see Center for Climate and Energy Solutions (2012).

APPENDIX 8.3

Policy recommendations and quotes from Republicans for Environmental Protection[a]

Section A. Policy recommendations

The evidence is clear that fossil fuel combustion is increasing the atmosphere's carbon dioxide load. Prudence demands that we acknowledge the facts and act. Despite the daunting nature of the challenge, taking it on will create large opportunities in America to reduce energy costs, build new industries, revitalize rural economies, and carry out a constructive foreign policy free from the corrosive influence of petroleum politics.

Here is what the federal government must do:

Establish a market for carbon reductions

The most important step that Congress and the administration must take to reduce oil dependence on lower greenhouse gas emissions is to put a price on those emissions. Market-friendly cap-and-trade legislation has been considered a preferred approach because it would include enforceable emissions limits. If Congress chooses instead to enact a carbon tax, the legislation must include assurances that actual emissions reductions would be achieved, funds would not be diverted to expand government, and the tax would not be riddled with exemptions or other complex "loopholes." Whichever option Congress chooses, it's important that legislation be adopted as soon as it can be drafted, vetted carefully, and acted upon with bipartisan support. Because of the potentially high costs of doing nothing, passing an imperfect bill would be better than failing to act.

Increase funding for energy research and development

Reducing oil dependence and stabilizing greenhouse gas concentrations will require scaling up numerous advanced energy technologies. A strong research and development program is necessary for moving promising technologies out of the lab and into the marketplace.

Strengthen energy efficiency standards and incentives

Energy efficiency is consistent with conservative values of frugality and stewardship. As the cheapest, cleanest, and most secure energy resource available, efficiency has a strong track record. It's time to build on that record of success, through measures to increase efficiency in buildings, industry, and transportation.

Expand transportation fuels from renewable resources

Ethanol is a promising resource for displacing significant quantities of gasoline when combined with plug-in hybrid-electric drive trains. Research, standards, and incentives should be adopted to accelerate broader use of cellulosic ethanol.

Expand electric power from renewable resources

Diversifying our electric power system with renewable resources will result in numerous economic benefits, including reduced vulnerability to fuel price and supply risks, economic development opportunities for rural communities, and greater freedom of choice for energy consumers. A renewable portfolio standard, extension of production tax credits, and other incentives should be adopted.

Keep a place for nuclear energy at the table

Nuclear energy can deliver large amounts of carbon-free baseload electricity. It is in the nation's interest to develop promising technologies for improving plant security and economics, managing high-level nuclear wastes, and minimizing risks of theft and diversion of fissile materials.

Ensure responsible use of natural gas

Natural gas is a relatively clean fuel for power generation and transportation. Gas can serve as a bridge to a cleaner, more diverse, less carbon-intensive energy economy. Steps should be taken to ensure the most efficient use of this fuel and minimize the impacts of gas production in the Intermountain West.

Clean up coal

The United States has large coal reserves. Coal, however, is the most problematic of the fossil fuels, because of climate, air quality, and land impacts. Through research and standards, the federal government should speed the transition to cleaner coal technologies, including large-scale carbon sequestration.

Conclusion

America stands at the threshold of both immense risk and opportunity. Beyond the practical economic and security benefits of moving to a cleaner, more secure energy economy, good stewardship is a moral imperative that is central to traditional conservatism. What is needed now is the will to marshal our nation's considerable assets, develop a conservative energy strategy for the future, and put it to work today.

Source: Republicans for Environmental Protection (2011a). Used with permission.

Section B. Quotes from conservative Republican presidents and presidential candidates

Theodore Roosevelt: "Conservation is a great moral issue, for it involves the patriotic duty of insuring the safety and continuance of the nation." (New Nationalism speech, Osawatomie, Kansas, August 31, 1910)

Barry Goldwater: "While I am a great believer in the free enterprise system and all that it entails, I am an even stronger believer in the right of our people to live in a clean and pollution-free environment." (*The Conscience of a Majority* [1970])

Ronald Reagan: "If we've learned any lessons during the past few decades, perhaps the most important is that preservation of our environment is not a partisan challenge; it's common sense. Our physical health, our social happiness, and our economic well-being will be sustained only by all of us working in partnership as thoughtful, effective stewards of our natural resources." (Remarks on signing annual report of Council on Environmental Quality, July 11, 1984)

John McCain: "Some urge we do nothing because we can't be certain how bad the (climate) problem might become or they presume the worst effects are most likely to occur in our grandchildren's lifetime. I'm a proud conservative, and I reject that kind of live-for-today, 'me generation,' attitude. It is unworthy of us and incompatible with our reputation as visionaries and problem solvers. Americans have never feared change. We make change work for us." (Address at Center for Strategic and International Studies, April 23, 2007)

[a] The organization subsequently changed its name to ConservAmerica.

Source: Compiled by the author from Republicans for Environmental Protection (2011b). Accessed at www.rep.org on May 1, 2011. Used with permission of ConservAmerica.

APPENDIX 8.4

Statement by a group of leading scholars: Thinking Through the Climate Change Challenge – An Open Letter

In October 2010, . . . a group of leading thinkers on environmental policy met at the Sustainable Consumption Institute at the University of Manchester for a conference in honour of Nobel economics Laureate Tom Schelling. At the event we formulated guidance for policymakers which draws on work that Schelling . . . has done on climate change. The analysis here relies on his concept of identifying "focal points" on which agreements can be based, and his emphasis on designing policies that are credible as well as easily monitored and enforced.

Problem overview

Global climate change is one of the greatest problems facing mankind that requires collective action in order to be solved. Although there would be substantial long-term gains to all societies from working together to limit greenhouse gas emissions, many countries lack strong incentives to reduce their own emissions over the short term. It is therefore unlikely that an economically efficient outcome could be achieved in the best of circumstances.

What's more, climate negotiations such as the meetings in Copenhagen and Cancun are likely to fail to reach an effective agreement on reducing greenhouse gas emissions. The terms of the negotiations thus need to be radically changed.

The international community has arrived at a focal point (in Schelling's terms) of limiting global temperature increases to 2°C. But without agreement on enforceable action to achieve the target, this will have little impact. We therefore offer ten concrete points for policy makers.

Guidance for decision makers

1. Economic analysis suggests that governments have significantly underinvested in mitigation relative to the level of effort that would be economically efficient from a global perspective.

2. All realistic options for addressing climate change should be seriously considered. These include controlling greenhouse gas emissions, removing carbon dioxide from the atmosphere, adaptation to change, and geo-engineering.
3. International agreements are needed because coordination would help ensure that policies achieve climate policy goals at minimum cost to society. But this does not imply that agreements must cover all countries and all sectors. Nor does it imply that action on the part of individual states must wait until there is an international accord.
4. New approaches that pass a benefit-cost test should be tried. While comprehensive approaches are appealing in principle, they face serious political hurdles. An alternative is to address specific greenhouse gases and sectors in separate agreements. There could, for example, be an agreement on maintaining forests and planting trees, and another on regulating carbon dioxide emissions from fossil fuel use. A primary focus of negotiations should be on the practical measures needed to monitor and enforce whatever is agreed. To date, enforcement of climate agreements has been weak.
5. Putting a price on greenhouse gas emissions (by taxing them or limiting aggregate output with a cap-and-trade mechanism) would be desirable because it would help to get consumers, businesses, and government to account for the full social cost of their behaviour. Many countries already have explicit or implicit prices on greenhouse gas emissions, which will help to reduce the cost of greenhouse gas reductions in the future. A potential issue complicating pricing policies is that they can create large revenue streams. Such revenue should be used productively – for example, by reducing other taxes that distort economic activity.
6. Climate stabilization requires that net carbon dioxide emissions eventually decline significantly. Achieving that goal will require a technological revolution. This is one reason why research and development in energy technologies should be a priority, though policies should be carefully designed to ensure innovative efforts are socially productive.
7. Research and development is also needed in technologies for removing carbon dioxide from the atmosphere and for managing solar radiation, even though these technologies may not be deployed for decades, if ever. Efforts should begin now to develop strong norms and governance arrangements for determining the appropriate use of geo-engineering technologies.
8. Businesses need appropriate incentives for innovation, investment and behavioural change. Thus, policy commitments for R&D and pricing greenhouse gas emissions should strive to be credible and reasonably stable over long periods.

9. The incentives for consumers, firms, and governments to adapt to climate change are strong because they will bear most of the costs if they do not adapt. The poorest countries, however, are least able to adapt. The industrialized countries should help them. The most effective means of providing assistance requires careful study. It may include a portfolio of efforts targeted more toward economic development than to climate adaptation.
10. There are great uncertainties in how best to manage the various components of the climate change problem. These uncertainties should be acknowledged by adopting a flexible approach to decision making that responds to new knowledge about climate change. Uncertainty should not, however, be used as a rationale for inaction.

Instead of negotiating about targets and timetables that are strongly opposed by key parties and that cannot be easily enforced, policymakers should focus on concrete alternatives that can be monitored and enforced. Continued efforts to reach a comprehensive agreement that lack[s] these characteristics offer little prospect of success.

[Signed by]: David Anthoff, University of California, Berkeley; Elizabeth Baldwin, University of Oxford; Scott Barrett, Columbia University; Linda Cohen, University of California, Irvine; Diane Coyle, Enlightenment Economics; Partha Dasgupta, University of Cambridge; Simon Dietz, London School of Economics and Political Science; David J. Frame, University of Oxford; Robert Hahn, The University of Manchester; James K. Hammitt, Toulouse School of Economics; Geoffrey Heal, Columbia University; Cameron Hepburn, University of Oxford; Michael Hoel, University of Oslo; Charles D. Kolstad, University of California, Santa Barbara; Andreas Lange, University of Hamburg; Robert Mendelsohn, Yale University; Karine Nyborg, University of Oslo; Ian W.H. Parry, Resources for the Future; Peter Passell, Milken Institute; Kenneth Richards, Indiana University; Robert Ritz, University of Cambridge; Thomas C. Schelling, University of Maryland; Massimo Tavoni, FEEM (Fondazione Eni Enrico Mattei); Alistair Ulph, The University of Manchester; Herman R.J. Vollebergh, Tilburg University Sustainability Centre (TSC); Anastasios Xepapadeas, Athens University of Economics and Business.

Source: Vox (2011); with minor formatting changes by the author. Used with permission.

APPENDIX 8.5

Spectrum of government–market mixes

The degree of state involvement in delivering social outcomes (such as environmental protection) might be considered to be on a spectrum running from "free market" at one end, to "nationalized delivery" at the other end:

Free market: no government involvement; individuals and firms voluntarily acquire information on externalities and voluntarily and altruistically internalize those externalities.

Information provision: government assumes the role of aggregating and disseminating information about externalities and their shadow prices, but does nothing more.

Moral suasion: government provides information and may even seek to persuade people and firms to change their preferences and objectives. In its best form, this might constitute a form of "government by discussion." Another recently popular, but some would argue more sinister, notion is that government might influence people's decisions by "soft paternalism" or nudging people's decisions by careful design of the "choice architecture" (Sunstein and Thaler 2003).

Economy-wide relative prices: government determines the appropriate price or quantity of the social good or externality (e.g. carbon dioxide emissions . . .) and implements policy to correct relative prices (e.g. economy-wide taxes, trading schemes, etc.).

Output-based intervention: government specifies output standards for specific sectors of firms (e.g. CO_2/MW standards), but does not require the use of any particular method to deliver those standards.

Input- or technology-based intervention: government specifies or encourages or requires firms to employ particular technologies or inputs . . . either through explicit regulation or through taxes or subsidies.

Project-level intervention: government specifies or encourages particular projects to occur, through subsidy or other financial . . . support (e.g. EU carbon capture and storage . . . program).

State capitalism: state-owned enterprises follow guidance given by their (government) shareholder; some flexibility for implementation may be retained if targets are expressed and political incentives put in place, but often executives are given direct instructions.

Nationalized delivery: government finances and delivers on environmental protection directly through central government departments.

For the environment, unlike other areas of economic activity, relying on the "free market" or on "information provision" is highly unlikely to deliver satisfactory outcomes because firms have inadequate incentives to internalize externalities without government intervention.

Source: Hepburn (2010); with minor modifications by the author in punctuation and formatting. Used with permission.

APPENDIX 8.6

Design issues for a cap-and-trade system[a]

. . . Designing a cap-and-trade program to achieve [emission] reductions would include important decisions about whether to sell or give away allowances. Those rights to emit greenhouse gases would have substantial value, and policymakers' choices about how to allocate them could have significant effects on the federal budget and on how the gains and losses brought about by the program were distributed among US households. If policymakers chose to sell the allowances, they could use the revenue that would arise in many different ways, including to offset other taxes, to assist workers or low-income households that might be adversely affected by the cap, to support other legislative priorities, or to reduce the budget deficit. Policymakers would also need to decide whether to include provisions to help contain the cost of the policy by allowing firms flexibility as to when they reduced their emissions and whether to include provisions to address effects on international trade, particularly for energy-intensive goods. . . .

A cap-and-trade program could raise significant [government] revenue because the value of the allowances created under the program would probably be substantial. . . .

Issuing allowances to entities at no charge, provided that the recipients can readily convert the allowances into cash, is economically equivalent to selling the allowances and dedicating the revenue to those same entities. . . .

Policymakers' decisions about how to allocate the allowances could have significant effects on the overall economic cost of capping CO_2 emissions and on the distribution of gains and losses among US households. Giving away a large share of the allowances to companies that produce fossil fuels or energy-intensive goods could be more costly to the economy and more regressive than selling them. . . .

If the government chose to sell emission allowances, it could use some of the revenue to offset the disproportionate economic burden that higher prices would impose on low-income households. Selling allowances could also significantly lessen the overall economic impact of a CO_2 cap. . . .

Policymakers could help reduce the cost of achieving any given long-term target for reducing emissions if they included provisions in a cap-and-trade program that allowed firms some degree of flexibility about when emission reductions take place. Such provisions would augment the flexibility about where and how emission cuts are made that is intrinsic to a cap-and-trade program. Timing flexibility would allow firms to reduce emissions more when the cost of doing so was low and would provide firms leeway to reduce their efforts when costs were high. One method of providing timing flexibility would be to set a ceiling and a floor for allowance prices. The ceiling would limit firms' expenses when the cost of cutting emissions was high, and the floor would automatically tighten the cap (and thereby increase emission reductions) when the cost of cutting emissions was low. Policymakers could periodically adjust the speed at which the price ceiling and floor increased to ensure that emission reductions were on track for achieving a long-term target.

[a] N.B. This is only an excerpt from the original source and therefore does not reflect the totality of the original analysis.
Source: Orszag (2008).

APPENDIX 8.7

Discussion of international institutional venue issues[a]

... [T]o the extent that there is no meeting of the minds between the United States, China, and the European Union, then moving the negotiations into [a forum other than the UNFCCC] won't solve the problem [of the presence of obstructionist governments in the UNFCCC].

Letting a thousand flowers bloom

A more promising approach to moving forward would be to split the climate-change problem up into different pieces and address the more tractable pieces in more specialized forums. To some degree this is happening already. The International Maritime Organization (IMO), for example, is considering international shipping, the International Civil Aviation Organization (ICAO) is considering civil aviation, and the Montreal Protocol is considering HFCs. These other forums are not immune from the influence of the UNFCCC – and in particular, the principle of common but differentiated responsibilities, which developing countries interpret to give them a free pass. But more progress may be possible in specialized forums, which have long traditions of cooperation, than in the UNFCCC. This decentralized approach will not be sufficient in itself to solve the climate-change problem. But it offers a useful supplement to the UNFCCC process and will be all the more important if the UNFCCC continues to be stalemated.

In the International Maritime Organization, an existing treaty – the International Convention for the Prevention of Pollution from Ships (MARPOL) – already addresses vessel-source pollution, including air pollution in Annex VI. So including controls on greenhouse-gas emissions could be adopted by amending Annex VI, which requires only a three-quarter majority vote (as compared to the consensus decisionmaking rule within the UNFCCC). This could occur as early as next year, when the parties consider a proposed amendment specifying ship efficiency standards designed to reduce emissions. The development of market-based measures to reduce emissions is also being considered by an IMO working group. Although the big developing countries object to these new measures, the IMO has greater potential than the UNFCCC to move forward since it can make decisions by qualified majority vote.

ICAO has made less progress in addressing emissions from civil aviation. But a high-level meeting in Fall 2009 agreed to a goal of improving global fuel efficiency by 2 percent per year. It also made some progress on reporting and verification issues and agreed to establish a process focusing on the use of economic instruments to address civil aviation emissions, such as an allowance trading scheme. An important factor helping to spur ICAO action is the impending expansion of the European Union emissions trading scheme to include emissions from flights to and from Europe. Finally, the Montreal Protocol on ozone-depleting substances has already had a huge impact in mitigating climate change, since ozone-depleting substances are also greenhouse gases. The decision by the Montreal Protocol in 2008 to accelerate the phase-out of HCFCs did more by itself to reduce climate forcings than the Kyoto Protocol. And a number of countries, including the United States and Mexico, are now proposing that HFCs – substitutes for ozone-depleting substances that are also very potent greenhouse gases – be regulated under the Montreal Protocol rather than the UNFCCC regime.

Of course, even if these initiatives in IMO, ICAO, and the Montreal Protocol are successful, they will make only a modest dent in global greenhouse-gas emissions. So we need additional action as well. And since such action is unlikely to be forthcoming anytime soon under the UNFCCC, we need to contemplate a decentralized regime, in which smaller groups of like-minded countries address particular issues such as deforestation, and in which countries and regional groupings take parallel action on their own, as the European Union and some states within the United States are already doing. Recently a Norwegian initiative resulted in an agreement by more than fifty countries on a framework to reduce deforestation. This could potentially serve as a model for other initiatives, for example addressing efficiency standards.

These extra-UNFCCC initiatives raise two issues. First, how should the countries that are taking action deal with each other? And second, how should countries taking action deal with those that aren't? In the former case, the answer could involve some type of integration or linkage between different national and regional trading systems, as many have proposed. In the latter case, the answer is likely to involve some type of trade measures against countries that fail to take meaningful action to combat climate change. For example, countries with policies that establish a price on carbon could impose border tax adjustments on imports of carbon-intensive goods from countries that fail to take action. This would serve two purposes. First, it would internalize the climate externalities associated with the production of these goods. Second, it would prevent countries that fail to take action from gaining a competitive advantage.

Conclusion

Global warming is a classic example of a collective action problem. So unless a technological magic bullet is found, solving the problem will ultimately require a collective agreement among the key contributors, like the Montreal Protocol on Substances that Deplete the Ozone Layer.

But since an agreement among the major emitters is unlikely anytime soon, we should seek progress where we can, through whatever means and in any forums that are available. Indeed, even if the prospects for a global agreement are brighter than this paper suggests, it does not make sense to put all of our eggs in one basket. Better to diversify our portfolio of climate-change policies, including through work on adaptation and geoengineering. Such an approach lacks the clarity and simplicity of a single comprehensive policy, such as a global emissions trading system or tax. But, still, muddling through is better than simply treading water from one UNFCCC meeting to the next.

[a] N.B. This is only an excerpt from the original source and therefore does not reflect the totality of the original analysis. Supported by the Harvard Project on Climate Agreements.

Source: Bodansky (2010). Used with permission.

References

Abatzoglou, John, Joseph F.C. DiMento, Pamela Doughman, and Stefano Nespor. 2007. "A Primer on Global Climate Change and Its Likely Impacts," in Joseph F.C. DiMento and Pamela Doughman, editors, *Climate Change: What It Means for Us, Our Children, and Our Grandchildren*. Cambridge, MA: MIT Press.

Abbasi, Daniel R. 2006. "Americans and Climate Change, A Synthesis of Insights and Recommendations from the 2005 Yale F&ES Conference on Climate Change," Yale University School of Forestry and Environmental Studies.

ABC News. 2009. Transcript: Rahm Emanuel and Rep. John Boehner, This Week with George Stephanopolis, April 19. Accessed at www.abcnews.com on May 1, 2011.

ABC News/Gallup. 2009. National survey on November 20–22, 2009, reported at PollingReport.com, Environment. Accessed on December 30, 2009.

ABC News/Planet Green/Stanford University. 2008. National survey on July 23–28, 2008, reported at PollingReport.com, Environment. Accessed on December 30, 2009.

Agrawala, S., and S. Andresen. 2001. "US Climate Policy: Evolution and Future Prospects," *Energy and Environment*, 9: 117–139.

Air Resources Board [of California]. 2010. Assembly Bill 32: Global Warming Solutions Act. Accessed at www.arb.ca.gov on November 8, 2010.

Aldy, Joseph E., Scott Barrett, and Robert Stavins. 2003. "Thirteen Plus One: A Comparison of Global Climate Policy Architectures," Working Paper Series rwp03-012, Harvard University, John F. Kennedy School of Government.

Aldy, Joseph E., and Robert N. Stavins, editors. 2007a. *Architectures for Agreement: Addressing Global Climate Change in the Post-Kyoto World*. Cambridge: Cambridge University Press.

Aldy, Joseph E., and Robert N. Stavins. 2007b. "Architectures for an International Global Climate Change Agreement: Lessons for the Policy Community," in Joseph E. Aldy and Robert N. Stavins, editors, *Architectures for Agreement: Addressing Global Climate Change in the Post-Kyoto World*. Cambridge: Cambridge University Press.

Aldy, Joseph E., and Robert N. Stavins, editors. 2009. *Post-Kyoto International Climate Policy: Summary for Policymakers*. Cambridge: Cambridge University Press.

Aldy, Joseph E., and Robert N. Stavins, editors. 2010. *Post-Kyoto International Climate Policy: Implementing Architectures for Agreement*. Cambridge: Cambridge University Press.

Aldy, Joseph E., Matthew J. Kotchen, and Anthony A. Leiserowitz. 2012. "Willingness to Pay and Political Support for a US National Clean Energy Standard," *Nature Climate Change*, 2: 596–599.

Allcott, Hunt, and Michael Greenstone. 2012. "Is There an Energy Efficiency Gap?" MIT Department of Economics Working Paper No. 12–03. Accessed at www.papers.nber.com on March 11, 2013.

Alt, James E., and Alberto Alesina. 1998. "Political Economy: An Overview," in Robert E. Goodin and Hans-Dieter Klingemann, editors, *A New Handbook of Political Science*. Oxford: Oxford University Press.

Ambec, Stefan, Mark A. Cohen, Stewart Elgie, and Paul Lanoie. 2010. "The Porter Hypothesis at 20: Can Environmental Regulation Enhance Innovation and Competitiveness?" Chairs' Paper. Montreal: Sustainable Prosperity. Accessed at www.sustainableprosperity.ca on June 26, 2010.

American Council for an Energy Efficient Economy (ACEEE). 2007. "Adapting to Climate Change: Planning for Resilient Energy Systems and Maintaining Global Competitiveness." Accessed at www.aceee.org on November 11, 2010.

American Petroleum Institute. 2001. *Comparison of Findings: National Academy of Sciences Report and the Intergovernmental Panel on Climate Change Report*. Washington, DC: American Petroleum Institute.

Anderegg,William R.L., James W. Prall, Jacob Harold, and Stephen H. Schneider. 2010. "Expert Credibility in Climate Change," *Proceedings of the National Academy of Sciences*, 107: 12107–12109. Accessed at www.pnas.org on July 25, 2010.

Andersen, Stephen O., and Durwood Zaelke. 2003. *Industry Genius: Inventions and People Protecting the Climate and Fragile Ozone Layer*. Sheffield: Greenleaf.

Anderson, Kai S. 2000. "The Climate Policy Debate in the US Congress," in Stephen H. Schneider, Armin Rosencranz, and John O. Niles, editors, *Climate Change Policy: A Survey*. Washington, DC: Island Press.

Antholis, William, and Strobe Talbott. 2010. *Fast Forward: Ethics and Politics in the Age of Global Warming*. Washington, DC: The Brookings Institution.

Applegate, John S. 2010. "Embracing a Precautionary Approach to Climate Change," in David M. Driesen, editor, *Economic Thought and US Climate Change Policy*. Cambridge, MA: MIT Press.

Archer, David, and Stefan Rahmstorf. 2010. *The Climate Crisis: An Introductory Guide to Climate Change*. Cambridge: Cambridge University Press.

Arris, L., editor. 1997. *Global Warming Yearbook*. Arlington, MA: Cutter Information.

Arris, L., editor. 1998. *Global Warming Yearbook*. Arlington, MA: Cutter Information.

Arris, L., editor. 2000. *Global Warming Yearbook*. Arlington, MA: Cutter Information.

Asia-Pacific Partnership on Clean Development and Climate. 2008. "Flagship Projects." Accessed at www.asiapacificpartnership.org on July 30, 2008.

Atkinson, A.B. 1998. "Political Economy, Old and New," in Robert E. Goodin and Hans-Dieter Klingemann, editors, *A New Handbook of Political Science*. Oxford: Oxford University Press.

Baldassare, Mark, Dean Bonner, Jennifer Paluch, and Sonja Petek. 2009. "Californians and the Environment," Public Policy Institute of California. Accessed at www.ppic.org on January 20, 2010.

Barrett, Scott. 1994. "Self-Enforcing International Environmental Agreements," *Oxford Economic Papers*, 46: 878–894.

Barrett, Scott. 1997. "Towards a Theory of International Cooperation," in C. Carraro and D. Siniscalco, editors, *New Directions in the Economic Theory of the Environment*. Cambridge: Cambridge University Press.

Barrett, Scott. 1998. "On the Theory and Diplomacy of Environmental Treaty-Making," *Environmental and Resource Economics*, 11: 317–333.

Barrett, Scott. 2001. "Towards a Better Climate Treaty," *Policy Matters*, 01–29. Washington, DC: AEI-Brookings Joint Center for Regulatory Studies.

Barrett, Scott. 2003. *Environment and Statecraft: The Strategy of Environmental Treaty-Making*. Oxford: Oxford University Press.

Barrett, Scott. 2007. "A Multitrack Climate Treaty System," in Joseph E. Aldy and Robert N. Stavins, editors, *Architectures for Agreement: Addressing Global Climate Change in the Post-Kyoto World*. Cambridge: Cambridge University Press.

Barrett, Scott. 2010. "A Portfolio System of Climate Treaties," in J.E. Aldy and Robert N. Stavins, editors, *Post-Kyoto International Climate Policy: Implementing Architectures for Agreement*. Cambridge: Cambridge University Press.

Bauer, Raymond, Ithiel de Sola Pool, and Lewis Anthony Dexter. 1963. *American Business and Public Policy*. New York, NY: Atherton.

Betsill, Michele M. 2000. "The United States and the Evolution of International Climate Change Norms," in Paul G. Harris, editor, *Climate Change and American Foreign Policy*. New York, NY: St. Martin's.

Betsill, Michele M. 2009. "NAFTA as a Forum for CO2 Permit Trading?" in Henrik Selin and Stacy D. VanDeveer, editors, *Changing Climates in North American Politics: Institutions, Policymaking, and Multilevel Governance*. Cambridge, MA: MIT Press.

Betsill, Michele M., and Harriet Bulkeley. 2006. "Cities and the Multilevel Governance of Global Climate Change," *Global Governance*, 12: 141–159.

Beveridge & Diamond. 2007. "Massachusetts becomes RGGI Signatory," February 7.

Bianco, Nicholas M., Franz T. Litz, Kristin Igusky Meek, Rebecca Gasper. 2013. *Can The US Get There From Here? Using Existing Federal Laws and State Action to Reduce Greenhouse Gas Emissions*. Washington, DC: WRI. Accessed at www.wri.org on February 21, 2013.

Biermann, Frank, and Rainer Brohm. 2005. "Implementing the Kyoto Protocol Without the United States: The Strategic Role of Energy Tax Adjustments at the Border," *Climate Policy*, 4: 289–302.

Bipartisan Policy Center. 2011. "Reassessing Renewable Energy Subsidies." Accessed at www.bipartisanpolicy.org on April 10, 2011.

Bloomberg, Michael. 2007. Quoted in "Cities Take Lead On Environment As Debate Drags At Federal Level: 522 Mayors Have Agreed To Meet Kyoto Standards." By Anthony Faiola and

Robin Shulman. *Washington Post*, June 9. Accessed at www.washingtonpost.com on June 12, 2007.

Bloomberg New Energy Finance. 2011. "Global Renewable Energy Market Outlook. Executive Summary." Accessed at www.bnef.com on January 28, 2013.

Blue Green Alliance. 2010a. "About the Blue Green Alliance." Accessed at www.bluegreenalliance.org on August 24, 2010.

Blue Green Alliance. 2010b. "Policy Statement on Climate Change Legislation, March 2009." Accessed at www.bluegreenalliance.org on August 24, 2010.

Bodansky, Daniel. 2010. "The International Climate Change Regime: The Road from Copenhagen. Harvard Project on International Climate Agreements." Accessed at www.belfercenter.ksg.harvard.edu on April 10, 2013.

Bonn Center for Local Climate Action. 2011. "City Commitments." Accessed at www.carbonn.org on January 4, 2011.

Borick, Christopher P., and Barry G. Rabe. 2008. "A Reason to Believe: Examining the Factors that Determine Americans' Views on Global Warming," *Issues in Governance Studies*, 18. Washington, DC: The Brookings Institution.

Borick, Christopher, Erick Lachapelle, and Barry Rabe. 2011. *Climate Compared: Public Opinion on Climate Change in the United States & Canada.* Key Findings Report for the National Survey of American Public Opinion on Climate Change and Public Policy Forum – Sustainable Prosperity Survey of Canadian Public Opinion on Climate Change.

Bowen, Mark. 2008. *Censoring Science: Inside the Political Attack on Dr James Hansen and the Truth of Global Warming.* New York, NY: Dutton.

Boykoff, Maxwell T., and Jules M. Boykoff. 2004. "Balance as Bias: Global Warming and the US Prestige Press," *Global Environmental Change*, 14: 125–136. Newspapers covered: *New York Times, Washington Post, Wall Street Journal, Los Angeles Times.*

Bradley, Raymond S. 2011. *Global Warming and Political Intimidation: How Politicians Cracked Down on Scientists as the Earth Heated Up.* Amherst and Boston, MA: University of Massachusetts Press.

Bravender, Robin. 2010. "Climate: Manufacturers Worry of Pitfalls from Utility-Only Carbon Cap," *Environment & Energy Daily*, July 2. Accessed at www.eenews.net on July 2, 2010.

Breitmeier, Helmut, Oran R. Young, and Michael Zürn. 2006. *Analyzing International Environmental Regimes: From Case Study to Database.* Cambridge, MA: MIT Press.

Brewer, Thomas L. 2003. "The Trade Regime and the Climate Regime: Institutional Evolution and Adaptation," *Climate Policy*, 3: 329–341.

Brewer, Thomas L. 2004a. "The WTO and the Kyoto Protocol: Interaction Issues," *Climate Policy*, 4: 3–12.

Brewer, Thomas L. 2004b. "Multinationals, the Environment and the WTO: Issues in the Environmental Goods and Services Industry and in Climate Change Mitigation," in Sarianna Lundan, Alan Rugman, and Alain Verbecke, editors, *Multinationals, the Environment and Global Competition.* Amsterdam: Elsevier.

Brewer, Thomas L. 2005a. "Global Warming and Climate Change in the 21st Century: New Issues for Firms, Governments and Research on Business-Government Relations," in Robert

Grosse, editor, *International Business-Government Relations in the 21st Century*. Cambridge: Cambridge University Press.

Brewer, Thomas L. 2005b. "Industry Location and Climate Change Policy-Making in the United States," Policy Brief. Brussels: Centre for European Policy Studies. Available at www.ceps.eu.

Brewer, Thomas L. 2005c. "US Public Opinion on Climate Change Issues: Implications for Consensus-Building and Policymaking," *Climate Policy*, 4: 359–376.

Brewer, Thomas L. 2007. "Border Measures To Address Climate Change-Related Competitiveness Concerns," *Bridges Trade BioRes: Trade & Environment Review*, 1: 6–7.

Brewer, Thomas L. 2008a. "Climate Change Technology Transfer: A New Paradigm and Policy Agenda," *Climate Policy*, 8: 516–526.

Brewer, Thomas L. 2008b. "International Energy Technology Transfers for Climate Change Mitigation: What, Who, How, Why, When, Where, How Much . . . and the Implications for International Institutional Architecture," Paper prepared for CESifo Venice Summer Institute Workshop: Europe and Global Environmental Issues. Venice, Italy, July 14–15; CESifo Working Paper No. 2408.

Brewer, Thomas L. 2008c. "The Technology Agenda for International Climate Change Policy: A Taxonomy for Structuring Analyses and Negotiations," in Christian Egenhofer, editor, *Beyond Bali: Strategic Issues for the Post-2012 Climate Change Regime*. Brussels: Centre for European Policy Studies.

Brewer, Thomas L. 2009a. "Pluralistic Politics and Public Choice: Theories of Business and Government Responses to Climate Change," in Paul G. Harris, editor, *Theory and Praxis of Environmental Foreign Policy*. London and New York, NY: Routledge.

Brewer, Thomas L. 2009b. "Technology Transfer and Climate Change: International Flows, Barriers and Frameworks," in Lael Brainard and Isaac Sorkin, editors, *Climate Change, Trade and Competitiveness*. Washington, DC: The Brookings Institution.

Brewer, Thomas L. 2010. "Trade Policies and Climate Change Policies: A Rapidly Expanding Joint Agenda," Introduction to a symposium on Trade Issues and Climate Change Issues, *The World Economy*, 33: 799–809.

Brewer, Thomas L. 2012. *International Technology Diffusion in a Sustainable Energy Trade Agreement (SETA): Issues and Options for Institutional Architectures*. Geneva: International Centre for Trade and Sustainable Development (ICTSD). Accessible at www.ictsd.org.

Brewer, Thomas L. 2013a. "Overview of Trade-Climate-Technology Transfer Cases." Accessible at www.TradeAndClimate.net.

Brewer, Thomas L. 2013b. *The Shale Gas "Revolution": Implications for Sustainable Development and International Trade*. Geneva: International Centre for Trade and Sustainable Development (ICTSD). Accessible at www.ictsd.org.

Brewer, Thomas L. 2014. "Shale Gas: New Issues on the Trade and Sustainable Development Agendas," *Bridges Trade BioRes: Trade & Environment Review*, 7. Accessible at www.ictsd.org.

Brewer, Thomas L., and Sarianna Lunden. 2006. "Environmental Policy and Institutional Transparency in Europe," in Lars Oxelheim, editor, *Corporate and Institutional Transparency for Economic Growth in Europe*. Amsterdam: Elsevier.

Brewer, Thomas L., and Michael Mehling. 2014. "Transparency Issues in Carbon Markets, Corporate Disclosure Practices and Government Climate Change Policies," in Jens Forssbæck and Lars Oxelheim, editors, *Transparency in International Markets.* Oxford: Oxford University Press.

Brown, Marilyn A. 2001. "Market Failures and Barriers as a Basis for Clean Energy Policies," *Energy Policy*, 29: 1197–1207.

Brown, Marilyn A., Andrea Sarzynski, and Frank Southworth. 2008. *Shrinking the Carbon Footprint of Metropolitan America.* Washington, DC: Brookings Institution. Accessed at www.brookings.edu on March 11, 2013.

Browne, John. 1997. Speech. Accessed at www.gsb.stanford.edu/community/bmag/sbsm0997/feature_ranks.html on March 2, 2010.

Bryner, Gary. 2000. "Congress and the Politics of Climate Change," in Paul G. Harris, editor, *Climate Change and American Foreign Policy.* New York, NY: St. Martin's.

Buchner, Barbara, Carlo Carraro, Igo Cersosimo, and Carmen Marchiori. 2002. "Back to Kyoto? US Participation and the Linkage between R&D and Climate Cooperation," Fondazione Eni Enrico Mattei. Accessed at www.feem.it on May 4, 2005.

Bush, George W. 2006. State of the Union Speech. Accessed at www.whitehouse.gov on March 3, 2006.

Business Roundtable. 2001. "Task Force Statement." Accessed at www.businessroundtable.org on October 25, 2004.

Business Roundtable. 2004. *Every Sector, One RESOLVE – 2004 Survey Results and Annual Report.* Accessed at www.businessroundtable.org on August 24, 2010.

Business Roundtable. 2009. "Climate RESOLVE." Accessed at www.businessroundtable.org on August 24, 2010.

Byrne, John, Kristen Hughes, Wilson Rickerson, and Lado Kurdgelashvili. 2007. "American Policy Conflict in the Greenhouse: Divergent Trends in Federal, Regional, State and Local Green Energy and Climate Change Policy," *Energy Policy* 35: 4555–4573.

Carbon Disclosure Project. 2006. *Annual Report.* Accessed at www.cdproject.int on April 22, 2008.

Carbon Disclosure Project. 2012. *Annual Report.* Accessed at www.cdproject.int on January 13, 2013.

Carbon Disclosure Project. 2013. "CDP Signatories and Members." Accessed at www.cdproject.int on January 28, 2013.

Carraro, Carlo. 2007. "Incentives and Institutions. A Bottom-up Approach to Climate Policy," in J. Aldy and R. Stavins, editors, *Architectures for Agreement: Addressing Global Climate Change in the Post-Kyoto World.* Cambridge: Cambridge University Press.

CBS News/*New York Times.* 2009. National survey, "Americans' View on the Environment," April 20–24, 2007, released on April 26, 2009. Accessed at PollingReport.com on December 30, 2009.

Center for Climate and Energy Solutions (C2ES). 2012. "The Candidates on Climate and Energy: A Guide to the Key Policy Positions of President Obama and Governor Romney."

Accessed at www.c2es.org on September 4, 2012. The Center for Climate and Energy Solutions (C2ES) was previously the Pew Center on Global Climate Change.

Center for Climate and Energy Solutions (C2ES). 2013a. "US States and Regions: Climate Action." Version 01/24/2013. Accessed at www.c2es.org on February 22, 2013.

Center for Climate and Energy Solutions (C2ES). 2013b. "Business Environmental Leadership Council (BELC) Member Companies." Accessed at www.c2es.org on January 27, 2013.

Center for Integrative Environmental Research. 2007. "The US Economic Impacts of Climate Change and the Costs of Inaction: A Review and Assessment," CIER at the University of Maryland.

Center for Climate Strategies. 2010. "Map of States' Climate Plans." Accessed at www.climatestrategies.us on December 28, 2010.

Center for Public Integrity. 2009a. "Methodology." February 24. Accessed at www.publicintegrity.org on August 24, 2010.

Center for Public Integrity. 2009b. "Number of Lobbyists on Climate Change by Sector, 2003 and 2008." Accessed at www.publicintegrity.org on August 24, 2010.

Center for Public Integrity. 2009c. "Tally of Interests on Climate Bill Tops a Thousand." Accessed at www.publicintegrity.org on August 28, 2010.

Center for Responsive Politics. 2011. "Energy/Natural Resources: Long-Term Contribution Trends." Accessed at www.opensecrets.org on January 20, 2011.

Center for Responsive Politics. 2012. "Energy/Natural Resources: Long-Term Contribution Trends." Accessed at www.www.opensecrets.org on September 10, 2012.

Chicago Council on Foreign Relations. 1999. *American Public Opinion and US Foreign Policy.* Accessed at www.ccfr.org on December 12, 2002.

Chicago Council on Foreign Relations. 2004. *Global Views 2004: American Public Opinion and Foreign Policy.* Accessed at www.ccfr.org on October 4, 2004.

Chicago Council on Foreign Relations. 2012. *Plurality of Americans Think US Government Neglecting Action on Climate Change.* Accessed at thechicagocouncil.org on April 20, 2014.

Chicago Council on Foreign Relations and German Marshall Fund of the United States. 2002a. *Worldviews 2002: American and European Public Opinion and Foreign Policy.* Accessed at www.worldviews.org on December 27, 2002.

Chicago Council on Foreign Relations and German Marshall Fund of the United States. 2002b. *Worldviews 2002 Survey,* Chapters 1, 4, 8. Accessed at www.worldviews.org on December 27, 2002.

Chicago Council on Foreign Relations and German Marshall Fund of the United States. 2002c. *Worldviews 2002: Comparing American and European Public Opinion on Foreign Policy: Transatlantic Key Findings Topline Data, Full Release.* Accessed at www.worldviews.org on December 27, 2002.

Chicago Council on Foreign Relations and German Marshall Fund of the United States. 2002d. *Worldviews 2002: US Leaders Topline Report.* Accessed at www.worldviews.org on December 27, 2002.

Chicago Council on Foreign Relations and German Marshall Fund of the United States. 2002e. *Worldviews 2002: Project Methodology.* Accessed at www.worldviews.org on December 27, 2002.

Chicago Council on Global Affairs. 2012. *Foreign Policy in the New Millennium.* Accessed at www.thechicagocouncil.org on September 18, 2012.

Christianson, G.E. 1999. *Greenhouse: The 200-Year Story of Global Warming.* New York, NY: Penguin.

Chu, Stephen. 2009. Speech at Harvard University. Quoted in Howell (2009). Accessed at www.eedaily.com on February 10, 2010.

Citigroup. 2007. "Statement on Climate Change." Accessed at www.citigroup.com on May 8, 2007.

Climate Strategies. 2010. "Reports: Industries and Sectors." Accessed at www.climatestrategies.org on April 22, 2010.

Climatebiz.com. 2010. "Cases." Accessed at www.climatebiz.com on April 22, 2010.

Climate Interactive. 2010. "C-SPAN." Accessed at www.climateinteractive.org on December 10, 2010. Used with permission.

ClimateWire. 2009. "Senators Prep for Cap-and-Trade Debate with Hearings, High-Level Meetings." By Darren Samuelsohn. July 7. Accessed at www.ee.net on August 5, 2009.

ClimateWire. 2010a. "Insurance: Regulators in Ind. Reject Climate Regulation; Some Others Appear Lukewarm." By Evan Lehmann. February 1. Accessed at www.eenews.net on January 5, 2011.

ClimateWire. 2010b. "Mann Cleared of Misconduct as Climategate Probe Ends." July 2. Accessed at www.eenews.net on July 5, 2010.

ClimateWire. 2010c. "Researchers Fear 'Climategate' Damage Is Done Despite Exoneration." July 20. Accessed at www.eenews.net on July 21, 2010.

ClimateWire. 2011. "Business: Corporations Face Record Number of Climate Shareholder Resolutions." February 17. Accessed at www.eenews.net on February 18, 2011.

ClimateWire. 2013. "Republican Mayor Pushes Climate Action, Wants Solar on New Homes." By Anne C. Mulkern. Accessed at www.eenews.net on March 8, 2013.

Cline, William R. 1992. *The Economics of Global Warming.* Washington, DC: Institute for International Economics.

CNN. 2001. Global Warming Poll. Accessed at www.cnn.com on April 3, 2002.

CNN/Opinion Research Corporation. 2009. CNN|ORC International Poll Results. Accessed at www.orcinternational.com on October 1, 2010.

Coalition for Environmentally Responsible Economics (CERES). 2003. *Corporate Governance and Climate Change: Making the Connection.* Boston: CERES.

Coalition for Environmentally Responsible Economics (CERES). 2006. *Corporate Governance and Climate Change: Making the Connection.* Boston: CERES.

Coalition for Environmentally Responsible Economics (CERES). 2009. *Fast Forward: Ceres Annual Report 2008–2009.* Accessed at www.ceres.org on September 10, 2012.

Coalition for Environmentally Responsible Economics (CERES). 2011a. *Climate Risk Disclosure by Insurers: Evaluating Insurer Responses to the NAIC Climate Disclosure Survey.* Accessed at www.ceres.org on September 10, 2012.

Coalition for Environmentally Responsible Economics (CERES). 2011b. *Global Investor Survey on Climate Change.* Accessed at www.ceres.org on September 20, 2012.

Coalition for Environmentally Responsible Economics (CERES). 2012a. *Insurer Climate Risk Disclosure Survey: 2012. Findings and Recommendations.* By Sharlene Leurig and Dr. Andrew Dlugolecki. Accessed at www.ceres.org on March 9, 2013.

Coalition for Environmentally Responsible Economics (CERES). 2012b. *Stormy Future for US Property/Casualty Insurers: The Growing Costs and Risks of Extreme Weather Events.* Accessed at www.ceres.org on September 20, 2012.

Coalition for Environmentally Responsible Economics (CERES). 2013a. "Who We Are." Accessed at www.ceres.org on January 28, 2013.

Coalition for Environmentally Responsible Economics (CERES). 2013b. "Shareholder Resolutions." Accessed at www.ceres.org on January 28, 2013.

Conference Board. 2008. "About Us" [from its *Annual Report,* 2008]. Accessed at www.conference-board.org on August 24, 2010.

Conference Board. 2009. *Shift in US Policy on Climate Change Opens a New Era for Business.* Accessed at www.conference-board.org on August 24, 2010.

Congressional Quarterly. 2003. "The Budget Process in Brief." In *Guide to Current American Government.* Washington, DC: Congressional Quarterly; pp. 119–120.

Conrad, Kevin. 2007. Quoted in "Issuing a Bold Challenge to the US Over Climate," *New York Times,* January 22, 2008. Accessed at http://www.nytimes.com/2008/01/22/science/earth/22conv.html on November 3, 2010.

Cook, John. 2013. "Quantifying the Consensus on Anthropogenic Global Warming in the Scientific Literature," *Environmental Research Letters,* 8: 024024. Accessed at http://iopscience.iop.org on April 22, 2014.

Cornes, Ricard, and Todd Sandler. 1996. *The Theory of Externalities, Public Goods, and Club Goods,* 2nd edition. Cambridge: Cambridge University Press.

Council on Foreign Relations. 2008. *Confronting Climate Change: A Strategy for US Foreign Policy.* Report of an Independent Task Force. By George E. Pataki, Thomas J. Vilsack, Michael A. Levi, and David G. Victor. New York, NY: Council on Foreign Relations.

Council on Foreign Relations. 2009a. "US Opinion on the Environment." November 19, p. 28. Accessed at www.cfr.org on December 3, 2009.

Council on Foreign Relations. 2009b. "Chapter 5a: World Opinion on the Environment," in *Public Opinion on Global Issues.* Accessed at www.cfr.org/public_opinion on April 20, 2014.

Council on Foreign Relations. 2012. "Joseph R. Biden, Jr." Accessed at www.cfr.org on September 4, 2012.

Dahl, Robert. 1961. *Who Governs? Democracy and Power in an American City.* New Haven, CT: Yale University Press. 2nd edition by Dahl and Douglas W. Rae (2005).

Dahl, Robert A. 2005. *Who Governs?: Democracy and Power in an American City,* 2nd edition. New Haven, CT: Yale University Press.

Dahl, Robert A., and Bruce Stinebrickner. 2002. *Modern Political Analysis*, 6th edition. Englewood Cliffs, NJ: Prentice-Hall.

Database of State Incentives for Renewables and Efficiency. 2011. "Texas." Accessed at www.dsireusa.org on January 28, 2011.

Democratic Party Convention. 2012. "Democratic National Platform." Accessed at www.democrats.org on September 4, 2012.

Derthick, Martha. 2010. "Compensatory Federalism," in Barry G. Rabe, editor, *Greenhouse Governance: Addressing Climate Change in America*. Washington, DC: Brookings Institution.

Dessler, Andrew, and Edward A. Parson. 2010. *The Science and Politics of Global Climate Change: A Guide to the Debate*, 2nd edition. Cambridge: Cambridge University Press.

Deutsch, Karl W. 1963. *The Nerves of Government*. Glencoe, IL: Free Press.

Development Bank of Japan. 2003. "Promoting Corporate Measure to Combat Global Warming: An Analysis of Innovative Activities in the Field," Research Report No. 42.

Dijkstra, B.R. 1999. *The Political Economy of Environmental Policy: A Public Choice Approach to Market Instruments*. Cheltenham, UK, and Northampton, MA: Elgar.

Dillard, Kirk. 2009. Quoted in "Global Warming Heats Up Ill. Governor Race." By Christopher Wills, The Associated Press, November 11.

DiMento, Joseph F.C., and Pamela Doughman, editors. 2007. *Climate Change: What It Means for Us, Our Children, and Our Grandchildren*. Cambridge, MA: MIT Press.

Dingell, John. 2008. Quoted in US Congress, House Committee on Energy and Commerce, Subcommittee on Energy and Air Quality hearing entitled, "Chairman Dingell, Legislative Proposals to Reduce Greenhouse Gas Emissions: An Overview," 110th Cong. H. Doc. Committee on Energy and Commerce, June 10, 2008. Accessed at www.energycommerce.house.gov on October 5, 2009.

DiPaola, M., and Arris, L., editors. 2001. *Global Warming Yearbook: 2001*. Arlington, MA: Cutter Information.

Diringer, Elliot. 2010. "Why Cancun Delivered," *National Journal*, December 12. Accessed at www.climate.nationaljournal.com on May 27, 2011.

Dizikes, Peter. 2012. "Presidential Campaigns Offer Energetic Energy Debate at MIT: Representatives of Obama, Romney Camps Lay Out Differences in Crucial Policy Domains of Energy and the Environment," MIT News. Accessed at www.mit.edu on March 26, 2013.

Dlugolecki, A., and M. Keykhah. 2002. "Climate Change and the Insurance Sector: Its Role in Adaptation and Mitigation," *Greener Management International*, 39: 83–98.

Driesen, David M., editor. 2010a. *Economic Thought and US Climate Change Policy*. Cambridge, MA: MIT Press.

Driesen, David M. 2010b. "Conclusion," in David M. Driesen, editor, *Economic Thought and US Climate Change Policy*. Cambridge, MA: MIT Press.

Driesen, David M. 2010c. "Introduction," in David M. Driesen, editor, *Economic Thought and US Climate Change Policy*. Cambridge, MA: MIT Press.

Dunn, S. 2002. "Down to Business on Climate Change: An Overview of Corporate Strategies," *Greener Management International*, 39: 27–41.

Ebi, Kristie L., Gerald A. Meehl, Dominique Bachelet, Robert R. Twilley, and Donald F. Boesch, et al. 2007. *Regional Impacts of Climate Change: Four Case Studies in the United States.* Washington, DC: Pew Center on Global Climate Change.

Ecofys. 2009. "G8 Climate Scorecards 2009." Accessed at www.wwf.org on June 14, 2010.

E&E Daily. 2004. "Senator Landrieu of Louisiana, an Oil and Gas State, Voted Against the McCain-Lieberman Bill in 2003." 6 July. Accessed at www.eenews.net on June 30, 2010.

E&E Daily. 2010a. "Climate: Obama Stresses Need for Price on Carbon as Dems Look to Compromise." June 29. Accessed at www.eenews.net on June 30, 2010.

E&E Daily. 2010b. "Climate: White House Meeting Keeps Carbon Caps Alive in Senate." June 30. Accessed at www.eenews.net on June 30, 2010.

E&E Daily. 2010c. "Climate: EDF Urges Obama to Roll up His Sleeves." July 1. Accessed at www.eenews.net on July 2, 2010.

E&E Daily. 2010d. "Senate Climate Debate: The 60-Vote Climb." Updated January 20. Accessed at www.E%26;Edaily.com on January 30, 2010.

E&E News. 2010. "ConocoPhillips, Caterpillar Inc. and BP America left. Lobbying: Influence Spending by Wind, All Renewables Soared in 2009." February 22. Accessed at www.eenews.net on June 14, 2010.

E&E News. 2012. "Exxonmobil CEO Pivots on Climate, Slams 'Manufactured Fear' on Fracking." June 28. Accessed at www.eenews.net on July 5, 2012.

Eggleton, Tony. 2013. *A Short Introduction to Climate Change.* Cambridge: Cambridge University Press.

Einstein, Albert. No date. Accessed at www.rescomp.stanford.edu on November 27, 2009.

Ellerman, A. Denny. 2008. *The European Union's Emissions Trading System in Perspective.* Washington, DC: Pew Center on Global Climate Change.

Ellerman, A. Denny, Frank J. Convery, and Christian de Perthuis. 2010. *Pricing Carbon: The European Union Emissions Trading Scheme.* Cambridge: Cambridge University Press.

Emanuel, Kerry. 2012. *What We Know About Climate Change*, 2nd edition. Cambridge, MA: MIT Press.

Energy Efficiency News. 2012. "WWF Corporate Climate Savers' Cut 100 Million Tonnes of Emissions." May 11. Accessed at www.energyefficiencynews.com on May 15, 2012.

Engel, Kirsten H. 2010. "Courts and Climate Policy: Now and in the Future," in Barry G. Rabe, editor, *Greenhouse Governance: Addressing Climate Change in America.* Washington, DC: Brookings Institution.

Environmental Defense Fund. 2010. *Our Approach: Corporate Partnership Guide.* Accessed at www.edf.org on November 6, 2010.

Erlandson, Dawn. 1994. "The Btu Tax Experience: What Happened and Why It Happened," *Pace Environmental Law Review*, 12: 172–184.

Esty, Daniel C., and Richard E. Caves. 1983. "Market Structure and Political Influence: New Data On Political Expenditures, Activity, and Success," *Economic Inquiry*, 21: 24–38.

Falke, Andreas. 2011. "Business Lobbying and the Prospect for American Climate Change Legislation," *GAIA*, 20: 20–25.

Farrell, Alexander E., and W. Michael Hanemann. 2009. "Field Notes on the Political Economy of California Climate Policy," in Henrik Selin and Stacy D. VanDeveer, editors, *Changing Climates in North American Politics.* Cambridge, MA: MIT Press.

Feldman, L., M.C. Nisbet, A. Leiserowitz, and E. Maibach. 2010. *The Climate Change Generation? Survey Analysis of the Perceptions and Beliefs of Young Americans.* Joint Report of American University's School of Communication, The Yale Project on Climate Change, and George Mason University's Center for Climate Change Communication. Accessed at http://www.climatechangecommunication.org on May 9, 2010.

Fickling, Meera. 2010. "US and Canadian Climate Legislation by State and Province," Peterson Institute for International Economics. Accessed at www.piie.com on December 22, 2010.

Financial Times. 2010a. "Climate Change: Not Necessarily a Threat but Insurers Can't Just Muddle Through." By Paul J. Davies. April 26. Accessed at www.ft.com on January 5, 2011.

Financial Times. 2010b. "Inquiry Backs Climate Scientists." By Fiona Harvey. July 7. Accessed at www.ft.com on July 8, 2010.

Financial Times. 2012a. "GE Rejects Republicans' Climate Change Doubts." March 11. Accessed at www.ft.com on March 12, 2012.

Financial Times. 2012b. "US Oil Jumps on the Campaign Trail." June 5. Accessed at www.ft.com on June 5, 2012.

Financial Times. 2012c. "Global Warming 'Manageable', Says Exxon Chief." June 27. Accessed at www.ft.com on January 27, 2013.

Fischer-Vanden, Karen. 2000. "International Policy Instrument Prominence in the Climate Change Debate," in Paul G. Harris, editor, *Climate Change and American Foreign Policy.* New York, NY: St. Martin's.

Fox News/Opinion Dynamics Poll. 2009. Fox News Poll: "Where Americans Stand on the Issues." Accessed at www.foxnews.com on October 24, 2013.

Frankel, Jeffrey. 2007. "Formulas for Quantitative Emission Targets," in Joseph E. Aldy and Robert N. Stavins, editors, *Architectures for Agreement: Addressing Global Climate Change in the Post-Kyoto World.* Cambridge: Cambridge University Press.

Fransen, Taryn. 2009. "Enhancing Today's MRV Framework to Meet Tomorrow's Needs: The Role of National Communications and Inventories," WRI Working Paper. Washington, DC: World Resources Institute. Available at www.wri.org.

Fransen, Taryn, Kirsten Stasio, and Smita Nakhooda. 2012. *The US Fast-Start Finance Contribution.* Washington, DC: World Resources Institute and Overseas Development Institute.

Friedman, Milton. 1955. "The Role of Government in Education," in Robert A. Solo, editor, *Economics and the Public Interest.* New Brunswick, NJ: Rutgers University Press.

Friedman, Thomas. 2010. "How the G.O.P. Goes Green," *New York Times,* February 28.

Gallagher, Kelly Sims and Laura D. Anadon. 2012. "DOE Budget Authority for Energy Research, Development, and Demonstration Database." Energy Technology Innovation Policy research group, Belfer Center for Science and International Affairs, Harvard Kennedy

School. February 29. Accessed at www.belfercenter.ksg.harvard.edu on March 11, 2013.
Gallup. 2001a. "Americans Consider Global Warming Real, but Not Alarming." By Frank Newport and Lydia Saad. April 9. Accessed at www.Gallup.com on August 29, 2002.
Gallup. 2001b. "Scientists Deliver Serious Warning About Effects of Global Warming." By Karren K. Carlson. January 23. Accessed at www.Gallup.com on December 27, 2002.
Gallup. 2002a. "Poll Topics and Trends: Environment." Accessed at www.Gallup.com on December 27, 2002.
Gallup. 2002b. "Americans Sharply Divided on Seriousness of Global Warming." By Lydia Saad. March 25. Accessed at www.Gallup.com on December 27, 2002.
Gallup. 2002c. "Poll Topics and Trends: Environment." Accessed at www.Gallup.com on April 19, 2002.
Gallup. 2002d. "Americans Want to Breathe Easier." By Kelly Maybury. April 16. Accessed at www.Gallup.com on April 19, 2002.
Gallup. 2004a. "Global Warming on Public's Back Burner." By Lydia Saad. April 20. Accessed at www.Gallup.com on September 16, 2004.
Gallup. 2004b. "Americans Tepid on Global Warming." By David W. Moore. April 13. Accessed at www.Gallup.com on April 16, 2004.
Gallup. 2004c. "Bush and the Environment." By Riley E. Dunlap. April 5. Accessed at www.Gallup.com on April 13, 2004.
Gallup. 2008. National survey on March 6–9, 2008, reported at PollingReport.com on Environment. Accessed on December 30, 2009.
Gallup. 2009. "Increased Number Think Global Warming Is 'Exaggerated.'" By Lydia Saad. March 11. Accessed at www.gallup.com on April 8, 2013.
Gallup. 2010a. "American's Global Warming Concerns Continue to Drop." By Frank Newport. Accessed at www.Gallup.com on April 13, 2010.
Gallup. 2010b. "Conservative's Doubts about Global Warming Grow." By Jeffrey M. Jones. Accessed at www.Gallup.com on May 8, 2010.
Gallup. 2010c. "In US, Many Environmental Issues at 20-Year Low Concern." By Jeffrey M. Jones. Accessed at www.Gallup.com on May 8, 2010.
Gallup. 2011. "In US, Concerns About Global Warming Stable at Lower Levels." By Jeffrey M. Jones. Accessed at www.Gallup.com on May 9, 2011.
Gallup. 2012. "In US Global Warming Views Steady Despite Warm Winter." By Lydia Saad. Accessed at www.Gallup.com on April 16, 2012.
Gallup. 2013. "Americans' Concerns About Global Warming on the Rise." By Lydia Saad. Accessed at www.Gallup.com on April 8, 2013.
Gardiner, Stephen M., Simon Caney, Dale Jamieson, and Henry Shue, editors. 2010. *Climate Ethics: Essential Readings*. Oxford and New York, NY: Oxford University Press.
Geithner, Timothy F. 2010. "Secretary of the Treasury Timothy F. Geithner Written Testimony before the House Committee on Appropriations Subcommittee on State, Foreign Operations, and Related Programs on the Fiscal Year 2011 International Programs Budget Request." March 25. Accessed at www.treas.gov on November 22, 2010.

German Marshall Fund of the United States. 2009. "Transatlantic Trend 2008." Reported in Council on Foreign Relations, World Opinion on the Environment, November 19, p. 12. Accessed at http://www.cfr.org/publication/20135 on December 3, 2009.

Gerrard, Michael B., editor. 2007. *Global Climate Change and US Law.* Chicago, IL: American Bar Association.

Gerrard, Michael B., and J. Cullen Howe. 2011. "Climate Change Litigation in the US." Accessed at www.climatecasechart.com on January 2, 2011.

Gimpel, J.G., K.M. Kaufmann, and S. Pearson-Merkowitz. 2007. "Battleground States versus Blackout States: The Behavioral Implications of Modern Presidential Campaigns," *Journal of Politics*, 69: 786–797.

Glicksman, Robert L. 2010. "Anatomy of Industry Resistance to Climate Change: A Familiar Litany," in David M. Driesen, editor, *Economic Thought and US Climate Change Policy.* Cambridge, MA: MIT Press.

Goldman Sachs. 2009. *Change Is Coming: A Framework for Climate Change.* Accessed at www.goldmansachs.com on November 11, 2010.

Goodin, Robert E. 2009. "The State of the Discipline, the Discipline of the State," in Robert E. Goodin, editor. *The Oxford Handbook of Political Science.* Oxford and New York, NY: Oxford University Press.

Goodin, Robert E., and Hans-Dieter Klingemann, editors. 1998a. *A New Handbook of Political Science.* Oxford: Oxford University Press.

Goodin, Robert E., and Hans-Dieter Klingemann. 1998b. "Political Science: The Discipline," in Robert E. Goodin and Hans-Dieter Klingemann, editors. *A New Handbook of Political Science.* Oxford: Oxford University Press.

Gore, Christopher, and Pamela Robinson. 2009. "Local Government Response to Climate Change: Our Last, Best Hope?" in Henrik Selin and Stacy D. VanDeveer, editors, *Changing Climates in North American Politics.* Cambridge, MA: MIT Press.

Goulder, Lawrence H., and Robert N. Stavins. 2011. "Challenges from State-Federal Interactions in US Climate Change Policy," *American Economic Review: Papers and Proceedings*, 101: 253–257.

Graham, Lindsey. 2010. Quoted in Thomas L. Friedman, "How the G.O.P. Goes Green," *New York Times*, February 28.

Greenhouse Gas Protocol [WRI and WBCSD]. 2010. "The Greenhouse Gas Protocol Initiative." Accessed at www.ghgprotocol.org on November 6, 2010.

Greenwire. 2010a. "Lobbying: Millions Spent to Sway Calif. Climate Rules." Accessed at www.eenews.net on July 13, 2010.

Greenwire. 2010b. "Oil and Gas Interests Set Spending Record in 2009." February 2. Accessed at www.eenews.int on August 24, 2010.

Greenwire. 2011. "CLIMATE: Texas Faces Uphill Legal Battle against EPA." January 5. Accessed at www.eenews.net on January 6, 2011.

Grofman, Bernard. 1998. "Political Economy: Downsian Perspectives," in Robert E. Goodin and Hans-Dieter Klingemann, editors. *A New Handbook of Political Science.* Oxford: Oxford University Press.

Gros, Daniel, Christian Egenhofer, Noriko Fujiwara, Anton Georgiev, and Selen Sarisoy Guerin. 2010. *Climate Change and Trade: Taxing Carbon at the Border?* Brussels: Centre for European Policy Studies.

Grubb, Michael, and Karston Neuhoff, editors. 2006. "Emissions Trading and Competitiveness." Special Issue of *Climate Policy.*

Grubb, Michael, Thomas L. Brewer, Misato Sato, Robert Heilmayr, and Dora Fazekas. 2010. *Climate Policy and Industrial Competitiveness: Ten Insights from Europe on the EU Emissions Trading System.* Washington, DC: German Marshall Fund.

Hahn, Robert, and A. Ulph, editors. 2012. *Climate Change and Common Sense: Essays in Honour of Tom Schelling.* Oxford: Oxford University Press.

Hall, Richard L., and Alan V. Deardorff. 2006. "Lobbying as Legislative Subsidy," *American Political Science Review*, 100: 69–84.

Hansen, James. 1988. Statement of Dr James Hansen. Reprinted in Bill McKibben, editor, *The Global Warming Reader.* New York, NY: Penguin, 2012.

Harris, Paul G., editor. 2000. *Climate Change and American Foreign Policy.* New York, NY: St. Martin's.

Harris, Paul G. 2007. "Collective Action on Climate Change: The Logic of Regime Failure," *Natural Resources*, 47: 195–202.

Harris, Paul G., editor. 2009. *Climate Change and Foreign Policy: Case Studies from East to West.* London: Routledge.

Harris, Richard. 2011. "Romney Seemingly Shifts on Climate Change." NPR, All Things Considered, October 28.

Harris Poll. 2001a. The Harris Poll #45, "Large Majority of Public Now Believes In Global Warming And Supports International Agreements To Limit Greenhouse Gases." Accessed at www.harrisinteractive.com on October 10, 2004.

Harris Poll. 2001b. Results of survey of August 15–22. Accessed at www.harrisinteractive.com on November 28, 2004.

Harris Poll. 2002. The Harris Poll #56, "Majorities Continue to Believe in Global Warming and Support Kyoto Protocol." Accessed at www.harrisinteractive.com on October 10, 2004.

Harrison, Kathryn, and Lisa McIntosh, editors. 2010. *Global Commons, Domestic Decisions: The Comparative Politics of Climate Change.* Cambridge, MA: MIT Press.

Harrison, Neil E. 2000. "From the Inside Out: Domestic Influences on Global Environmental Policy," in Paul G. Harris, editor, *Climate Change and American Foreign Policy.* New York, NY: St. Martin's.

Haufler, Virginia. 2009. "Insurance and Reinsurance in a Changing Climate," in Henrik Selin and Stacy D. VanDeveer, editors, *Changing Climates in North American Politics: Institutions, Policymaking, and Multilevel Governance.* Cambridge, MA: MIT Press.

Helm, Carsten, and Detlef Sprinz. 2000. "Measuring the Effectiveness of International Environmental Regimes," *Journal of Conflict Resolution*, 44: 630–652.

Helm, Dieter. 2010. "Government Failure, Rent-Seeking, and Capture: The Design of Climate Change Policy," *Oxford Review of Economic Policy*, 26: 182–196.

Helm, Dieter, and Cameron Hepburn. 2009. *The Economics and Politics of Climate Change*. Oxford: Oxford University Press.

Hepburn, Cameron. 2010. "Environmental Policy, Government, and the Market," *Oxford Review of Economic Policy*, 26: 117–136.

Hibbard, Paul J., and Susan F. Tierney. 2011. Carbon Control and the Economy: Economic Impacts of RGGI's First Three Years," *The Electricity Journal*, 24: 30–40.

Hoffman, Andrew. 2005. "Climate Change Strategy: The Business Logic Behind Voluntary Greenhouse Gas Reductions," *California Management Review*, 47: 21–46.

Hofman, P.S. 2002. "Becoming a First Mover in Green Electricity Supply: Corporate Change Driven by Liberalization and Climate Change," *Greener Management International*, 39: 99–108.

Hoggan, James. 2009. *Climate Cover-Up: The Crusade to Deny Global Warming*. Vancouver: Greystone Books.

Houghton, John. 2009. *Global Warming: The Complete Briefing*, 4th edition. Cambridge: Cambridge University Press.

Howell, Katie. 2009. "Tech Revolution Key to US Prosperity," E&E News PM, August 6.

Hoyos, C.D., P.A. Agudelo, P.J. Webster, and J.A. Curry. 2006. "Deconvolution of the Factors Contributing to the Increase in Global Hurricane Intensity, *Science*, 312: 94–97.

HSBC. 2009. "Building a Green Recovery. Summary." Accessed at www.hsbc.com on February 26, 2011.

Hufbauer, Gary, Steve Charnovitz, and J. Kim. 2009. *Global Warming and the World Trade System*. Washington, DC: Peterson Institute for International Economics and the World Resources Institute.

Immelt, Jeffrey. 2005. Speech. Reported at www.bloomberg.com. Accessed on March 2, 2010.

Institutional Investors Group on Climate Change, Investor Network on Climate Change, and Investor Group on Climate Change. 2012. *Global Investor Survey on Climate Change*. Available at www.ceres.org.

InterAcademy Council. 2010. *Climate Change Assessments: Review of the Processes and Procedures of the IPCC*. Accessed at www.reviewipcc.net on January 28, 2011.

International Emissions Trading Association (IETA). 2003. *Greenhouse Gas Market 2003*. Geneva: IETA.

International Energy Agency (IEA). 2008. *Energy Technology Perspectives, 2008*. Paris: IEA.

International Energy Agency (IEA). 2011. *Are We Entering a Golden Age of Gas? World Energy Outlook 2011*, Special Report. Accessed at www.iea.org on April 1, 2013.

International Energy Agency (IEA). 2009. *Energy Technology Perspectives, 2009*. Paris: IEA.

International Energy Agency, Organization for Economic Cooperation and Development and World Bank. 2010. *The Scope of Fossil-Fuel Subsidies in 2009 and a Roadmap for Phasing Out Fossil-Fuel Subsidies*. Accessed at www.iea.org on January 15, 2011.

International Herald Tribune. 2001. Survey on US attitudes. Accessed at www.iht.com/poll/bushpoll.htm on August 16, 2001.

Investor Network on Climate Risk. 2010. "Climate Resolution Tracker for 2010" as of July 19, 2010. Accessed at www.incr.com on August 24, 2010.

Jacobsson, Staffan, and Anna Johnson. 2000. "The Diffusion of Renewable Energy Technology: An Analytical Framework and Key Issues for Research," *Energy Policy*, 28: 625–640.

Jaffe, Adam B., Richard G. Newell, and Robert N. Stavins. 2005. "A Tale of Two Market Failures: Technology and Environmental Policy," *Ecological Economics*, 54: 164–174.

Jepma, Catrinus J., and Mohan Munasinghe. 1998. *Climate Change Policy: Facts, Issues, and Analyses*. Cambridge: Cambridge University Press.

Jones, Charles A., and David L. Levy. 2009. "Business Strategies and Climate Change," in Henrik Selin and Stacy D. VanDeveer, editors, *Changing Climates in North American Politics: Institutions, Policymaking, and Multilevel Governance*. Cambridge, MA: MIT Press.

Kamieniecki, Sheldon. 2006. *Corporate America and Environmental Policy: How Often Does Business Get Its Way?* Stanford, CA: Stanford University Press.

Kamieniecki, Sheldon, and Michael Kraft. 2009. "Series Forward," in Henrik Selin and Stacy D. VanDeveer, editors. *Changing Climates in North American Politics: Institutions, Policymaking, and Multilevel Governance*. Cambridge, MA: MIT Press.

Kaplan, Abraham, and Harold D. Lasswell. 1950. *Power and Society: A Framework for Political Inquiry*. New Haven, CT: Yale University Press.

Keeling, Charles D. 1960. "The Concentration and Isotopic Abundances of Carbon Dioxide in the Atmosphere," *Tellus*, 12: 200–203.

Kelman, Steven. 1981. "Cost-Benefit Analysis: An Ethical Critique," *AEI Journal on Government and Society Regulation*, January/February: 33–40.

Kennedy, Craig, and Marshall M. Bouton. 2002. "The Real Trans-Atlantic Gap," *Foreign Policy*, November/December: 65–74.

Kenney, Robyn. 2008. "Energy Policy Act of 1992, United States," in Cutler J. Cleveland, editor, *Encyclopaedia of the Earth*. Washington, DC: Environmental Information Coalition, National Council for Science and the Environment. First published in the *Encyclopaedia of the Earth* August 29, 2006; last revised November 18, 2008. Accessed at www.eoearth.org on February 10, 2010.

Kolk, Ans. 2000. *Economics of Environmental Management*. London: *Financial Times* / Prentice Hall, Harlow.

Kolk, Ans, and David Levy. 2001. "Winds of Change: Corporate Strategy, Climate change and Oil Multinationals," *European Management Journal*, 19: 501–509.

Kolk, Ans, and Jonatan Pinkse. 2005. "Business Responses to Climate Change: Identifying Emergent Strategies," *California Management Review*, 47: 6–20.

Kopp, Ray J., and Michael A. Toman. 2000. "International Emissions Trading: A Primer," in Raymond J. Kopp and Jennifer B. Thatcher, editors, *The Weathervane Guide to Climate Policy: An RFF Reader*. Washington, DC: Resources for the Future. Accessed at www.rff.org on March 15, 2008.

Kopp, Raymond J., and William A. Pizer, editors. 2007. *Assessing US Climate Policy Options*. Resources for the Future, and US Climate Policy Forum. Accessed at www.rff.org on January 15, 2011.

Kraft, Michael E. 2007. *Environmental Policy and Politics*, 4th edition. New York, NY: Pearson-Longman.

Kraft, Michael E., and Scott R. Furlong. 2010. *Public Policy: Politics, Analysis, and Alternatives*, 3rd edition. Washington, DC: CQ Press.

Kraft, Michael E., and Sheldon Kamieniecki. 2007a. "Analyzing the Role of Business in Environmental Policy," in Michael E. Kraft and Sheldon Kamieniecki, editors, *Business and Environmental Policy: Corporate Interests in the American Political System*. Cambridge, MA: MIT Press.

Kraft, Michael E., and Sheldon Kamieniecki, editors. 2007b. *Business and Environmental Policy: Corporate Interests in the American Political System*. Cambridge, MA: MIT Press.

Kriegler, Elmar, Jim W. Hall, Hermann Held, Richard Dawson and Hans Joachim Schellnhuber. 2009. "Imprecise Probability Assessment of Tipping Points in the Climate System," *Proceedings of the National Academy of Sciences*, 106: 5041–5046. Accessed at www.pnas.org on July 25, 2010.

Krosnick, Jon. 2010. "Large Majority of Americans Support Government Solutions to Address Global Warming." Accessed at www.woods.stanford.edu on June 3, 2011.

Krosnick, Jon. 2013. "Public Perceptions about Global Warming and Government Involvement in the Issue." Presentation for the Environmental and Energy Studies Institute, March 28. Accessible at www.eesi.org.

Krosnick, Jon, Allyson L. Holbrook and Penny S. Visser. 2000. "The Impact of the Fall 1997 Debate about Global Warming on American Public Opinion," *Public Understanding of Science*, 9: 239–260.

Krosnick, Jon, and Ana Villar. 2010. Global Warming Poll: Stanford University. Released June 9. Accessed at www.woods.stanford.edu on June 2, 2011.

Labor Strategies. 2010. Statement by Richard Trumka. Accessed at www.laborstrategies.blogs.com on April 22, 2010.

Lasswell, Harold D. 1936. *Politics: Who Gets What, When, and How*. New York, NY: McGraw-Hill.

Layzer, Judith A. 2007. "Deep Freeze: How Business Has Shaped the Global Warming Debate in Congress," in Michael E. Kraft and Sheldon Kamieniecki, editors, *Business and Environmental Policy: Corporate Interests in the American Political System*. Cambridge, MA: MIT Press.

Layzer, Judith. 2011. *The Environmental Case: Translating Values Into Policy*, 3rd edition. Washington, DC: CQ Press.

Lee, Henry. 2001. "US Climate Policy: Factors and Constraints," in Eileen Clausen, Vicki Arroyo Cochran, and Debra P. Davis, editors, *Climate Change: Science, Strategies, and Solutions*. Boston, MA: Brill.

Lehman Brothers. 2007. *The Business of Climate Change: Opportunities and Challenges*. By John Lewellwyn. London: Lehman Brothers.

Lehman Brothers. 2009. *The Business of Climate Change*, II. By John Llewellwyn and Camille Chaix. London: Lehman Brothers.

Leiserowitz, Anthony. 2003. "American Opinions on Global Warming." University of Oregon Survey Research Laboratory. Accessed at www.osrl.uoregon.edu/projects/globalwarm on August 8, 2003.

Leiserowitz, Anthony, 2007a. "International Public Opinion, Perception, and Understanding of Global Climate Change." Yale University. Accessed at http://environment.yale.edu on April 5, 2010.

Leiserowitz, Anthony. 2007b. "American Support for Local Action on Climate Change." Yale University School of Forestry and Environmental Studies. Accessed at http://environment.yale.edu on April 5, 2010.

Leiserowitz, Anthony, 2007c. "American Opinions on Global Warming: Summary." Yale University School of Forestry and Environmental Studies. Accessed at http://environment.yale.edu on April 5, 2010.

Leiserowitz, Anthony, Edward Maibach, and Connie Roser-Renouf. 2009. *Climate Change in the American Mind: Americans' Climate Change Beliefs, Attitudes, Policy Preferences, and Actions.* George Mason University Center for Climate Change Communication. Accessed at http://www.climatechangecommunication.org on May 9, 2010.

Leiserowitz, Anthony, Edward Maibach, and Connie Roser-Renouf. 2010. *Americans' Actions to Conserve Energy, Reduce Waste, and Limit Global Warming.* Yale University and George Mason University. New Haven, CT: Yale Project on Climate Change. Accessed at http://environment.yale.edu on April 5, 2010.

Leiserowitz, Anthony, Edward Maibach, Connie Roser-Renouf, Geoff Feinberg, and Peter Howe. 2013. *Global Warming's Six Americas in September 2012.* Yale University and George Mason University. New Haven, CT: Yale Project on Climate Change Communication. Accessed at http://environment.yale.edu/climate/publications/Six-Americas-September-2012 on December 15, 2013.

Leiserowitz, Anthony, Edward Maibach, Connie Roser-Renouf, and Nicholas Smith. 2010. *Global Warming's Six Americas.* Yale University and George Mason University. New Haven, CT: Yale Project on Climate Change. Accessed at http://environment.yale.edu on May 1, 2011.

Leiserowitz, Anthony, Edward Maibach, Connie Roser-Renouf, and Jay D. Hmielowski. 2011. *Politics and Global Warming: Democrats, Republicans, Independents, and the Tea Party.* Yale University and George Mason University. New Haven, CT: Yale Project on Climate Change. Accessed at http://environment.yale.edu on March 9, 2012.

Levy, David L. 2005. "Business and the Evolution of the Climate Regime: The Dynamics of Corporate Strategies," in David L. Levy and Peter J. Newell, editors, *The Business of Global Environmental Governance.* Cambridge, MA: MIT Press.

Levy, David L., and D. Egan. 1998. "Capital Contests: National and Transnational Channels of Corporate Influence on the Climate Change Negotiations," *Politics and Society*, 26: 335–359.

Levy, David L., and Peter J. Newell, editors. 2005. *The Business of Global Environmental Governance.* Cambridge, MA: MIT Press.

Lieberman, Bruce. 2013. "SCC – Social Cost of Carbon: A Continuing Little-Told Story," The Yale Forum on Climate Change and The Media. Accessed at www.yaleclimatemediaforum.org on September 23, 2013.

Lindblom, Charles E. 1980. *Politics and Markets: The World's Political Economic Systems.* New York, NY: Basic Books.

Litz, Franz T. 2008. "Toward a Constructive Dialogue on Federal and State Roles in US Climate Policy," Pew Center on Global Climate Change. Accessed at www.pewclimate.org on December 28, 2010.

Lizza, Ryan. 2010. "As the World Burns: How the Senate and the White House Missed Their Best Chance to Deal with Climate Change," *New Yorker*, October 11.

Lofgren, K.-G. 1995. "Markets and Externalities," in H. Folmer, H. L. Gabel, and H. Opschoor, editors, *Principles of Environmental Economics*. Cheltenham: Elgar.

Lowi, Theodore. 1964. "American Business, Public Policy, Case-Studies, and Political Theory," *World Politics*, 16: 677–715.

Luce, Edward, and Fiona Harvey. 2010. "Senators Urge Obama on Climate," *Financial Times*, July 23. Accessed at www.ft.com on July 24, 2010.

Luterbacher, Urs, and Detlef F. Sprinz, editors. 2001. *International Relations and Global Climate Change*. Cambridge, MA: MIT Press.

Maibach, Edward, Anthony Leiserowitz, Connie Roser-Renouf, Karen Akerlof, and Matthew Nisbet. 2010. "Saving Energy is a Value Shared by All Americans: Results of a Global Warming Audience Segmentation Analysis." For publication in *Human Resources for Climate Solutions: Energy-Smart Behaviors, People-Centered Policies, and Public Engagement*. Edited by Karen Ehrhardt-Martinez. Washington, DC: American Council for an Energy-Efficient Economy.

Mann, Michael E. 2012. *The Hockey Stick and the Climate Wars*. New York, NY: Columbia University Press.

Mann, Michael E., Raymond S. Bradley, and Malcolm K. Hughes. 1999. "Northern Hemisphere Temperatures During the Past Millenium: Inferences, Uncertainties, and Limitations," *Geophysical Research Letters*, 26: 759–762.

March, James G., and Johan P. Olsen. 1984. "The New Institutionalism: Organizational Factors in Political Life," *The American Political Science Review*, 78: 734–749.

March, James G., and Johan P. Olsen. 1998. "The Institutional Dynamics of International Political Orders," *International Organization*, 52: 943–969.

Marshall, Alfred. 1890. *Principles of Economics*. London: Macmillan.

Marshall, Christa. 2008. "Lobbying: K Street Getting Hot for Climate Change," *ClimateWire*, May 27. Accessed at www.eenews.net on June 14, 2010.

Marshall, Christa. 2010. "Politics: Cantwell-Collins bill generates lobbying frenzy," *ClimateWire*, February 15. Accessed at www.eenews.net on June 14, 2010.

Mathez, Edmond A. 2009. *Climate Change: The Science of Global Warming and Our Energy Future*. New York, NY: Columbia University Press.

McGarity, Thomas O. 2010. "The Cost of Greenhouse Gas Reductions," in David M. Driesen, editor, *Economic Thought and US Climate Change Policy*. Cambridge, MA: MIT Press.

McKibben, Bill, editor. 2012. *The Global Warming Reader*. New York, NY: Penguin.

McKibbin, Warwick J., and Peter J. Wilcoxen. 2007. "A Credible Foundation for Long-term International Cooperation on Climate Change," *Brookings Discussion Papers in International Economics*, 171.

McKinsey. 2007. *Reducing US Greenhouse Gas Emissions: How Much at What Cost?* Accessed at www.mckinsey.com on February 26, 2011.

Meckling, Jonathan. 2012. *Carbon Coalitions: Business, Climate Politics, and the Rise of Emissions Trading.* Cambridge, MA: MIT.

Michaelowa, Axel. 2007. "Graduation and Deepening," in Joseph E. Aldy and Robert N. Stavins, editors, *Architectures for Agreement: Addressing Global Climate Change in the Post-Kyoto World.* Cambridge: Cambridge University Press.

Midwestern Greenhouse Gas Reduction Accord. 2010. Midwestern Greenhouse Gas Accord. Accessed at www.midwesternaccord.org on December 3, 2010.

Mills, Evan. 2007. *From Risk to Opportunity: Insurer Responses to Climate Change.* Boston: CERES.

Mills, Evan. 2012. "The Greening of Insurance," *Science*, 338. Accessed at www.sciencemag.org on January 28, 2013.

Minar, David M. 1961. "Ideology and Political Behavior," *Midwest Journal of Political Science*, 5: 317–331.

Misbach, Andreas. 2000. "Regulation Theory and Climate Change Policy," in Paul G. Harris, editor, *Climate Change and American Foreign Policy.* New York, NY: St. Martin's.

Monitz, Ernest J., et al. 2012. *The Future of Natural Gas.* Cambridge, MA: MIT Press.

Mount, Timothy D. 1999. "Redirecting Energy Policy in the USA to Address Global Warming," in Mohanned H.I. Dore and Timothy D. Mount, editors, *Global Environmental Economics.* Oxford: Blackwell.

Mulkern, Anne C. 2010. "Lobbying: Ad Blitz Accompanies Senate Push on Climate, Energy," *Greenwire*, July 16. Accessed at www.eenews.net on July 26, 2010.

Muller, Richard A. 2009. *Physics for Future Presidents.* New York, NY: W.W. Norton.

Narayanamurti, Venkatesh, Laura D. Anadon, and Ambuj D. Sagar. 2009. "Transforming Energy Innovation," *Issues In Science and Technology*, Fall: 57–64.

National Association of Insurance Commissioners. 2010. "Insurer Climate Risk Disclosure Survey," Adopted Version, March 28, 2010; as reported by National Association of Mutual Insurance Companies 2011. Accessed at www.namic.org on January 5, 2011.

National Association of Manufacturers. 2001a. "Statement on Climate Change." Accessed at www.nam.org on August 24, 2006.

National Association of Manufacturers. 2001b. "Press Release on Climate Change." Accessed at www.nam.org on August 24, 2006.

National Association of Manufacturers. 2009. "NAM Principles on Climate Change." Accessed at www.nam.org on August 24, 2010.

National Association of Mutual Insurance Companies 2011. "Insurer Climate Risk Disclosure Survey," Adopted Version, March 28, 2010. Accessed at www.namic.org on January 5, 2011.

National Commission on Energy Policy (NCEP). 2004. *Ending the Energy Stalemate: A Bipartisan Strategy to Meet America's Energy Challenges.* Washington, DC: NCEP.

National Journal. 2011. "Upton Takes Aim at Obama Agenda," *National Journal*, February 8. Accessed at www.nationaljournal.com on May 1, 2011.

Natural Resources Defense Council (NRDC). 2009. NRDC's second and updated summary of Waxman–Markey. Accessed at www.blog.climateandenergy.org on January 28, 2011.

Natural Resources Defense Council (NRDC). 2010. "Center for Market Innovation." Accessed at www.nrdc.org on November 6, 2010.

Natural Resources Defense Council (NRDC). 2012. "Ready or Not: An Evaluation of State Climate and Water Preparedness Planning," Issue Brief, 12–03-A. Accessed at www.nrdc.org on May 15, 2012.

New York Times. 2010. "States Take Sides in Greenhouse Gas 'Endangerment' Brawl." By Robin Bravender of *Greenwire.* March 19. Accessed at www.nytimes.com on January 27, 2011.

Newell, Peter, and Matthew Patterson. 2010. *Climate Capitalism: Global Warming and the Transformation of the Global Economy.* Cambridge: Cambridge University Press.

Newell, Richard G. 2010. "The Role of Markets and Policies in Delivering Innovation for Climate Change Mitigation," *Oxford Review of Economic Policy*, 26: 253–269.

Nordhaus, William D. 1994. *Managing the Global Commons: The Economics of Climate Change.* Cambridge, MA: MIT Press.

Nordhaus, William D. 2013. *The Climate Casino: Risk, Uncertainty, and Economics for a Warming World.* New Haven, CT: Yale University Press.

Nordquist, J., C. Boyd, and H. Klee. 2002. "Three big Cs: Climate, Cement and China," *Greener Management International*, 39: 69–82.

Norgaard, Kari Marie. 2011. *Living in Denial: Climate Change, Emotions, and Everyday Life.* Cambridge, MA: MIT Press.

Northrop, Michael, and David Sassoon. 2008. "States Take the Lead on Climate." Yale Environment 360, April 3. Accessed at www.e360.yale.edu on March 7, 2011.

Oak Ridge Center for Advanced Studies. 2005. "ORCAS/Harris Interactive ® Survey of Public Perceptions of Climate Change." Accessed at http://orcas.orau.org on October 14, 2005.

Obama, President Barack. 2009. Remarks at the Copenhagen International Conference on Climate Change. Accessed at http://www.whitehouse.gov/blog/2009/12/18/a-meaningful-and-unprecedented-breakthrough-here-copenhagen on March 2, 2010.

Obama, President Barack. 2010. Remarks by the President to the Nation on the BP Oil Spill. Accessed at www.whitehouse on April 20, 2014.

Obama, President Barack. 2011. Remarks by the President in the State of the Union Address. Accessed at www.whitehouse.gov on January 29, 2011.

Obama, President Barack. 2013. Remarks by the President in the State of the Union Address. Accessed at www.whitehouse.gov on April 21, 2014.

Obama, President Barack. 2014. President Barack Obama's State of the Union Address. Accessed at www.whitehouse.gov on February 4, 2014.

Offe, Claus. 1998. "Political Economy: Sociological Perspectives," in Robert E. Goodin and Hans-Dieter Klingemann, editors, *A New Handbook of Political Science.* Oxford: Oxford University Press.

Olson, Mancur. 1971. *The Logic of Collective Action: Public Goods and the Theory of Groups*, second printing with new preface and appendix. Cambridge, MA: Harvard University Press.

O'Neill, Karen, and Forrest Reinhardt. 2000. "What Every Executive Needs to Know about Global Warming," *Harvard Business Review*, 78: 128–135.

Oppenheimer, Michael. 2009. "Forward," in Edmond A. Mathez, *Climate Change: The Science of Global Warming and Our Energy Future.* New York, NY: Columbia University Press.

Oreskes, Naomi. 2007. "The Scientific Consensus on Climate Change: How Do We Know We're Not Wrong?" in Joseph F.C. DiMento and Pamela Doughman, editors, *Climate Change: What It Means for Us, Our Children, and Our Grandchildren.* Cambridge, MA: MIT Press.

Oreskes, Naomi, and Erik M. Conway. 2010. *Merchants of Doubt.* New York, NY: Bloombury Press.

Organisation for Economic Cooperation and Development (OECD) and International Energy Agency (IEA). 2003. *Technology Innovation, Development and Diffusion.* Paris: OECD and IEA.

Orszag, Peter R. 2008. "Issues in Designing a Cap-and-Trade Program for Carbon Dioxide Emissions," Statement before the Committee on Ways and Means, US House of Representatives, September 18. Washington, DC: Congressional Budget Office. Accessed at www.cbo.gov on April 10, 2013.

Ostrom, Elinor. 1990. *Governing the Commons: The Evolution of Institutions.* Cambridge: Cambridge University Press.

Oye, Kenneth A., and James H. Maxwell. 1995. "Self-Interest and Environmental Management," in Robert O. Keohane and Elinor Ostrom, editors, *Local Commons and Global Interdependence.* London: Sage Publications.

Pacala, Stephen W., and Robert H. Socolow. 2004. "Stabilization Wedges: Solving the Climate Problem for the Next 50 Years with Current Technologies," *Science,* 305: 968–972.

Park, Jacob. 2000. "Governing Climate Change Policy: From Scientific Obscurity to Foreign Policy Prominence," in Paul G. Harris, editor, *Climate Change and American Foreign Policy.* New York, NY: St. Martin's.

Partnership for a Secure America. 2013. "Thirty-Eight Leading US National Security Experts Urge Action on International Climate Change Initiatives." February 25. Accessed at www.psaonline.org on April 10, 2013.

Peace, Janet, and Robert Stavins. 2010. "Meaningful and Cost-Effective Climate Policy: The Case for Cap-and-Trade." Washington, DC: Pew Center on Global Climate Change. Accessed at www.pewclimate.org on July 27, 2010.

Peck, Louis. 2010. "A Veteran of the Climate Wars Reflects On US Failure to Act." Yale Environment 360, 4 January. Accessed at www.e360.yale.edu on February 16, 2011.

Pennsylvania Insurance Department. 2010. "Insurance Company Climate Risk Disclosure Survey." Accessed at www.portal.state.pa.us on January 6, 2011.

Peters, B. Guy. 1998. "Political Institutions, Old and New," in Robert E. Goodin and Hans-Dieter Klingemann, editors, *A New Handbook of Political Science.* Oxford: Oxford University Press.

Pew Center on Global Climate Change. 2008. "The Candidates and Climate Change: A Guide to Key Policy Positions." Accessed at www.pewclimate.org on September 15, 2008. The Pew Center changed its name to Center for Climate and Energy Solutions (C2ES) with a website at www.c2es.org.

Pew Center on Global Climate Change. 2009a. "US States and Regions." Accessed at www.pewclimate.org on April 10, 2009.

Pew Center on Global Climate Change. 2009b. "Waxman–Markey Discussion Draft." Accessed at www.pewclimate.org on January 29, 2011.
Pew Center on Global Climate Change. 2009c. "US Department of Energy's Recovery Act Spending." Accessed at www.pewclimate.org on February 26, 2011.
Pew Center on Global Climate Change. 2010a. "Business Environmental Leadership Council (BELC)." Accessed at www.pewclimate.org on April 27, 2010.
Pew Center on Global Climate Change. 2010b. "Climate Change 101: State Action." Accessed at www.pewclimate.org on December 28, 2010.
Pew Center on Global Climate Change. 2010c. "In Brief: The Business Case for Climate Legislation." Accessed at www.pewclimate.org on June 30, 2010.
Pew Global Attitudes Project. 2009. "World Opinion on the Environment." November 19, pp. 8–9. Accessed at http://www.cfr.org/publication/20135 on December 3, 2009. [Note that the Pew Global Attitudes Project is not the same as the Pew Center on Global Climate Change – which has changed its name to Center for Climate and Energy Solutions (C2ES).]
Pigou, Arthur C. 1920. *The Economics of Welfare.* London: Macmillan.
Pinkse, Jonatan, and Ans Kolk. 2009. *International Business and Global Climate Change.* Oxon: Routledge.
Pinkse, Jonatan, and Ans Kolk. 2010. "Challenges and Trade-offs in Corporate Innovation for Climate Change," *Business Strategy and the Environment,* 19: 261–272.
Pizer, William A. 2007. "Practical Global Climate Policy," Joseph E. Aldy and Robert N. Stavins, editors, *Architectures for Agreement: Addressing Global Climate Change in the Post-Kyoto World.* Cambridge: Cambridge University Press.
PollingReport.com. 2009. "Environment." Accessed at http://www.pollingreport.com/enviro.htm on December 30, 2009.
Pomerance, Rafe. 1989. "The Dangers from Climate Warming: A Public Awakening," in Dean Edwin Abrahamson, editor, *The Challenge of Global Warming.* Washington, DC: Island Press.
Pooley, Eric. 2010. *The Climate War: True Believers, Power Brokers, and the Fight to Save the Earth.* New York, NY: Hyperion.
Porter, Michael E. 1991. "America's Green Strategy," *Scientific American,* 264: 96–101.
Porter, Michael E., and C. van der Linde. 1995. "Towards a New Conception of the Environment-Competitiveness Relationship," *Journal of Economic Perspectives,* 9: 97–118.
Posner, Paul L. 2010. "The Politics of Vertical Diffusion: The States and Climate Change," in Barry G. Rabe, editor, *Greenhouse Governance: Addressing Climate Change in America.* Washington, DC: Brookings Institution.
Powell, James Lawrence. 2011. *The Inquisition of Climate Science.* New York, NY: Columbia University Press.
Prickett, Glenn. 2010. Quoted in Thomas L. Friedman, "Want the Good News First?" *New York Times,* July 27. Accessed at www.nytimes.com on July 28, 2010.
Program on International Policy Attitudes. 2002. "Americans and the World: Public Opinion on World Affairs." Accessed at www.americans-world.org, April 22, 2003.
Proost, S. 1995. "Public Policies and Externalities," in H. Folmer, H.L. Gabel, and H. Opschoor, editors, *Principles of Environmental Economics.* Cheltenham: Elgar.

Public Policy Institute of California. 2005. "Special Surveys on Californians and the Environment." Accessed at www.ppic.org on January 30, 2005.

Public Policy Institute of California. 2010a. "Californians and the Environment." July 2010. Accessed at www.ppic.org on January 4, 2011.

Public Policy Institute of California. 2010b. "Special Surveys on Californians and the Environment." Accessed at www.ppic.org on April 17, 2010.

Public Policy Institute of California. 2012. "Californians and the Environment." July 2012. Accessed at www.ppic.org on August 27, 2012.

Pulver, Simone. 2007. "Making Sense of Corporate Environmentalism," *Organization and Environment*, 20: 44–83.

Putnam, Robert D. 2000. *Bowling Alone*. New York, NY: Simon and Schuster.

Rabe, Barry G. 2002. *Greenhouse and Statehouse: The Evolving State Government Role in Climate Change*. Washington, DC: Pew Center on Global Climate Change.

Rabe, Barry G. 2009. "Second-Generation Climate Policies: Proliferation, Diffusion, and Regionalization," in Henrik Selin and Stacy D. VanDeveer, editors, *Changing Climates in North American Politics*. Cambridge, MA: MIT Press.

Rabe, Barry G. 2010a. "Can Congress Govern the Climate?" in Barry G. Rabe, editor, *Greenhouse Governance: Addressing Climate Change in America*. Washington, DC: Brookings Institution.

Rabe, Barry G. 2010b. "Conclusion," in Barry G. Rabe, editor, *Greenhouse Governance: Addressing Climate Change in America*. Washington, DC: Brookings Institution.

Rabe, Barry G. 2010c. "Introduction: The Challenges of US Climate Governance," in Barry G. Rabe, editor, *Greenhouse Governance: Addressing Climate Change in America*. Washington, DC: Brookings Institution.

Rabe, Barry G. 2010d. "The 'Impossible Dream' of Carbon Taxes: Is the 'Best Answer' a Political Non-Starter?" in Barry G. Rabe, editor, *Greenhouse Governance: Addressing Climate Change in America*. Washington, DC: Brookings Institution.

Rabe, Barry G., and Christopher P. Borick. 2010. "The Climate of Belief: American Public Opinion on Climate Change," *Issues in Governance Studies*, 31. Washington, DC: The Brookings Institution.

Radio Netherlands (RNL). 2010. "Sea Level Blunder Enrages Dutch Minister." February 4. Accessed at www.rnw.nl on July 25, 2010.

Rapp, Tobias, Christian Schwägerl, and Gerald Traufette. 2010. "The Copenhagen Protocol: How China and India Sabotaged the UN Climate Summit," *Der Spiegel*, May 5. Accessed at www.spiegel.de on October 24, 2010. Used with permission. (The adaptation presented here includes changes that reflect more clearly the sequence of events during the meeting and that convey less emotive word choices in a few instances.)

Rasmussen. 2011a. "Only 33% Think Most Americans Blame Humans for Global Warming." Accessed at www.rasmussenreports.com on May 9, 2011.

Rasmussen. 2011b. "Energy Update." Accessed at www.rasmussenreports.com on May 9, 2011.

Regional Greenhouse Gas Initiative. 2010. "News and Updates." Accessed at www.rggi.org on December 3, 2010.

Reinhardt, Forrest L. 2000. "Global Climate Change and BP Amoco," Harvard Business School Case Study, N9–700–106.

Republican Party Convention. 2012. "Republican Party Platform." Accessed at www.gop.org on September 4, 2012.

Republicans for Environmental Protection. 2011a. "Energy and Climate Change," Policy Paper, Executive Summary. Accessed at www.rep.org on May 12, 2011.

Republicans for Environmental Protection. 2011b. "Why Conservation *is* Conservative As Expressed by History's Preeminent Conservative Minds: A Comprehensive Collection of Conservative Quotations." Accessed at www.rep.org on May 12, 2011.

Reuters. 2002. "US Voters Want Strict Greenhouse Gas Cuts, Says Survey." Accessed at www.enn.com on July 10, 2002.

Reuters. 2008. "GM Exec Stands by Calling Global Warming a 'Crock.'" February 22. Accessed at www.reuters.com on July 27, 2010.

Revelle, Roger, and Hans E. Suess. 1957. "Carbon Dioxide Exchange between Atmosphere and Ocean and the Question of an Increase of Atmospheric CO2 During the Past Decades," *Tellus*, 9: 18–27.

Revkin, Andrew C. 2010. "With No Obama Push, Senate Punts on Climate," *New York Times*, July 22. Accessed at www.nytimes.com on July 24, 2010.

Rhodes, R.A.W. 2009. "Old Institutionalisms: An Overview," in Robert E. Goodin, editor, *The Oxford Handbook of Political Science.* Oxford and New York, NY: Oxford University Press.

Richards, Kenneth, and Stephanie Richards. 2009. "US Senate Climate Change Bills in the 110th Congress: Learning by Doing," *Environs – Environmental Law and Policy Journal* 33: 1–110.

Richter, Burton. 2010. *Beyond Smoke and Mirrors: Climate Change and Energy in the 21st Century.* Cambridge: Cambridge University Press.

Rodden, Jonathan A. 2008. "Federalism," in Barry R. Weingast and Donald A. Wittman, editors, *The Oxford Handbook of Political Economy.* Oxford: Oxford University Press.

Rogers, James E. 2010. Speech reported in Eric Pooley, "The Smooth-Talking King of Coal – and Climate Change," *Bloomberg Businessweek*, June 3. Accessed at www.businessweek.com on August 24, 2010.

Romney, Mitt. 2010. *No Apologies.* New York, NY: Macmillan.

Rosencranz, Armin. 2002. "US Climate Change Policy," in Stephen H. Schneider, Armin Rosencranz, and John O. Niles, editors, *Climate Change Policy: A Survey.* Washington, DC: Island Press.

Rowlands, Ian H. 2009. "Renewable Electricity Politics across Borders," in Henrik Selin and Stacy D. VanDeveer, editors, *Changing Climates in North American Politics: Institutions, Policymaking, and Multilevel Governance.* Cambridge, MA: MIT Press.

Rowlands, Ian H. 2010. "Encouraging Renewable Electricity to Promote Climate Change Mitigation," in Barry G. Rabe, editor, *Greenhouse Governance: Addressing Climate Change in America.* Washington, DC: Brookings Institution.

Ruggie, John Gerard. 1972. "Collective Goods and Future International Collaboration," *American Political Science Review*, 66: 874–893.

Salamon, Lester M., and John J. Siegfried. 1977. "Economic Power and Political Influence: The Impact of Industry Structure on Public Policy," *The American Political Science Review*, 71: 1026–1043.

Salorio, Eugene M. 1991. "Trade Policy and Corporate Strategy: Why Some Firms Oppose Import Protection for Their Own Industry," Academy of International Business annual meeting, Miami.

Salorio, Eugene M. 1992. "Strategic Use of Import Protection: Seeking Shelter for Competitive Advantage," Conference on Global Strategic Management, European Institute for Advanced Studies in Management, Brussels.

Salorio, Eugene M. 1994. "Strategic Use of Import Protection: Seeking Shelter for Competitive Advantage," *Research in Global Strategic Management*, 4: 101–124.

Samuelsohn, Darren. 2010. "Climate: Power Companies Lie Back As Push Begins for Senate Bill," *Greenwire*, June 3. Accessed at www.eenews.net on June 4, 2010.

Samuelsohn, Darren. 2011. "Mitt Romney Hit for Past Emissions Stance," *Politico*, November 2.

Schelling, Thomas C. 1987. "Value of Life," *The New Palgrave: A Dictionary of Economics*, 4: 793–96.

Schelling, Thomas. 2007. "Epilogue: Architectures for Agreement," in Joseph E. Aldy and Robert N. Stavins, editors, *Architectures for Agreement: Addressing Global Climate Change in the Post-Kyoto World*. Cambridge, MA: Cambridge University Press.

Schick, Allen. 2007. *The Federal Budget: Politics, Policy, and Process*, 3rd edition. Washington, DC: Brookings Institution.

Schiller, Ben. 2012. "Insurance Companies Face Increased Risks from Warming." Yale Environment 360. Accessed at www.e360.yale.edu on April 26, 2012.

Schmidheiny, Stephan. 1992. *Changing Course: A Global Business Perspective on Development and the Environment*. Cambridge, MA: MIT Press.

Schmidt, Gavin, and Joshua Wolfe, editors. 2009. *Climate Change: Picturing the Science*. New York, NY: W.W. Norton.

Schneider, Stephen. 2009. *Science as a Contact Sport: Inside the Battle to Save Earth's Climate*. Washington, DC: National Geographic Society.

Schneider, Stephen H., Armin Rosencranz, and John O. Niles, editors. 2000. *Climate Change Policy: A Survey*. Washington, DC: Island Press.

Schrag, Daniel P. 2012. "Is Shale Gas Good for Climate Change?" *Daedalus*, 141: 72–80.

Schroeder, Christopher, and Robert L. Glicksman. 2010. "The United States' Failure to Act," in David M. Driesen, editor, *Economic Thought and US Climate Policy*. Cambridge, MA: MIT Press.

Schwarzenegger, Arnold. 2009. ABC News. *Good Morning America*. Accessed at www.abcnews.go.com on November 11, 2009.

ScienceDebate.org. 2012. "The Top American Science Questions." Results reported at www.scientificamerican.com. Accessed on September 5, 2012.

Scripps Institution of Oceanography. 2013. "Mauna Loa Record." Accessed at www.scrippsco2.ucsd.edu on August 26, 2013.

Selin, Henrik, and Stacy D. VanDeveer. 2007. "Political Science and Prediction: What Next for US Climate Policy?" *Review of Policy Research*, 24: 1–27.

Selin, Henrik, and Stacy D. VanDeveer, editors. 2009. *Changing Climates in North American Politics: Institutions, Policymaking, and Multilevel Governance.* Cambridge, MA: MIT Press.

Selin, Henrik, and Stacy D. VanDeveer. 2010. "Multilevel Governance and Transatlantic Climate Change Politics," in Barry G. Rabe, editor, *Greenhouse Governance: Addressing Climate Change in America.* Washington, DC: Brookings Institution.

Shaw, Daron R. 1999. "The Methods behind the Madness: Presidential Electoral College Strategies, 1988–1996," *The Journal of Politics*, 61: 893–913.

Sheppard, Kate. 2008. "A Voice in the Wilderness," *The Guardian*, June 23. Accessed at www.theguardian.com on August 26, 2013.

Shepsle, Kenneth A. 2008. "Old Questions and New Answers about Institutions," in Barry R. Weingast and Donald A. Wittman, editors, *The Oxford Handbook of Political Economy.* Oxford and New York, NY: Oxford University Press.

Shogren, J.F., and M.A. Toman. 2000. "Climate Change Policy," in P.R. Portney and R.N. Stavins, editors, *Public Policies for Environmental Protection*, 2nd edition. Washington, DC: Resources for the Future.

Shultz, George. 2013. George Shultz, Fmr. Sec. of State, Addresses Policymakers on Capitol Hill for the First Time in 20 Years. Transcript of speech on March 8, 2013. Partnership for a Secure America. Accessed at www.psaonline.org accessed on April 10, 2013.

Skolnikoff, Eugene B. 1990. "The Policy Gridlock on Global Warming," *Foreign Policy*, 79: 77–85.

Skolnikoff, Eugene B. 1999. "The Role of Science in Policy: The Climate Change Debate in the United States," *Environment*, 41: 16–21.

Smith, Jeffrey A., Matthew Morreale, and Kimberley Drexler. 2010. The SEC's Interpretive Release on Climate Change Disclosure. *Climate Change Law and Review*, 2: 147–153.

Smith, Joel B., Jason M. Vogel, Terri L. Cruce, Stephen Seidel, and Heather A. Holsinger. 2010. *Adapting to Climate Change: A Call for Federal Leadership.* Pew Center on Global Climate Change. Accessed at www.pewclimate.org on November 6, 2010.

Socolow, Robert H., and Stephen W. Pacala, 2006. "A Plan to Keep Carbon in Check," *Scientific American*, September: 50–57.

South, S.G., editor. 2000. *Corporate Leadership on Climate Change.* Arlington, MA: Cutter Information.

Spar, Debora L. 2009. "National Policies and Domestic Politics," in Alan M. Rugman, editor, *The Oxford Handbook of International Business.* Oxford: Oxford University Press.

Stabenow, Debbie, et al. 2009. Letter to Senate Majority Leader Harry Reid. June 6. Accessed at www.eenews.net on June 26, 2010.

Stavins, Robert. 2010a. "An Economic View of the Environment: The Real Options for US Climate Policy." July 23. Accessed at http://belfercenter.ksg.harvard.edu on July 24, 2010.

Stavins, Robert. 2010b. "What Happened (and Why): An Assessment of the Cancun Agreements." Accessed at www.belfercenter.ksg.harvard,edu on May 27, 2011.

Stavins, Robert N. 2012. "Cap-and-Trade, Carbon Taxes, and My Neighbor's Lovely Lawn. An Economic View of the Environment." October 21. Accessed at www.robertstavinsblog.org on April 10, 2013.

Stern, Nicholas. 2007. *The Economics of Climate Change.* Cambridge: Cambridge University Press.

Stoett, Peter. 2009. "Looking for Leadership: Canada and Climate Change Policy," in Henrik Selin and Stacy D. VanDeveer, editors, *Changing Climates in North American Politics: Institutions, Policymaking, and Multilevel Governance.* Cambridge, MA: MIT Press.

Stone, Brian Jr. 2012. *The City and the Coming Climate: Climate Change in the Places We Live.* Cambridge: Cambridge University Press.

Sugarman, Danielle. 2011. "The House Continuing Resolution: The Dismantling of Climate Regulation," Climate Law Blog, Center for Climate Change Law, March 7. Accessed at http://www.blogs.law.columbia.edu/climatechange on May 3, 2011.

Sunstein, Cass R. 2006. "On the Divergent American Reactions to Terrorism and Climate Change," AEI-Brookings Center for Regulatory Studies, Working Paper 06–13, May.

Sunstein, Cass R., and Richard H. Thaler. 2003. "Libertarian Paternalism Is Not an Oxymoron," *The University of Chicago Law Review,* 70: 1159–1202.

Suskind, Ron. 2004. *The Price of Loyalty: George W. Bush, the White House, and the Education of Paul O'Neill.* New York, NY: Simon and Schuster.

Tamiotti, Ludivine, Robert Teh, Vesile Kulacoglu, Anne Olhoff, Benjamin Simmons, and Hussein Abasa. 2009. *Trade and Climate Change.* A Report by the United Nations Environment Programme and the World Trade Organization. Geneva: WTO Publications.

The Project on Climate Science. 2011. Accessible at www.theprojectonclimatescience.org.

Telegraph. 2010. Articles on November 28, and December 20. Accessed at www.telegraph.co.uk on April 9, 2010.

Therborn, Goran. 2009. "Why and How Place Matters," in Robert E. Goodin, editor, *The Oxford Handbook of Political Science.* Oxford and New York, NY: Oxford University Press.

Time. 2001. Reports Results of a Poll Conducted for CNN and *Time* Magazine by Yankelovich/Harris on March 21–22, 2001. April 9, p. 32.

Times of India. 2010. Statement by Minister for Environment and Forests. Accessed at www.timesofindia.indiatimes.com on April 9, 2010.

Truman, David B. 1971. *The Governmental Process: Political Interests and Public Opinion,* 2nd edition. New York, NY: Knopf.

Tufte, Edward M. 1978. *Political Control of the Economy.* Princeton, NJ: Princeton University Press.

UK Royal Society. 2009. *Geoengineering and the Climate: Governance and Uncertainty.* London: The Royal Society.

UN. 2013. "Department of Economic and Social Affairs. Population Division. Data." Accessed at www.esa.un.org on February 21, 2013.

UN Framework Convention on Climate Change (UNFCCC). 2009. Copenhagen Accord. Accessed at www.unfccc.int on December 1, 2010.

UN Framework Convention on Climate Change (UNFCCC). 2010. "UNFCCC Publishes Reports Summing Up Results of 2009 UN Climate Change Conference in Copenhagen." Accessed at www.unfccc.int on April 22, 2014.

UN Framework Convention on Climate Change (UNFCCC). 2011a. Copenhagen Accord, Appendix I, "Quantified Economy-wide Emissions Targets for 2020." Accessed at www.unfccc.int on May 23, 2011.

UN Framework Convention on Climate Change (UNFCCC). 2011b. Copenhagen Accord, Appendix II, "Nationally Appropriate Mitigation Actions of Developing Country Parties." Accessed at www.unfccc.int on May 23, 2011.

UN Framework Convention on Climate Change (UNFCCC). 2011c. "Establishment of an Ad Hoc Working Group on the Durban Platform for Enhanced Action." Accessed at www.unfccc.int on March 11, 2012.

UN Framework Convention on Climate Change (UNFCCC). 2012. "Decisions adopted by COP 18 and CMP 8." Accessed at www.unfccc.int on January 13, 2013.

UN High-level Advisory Group on Climate Change Financing. 2010. "Report of the Secretary-General's High-level Advisory Group on Climate Change Financing." Accessed at www.un.org on May 31, 2011.

UN Intergovernmental Panel on Climate Change (IPCC). 1996. *Climate Change*. Cambridge: Cambridge University Press.

UN Intergovernmental Panel on Climate Change (IPCC). 2001. *Climate Change*. Cambridge: Cambridge University Press.

UN Intergovernmental Panel on Climate Change (IPCC). 2007. *Fourth Assessment Report*. Cambridge: Cambridge University Press.

UN Intergovernmental Panel on Climate Change (IPCC), Working Group I. 2013. *Climate Change 2013: The Physical Science Basis*. Cambridge: Cambridge University Press.

UN Intergovernmental Panel on Climate Change (IPCC), Working Group II. 2014. *Climate Change 2014: Impacts, Adaptation and Vulnerability*. Cambridge: Cambridge University Press.

UN Intergovernmental Panel on Climate Change (IPCC), Working Group III. 2014. *Mitigation of Climate Change*. Cambridge: Cambridge University Press.

Union of Concerned Scientists. 2002a. "New Poll Finds Little Support for Bush's Global Warming Policy." July 8. Accessed at www.ucsusa.org on July 10, 2002.

Union of Concerned Scientists. 2002b. "Zogby International Omnibus Polling Results: American Attitudes on Climate Change." Accessed at www.ucsusa.org on July 10, 2002.

Union of Concerned Scientists 2005. "Summary of the [2005] Energy Bill." Accessed at www.ucsusa.org on January 31, 2011.

Union of Concerned Scientists. 2007. *Smoke, Mirrors and Hot Air*. Accessed at www.ucsusa.org on August 24, 2010.

United Technologies Corporation (UTC). 2006. *Annual Report*. Accessed at www.utc.com on October 15, 2006.

United Technologies Corporation (UTC). 2009. *Annual Report*. Accessed at www.utc.com on June 10, 2010.

University of Minnesota, Center for the Study of Politics and Governance. 2009. Michele Bachmann on D.C.: "I'm a Foreign Correspondent On Enemy Lines." [Report of an interview on radio station WWTC AM 1280.] Accessed at www.blog.lib.umn.edu on May 11, 2011.

US Bureau of Economic Analysis. 2010a. "Coal Industry Data and Map." Data released in 2003. Accessed at http://www.bea.gov on March 11, 2010.

US Bureau of Economic Analysis. 2010b. "Oil and Gas Industry Data and Map." Data released on June 2, 2009. Accessed at http://www.bea.gov on March 11, 2010.

US Bureau of Economic Analysis. 2010c. "Auto Industry Data and Map." Data released on June 2, 2009. Accessed at http://www.bea.gov on March 11, 2010.

US Carbon Data Information Analysis Center. 2012. "Global Fossil-Fuel Carbon Emissions – Graphics." Accessed at www.cdiac.ornl.gov on April 13, 2013.

US Census Bureau. 2010. "Census Bureau Regions and Divisions." Accessed at www.censusbureau.gov on December 23, 2010.

US Census Bureau. 2013. "Population Estimates." Accessed at www.census.gov on February 21, 2013.

US Chamber of Commerce. 2001. "Statement on Climate Change." Accessed at www.uschamber.com on September 30, 2006.

US Chamber of Commerce. 2009. "About the US Chamber of Commerce." Accessed at www.uschamber.com on July 1, 2009.

US Chamber of Commerce. 2010. "Five Positions on Energy and the Environment." Accessed at www.uschamber.com on August 24, 2010.

US Climate Action Partnership (USCAP). 2007. "Policy Statements." Accessed at us-cap.org on November 1, 2010.

US Climate Change Technology Program (CCTP). 2003. *Strategic Plan.* Washington, DC: US DOE.

US Climate Change Technology Program (CCTP). 2005. *Strategic Plan.* Washington, DC: US DOE.

US Climate Change Technology Program (CCTP). 2010. "Member Agencies of the US Climate Change Technology Program." Accessed at www.climatetechnology.gov on January 26, 2010.

US Conference of Mayors. 2005. Climate Protection Agreement (as endorsed by the 73rd Annual US Conference of Mayors meeting, Chicago, 2005). Accessed at www.usmayors.org on March 10, 2010.

US Conference of Mayors. 2010. "Cities That Have Signed On." Accessed at www.usmayors.org on March 10, 2010.

US Congress. 2010. *Conference Committee Report of the Appropriations Committees.* Washington, DC: US Congress.

US Congressional Budget Office (CBO). 2001. *An Evaluation of Cap-and-Trade Programs for Reducing US Carbon Emissions.* Washington, DC: CBO.

US Congressional Budget Office (CBO). 2004. *Glossary of Budgetary and Economic Terms.* Washington, DC: CBO.

US Congressional Budget Office (CBO). 2008. *Issues in Designing a Cap-and-Trade Program for Carbon Dioxide Emissions.* Washington, DC: CBO.

US Congressional Budget Office (CBO). 2009. *Potential Impacts of Climate Change in the United States.* Available at www.cbo.gov.

US Congressional Research Service (CRS). 1999a. *Global Climate Change.* IB89005. Washington, DC: Congressional Research Service.

US Congressional Research Service (CRS). 1999b. *Global Climate Change Policy: From "No Regrets" to S. Res. 98.* Report for Congress. By Larry B. Parker and John E. Blodgett. Washington, DC: CRS.

US Congressional Research Service (CRS). 1999c. "*Global Climate Change: the Energy Tax Incentives in the President's FY 2000 Budget.* By Salvatore Lazzari. Washington, DC: CRS. Accessed at www.ncseonline.org on April 13, 2011.

US Congressional Research Service (CRS). 2000a. *Global Climate Change. CRS Issue Brief for Congress.* By Wayne E. Morrissey and John R. Justus. Washington, DC: CRS.

US Congressional Research Service (CRS). 2000b. *Global Climate Change Policy: From "No Regrets" to S. Res. 98. RL30024.* Washington, DC: Congressional Research Service.

US Congressional Research Service (CRS). 2000c. *Global Climate Change Briefing Book.* Washington, DC: Congressional Research Service.

US Congressional Research Service (CRS). 2000d. *Global Climate Change: A Survey of Scientific Research and Policy Reports.* RL30522. Washington, DC: Congressional Research Service.

US Congressional Research Service (CRS). 2001. *Climate Change Technology Initiative (CCTI): Research, Technology, and Related Programs.* By Michael M. Simpson. Washington, DC: CRS. Accessed at www.ncseonline.org on April 13, 2011.

US Congressional Research Service (CRS). 2003a. *The Congressional Budget Process Timetable.* CRS Report for Congress, by Heniff, Bill, Jr. Washington, DC: CRS.

US Congressional Research Service (CRS). 2003b. *Overview of the Executive Budget Process.* CRS Report for Congress, by Bill Heniff, Jr. Washington, DC: CRS.

US Congressional Research Service (CRS). 2004. *The Congressional Budget Process: A Brief Overview.* CRS Report for Congress, by James V. Saturno. Washington, DC: CRS.

US Congressional Research Service (CRS). 2007. *Energy Independence and Security Act of 2007: A Summary of Major Provisions.* Report for Congress by Fred Sissine. Accessed at www.crs.gov on July 30, 2008. Washington, DC: CRS.

US Congressional Research Service (CRS). 2009. *Greenhouse Gas Legislation Summary and Analysis of H.R. 2454 as Reported by the House Committee on Energy and Commerce.* By Mark Holt and Gene Whitney. Accessed at www.crs.gov on January 29, 2011.

US Congressional Research Service (CRS). 2010a. Memorandum. "Comparison of Selected Senate Energy and Climate Change Proposals." By Brent Yacobucci. Accessed at www.lugar.senate.gov on January 30, 2011.

US Congressional Research Service (CRS). 2010b. "A US-centric Chronology of the International Climate Change Negotiations." By Jane A. Leggett. Accessed at www.crs.gov on February 9, 2011.

US Council for International Business (USCIB). 2001. "Statement on Climate Change." Accessed at www.uscib.org on March 5, 2005.

US Council for International Business (USCIB). 2009. "USCIB Statement on the UN Global Leadership Forum on Climate Change." Accessed at www.uscib.org on August 24, 2010.

US Department of Energy (DOE). 2003. *Climate Change Technology Program: Research and Current Activities.* Washington, DC: DOE.

US Department of Energy (DOE). 2006. *Climate Change Technology Program: Strategic Plan.* Washington, DC: DOE.

US Department of Energy (DOE). 2008. P.L. 110–343, The Emergency Economic Stabilization Act of 2008: Energy Tax Incentives. Accessed at www.energy.gov on January 31, 2011.

US Department of Energy (DOE). 2009. "Consumer Energy Tax Incentives: What the American Recovery and Reinvestment Act Means to You." Accessed at www.energy.gov on January 31, 2011.

US Department of Energy (DOE). 2010. "Energy Efficiency and Renewable Energy. Fiscal Year 2010 Budget-in-Brief." Accessed at www.eere.energy.gov on April 27, 2011.

US Department of State (DOS). 2007. *Energy Needs, Clean Development and Climate Change: Partnerships in Action.* Washington, DC: DOS.

US Department of State (DOS). 2010. *US Climate Action Report.* Washington, DC: Global Publishing Services. Available at www.state.gov.

US Department of State (DOS). 2013. "US-China Climate Change Working Group Fact Sheet." Accessed at www.state.gov on July 24, 2013.

US Energy Information Administration (EIA). 2003. *Analysis of S. 139, the Climate Stewardship Act of 2003: Highlights and Summary.* Washington, DC: EIA.

US Energy Information Administration (EIA). 2008. "Federal Financial Intervention and Subsidies in Energy Markets. 2007." Accessed at www.eia.gov on February 1, 2011.

US Energy Information Administration (EIA). 2010a. "Monthly Energy Review." Available at http://www.eia.doe.gov.

US Energy Information Administration (EIA). 2010b. "US Census Regions and Divisions" [adapted from standard US Census Bureau groupings]. Accessed at www.iea.doe.gov on December 23, 2010.

US Energy Information Administration (EIA). 2011. *World Shale Gas Resources: An Initial Assessment of 14 Regions Outside the United States.* Report prepared by Advanced Resources International. Accessed at www.eia.gov on April 1, 2013.

US Environmental Protection Agency (EPA). 2010. *Inventory of US Greenhouse Gas Emissions and Sinks, 1990–2008.* Accessed at www.epa.gov on February 1, 2011.

US Environmental Protection Agency (EPA). 2011a. "Climate Change – Regulatory Initiatives." Accessed at www.epa.gov on February 26, 2011.

US Environmental Protection Agency (EPA). 2011b. "High Global Warming Potential Gases." Accessed at www.epa.gov on February 26, 2011.

US Environmental Protection Agency (EPA). 2012a. *Draft Inventory of US Greenhouse Gas Emissions and Sinks, 1990–2010.* Accessed at www.epa.gov on March 8, 2012.

US Environmental Protection Agency (EPA). 2012b. "New Source Review: Regulations and Standards." Accessed at www.epa.gov on March 28, 2012.

US Environmental Protection Agency (EPA). 2012c. "Obama Administration Finalizes Historic 54.5 mpg Fuel Efficiency Standards." August 28. Accessed at www.epa.gov on September 3, 2012.

US Environmental Protection Agency (EPA). 2013a. *Draft Inventory of US Greenhouse Gas Emissions and Sinks: 1990–2011* (February 2013). Accessed at www.epa.gov on February 22, 2013.

US Environmental Protection Agency (EPA). 2013b. "Emissions from Fossil Fuel Combustion – Million Metric Tons CO_2." Accessed at www.epa.gov on February 22, 2013.

US Federal Election Commission. 2010. "Federal Elections 2000." Accessed at www.fec.gov on November 8, 2010.

US Global Change Research Program. 2000. *Climate Change Impacts: The Potential Consequences of Climate Variability and Change.* Cambridge: Cambridge University Press.

US Global Change Research Program. 2009. *Global Climate Change Impacts in the United States.* Cambridge: Cambridge University Press.

US Government Accountability Office (GAO). 2004. "Alaska Native Villages: Villages Affected by Flooding and Erosion Have Difficulty Qualifying for Federal Assistance." Accessed at www.gao.gov on January 28, 2011.

US Government Accountability Office (GAO). 2010. "Climate Change: Preliminary Observations on Geoengineering Science, Federal Efforts, and Governance Issues." Statement of Frank Rusco, Director, Natural Resources and Environment, before the Committee on Science and Technology, House of Representatives.

US House of Representatives. 1993. 103rd Congress, 1st Session, Roll Call 199, May 27, 1993. Accessed at www.clerk.house.gov on February 20, 2010.

US House of Representatives. 2007. Report of the Committee on Oversight and Government Reform, Political Interference with Climate Change Science Under the Bush Administration. Washington, DC. Accessed at www.oversight.house.gov on August 23, 2010.

US House of Representatives. 2009a. 111th Congress, 1st Session, H.R. 2454, Roll Call 477, June 26, 2009. Accessed at www.clerk.house.gov on February 20, 2010.

US House of Representatives. 2009b. Floor debate, June 26, 2009; transcribed by the author. Accessed at www.c-span.org on May 9, 2011.

US House of Representatives. 2011. Appropriations Committee. "FY2011 Continuing Resolution Reductions."

US House of Representatives, Committee on the Budget. 2001. "Basics of the Budget: A Briefing Paper." Washington, DC: US House of Representatives.

US House of Representatives, Committee on Energy and Commerce. 2011. "Committee Votes." Accessed at www.energycommerce.house.gov on March 17, 2011.

US House of Representatives, Office of Congressman Joe Barton. 2011. *Climate Change and Policy Implications.* Accessed at www.joebarton.house.gov on May 10, 2011.

US Interagency Working Group on Social Cost of Carbon. 2013. "Technical Update of the Social Cost of Carbon for Regulatory Impact Analysis. Office of Management and Budget, and Ten Other Agencies." Accessed at www.whitehouse.gov/omb on February 4, 2014.

US Library of Congress. 2009a. Committee Reports. 111th Congress (2009–2010). House Report 111–137. American Clean Energy and Security Act of 2009. Accessed at www.thomas.loc.gov on January 28, 2011.

US Library of Congress. 2009b. American Clean Energy and Security Act of 2009 (House Of Representatives – June 26, 2009). Accessed at www.thomas.loc.gov on January 28, 2011.

US National Academy of Sciences (NAS). 1975. *Understanding Climatic Change: A Program for Action* (by the Committee for the Global Atmospheric Research Program – GARP). Washington, DC; Detroit, MI: National Academy of Sciences; Grand River Books.

US National Academy of Sciences (NAS). 1991. *Policy Implications of Greenhouse Warming.* Washington, DC: National Academy Press.

US National Academy of Sciences (NAS). 1999. *Our Common Journey.* Washington, DC: National Academy Press.

US National Academy of Sciences (NAS). 2001. *Climate Change Science: An Analysis of Some Key Questions.* Washington, DC: National Academy Press.

US National Academy of Sciences (NAS). 2010a. *Advancing the Science of Climate Change.* Washington, DC: National Academy Press.

US National Academy of Sciences (NAS). 2010b. *Limiting the Magnitude of Future Climate Change.* Washington, DC: National Academy Press.

US National Academy of Sciences (NAS). 2010c. *Adapting to the Impacts of Climate Change.* Washington, DC: National Academy Press.

US National Academy of Sciences (NAS). 2010d. *Informing an Effective Response to Climate Change.* Washington, DC: National Academy Press.

US National Academy of Sciences (NAS). 2010e. *Climate Stabilization Targets: Emissions, Concentrations, and Impacts Over Decades to Millennia.* Washington, DC: National Academy Press.

US National Academy of Sciences (NAS). 2011. *America's Climate Choices.* Washington, DC: National Academy Press. Accessed at www.nap.edu on June 2, 2011.

US National Aeronautics and Space Administration (NASA). 2011. "Research News. NASA Research Finds 2010 Tied for Warmest Year on Record." January 12. Accessed at www.giss.nasa.gov on April 13, 2013.

US National Economic Council. 2009. *A Strategy for American Innovation: Driving Towards Sustainable Growth and Quality Jobs.* Accessed at www.whitehouse.gov on March 24, 2011.

US National Economic Council, Council of Economic Advisers, and Office of Science and Technology Policy. 2011. *A Strategy for American Innovation: Securing Our Economic Growth and Prosperity.* Accessed at www.whitehouse.gov on March 24, 2011.

US National Oceanic and Atmospheric Administration (NOAA). 2010. *State of the Climate in 2009 National Climatic Data Center.* Available at www.noaa.gov. Also appears as the June 2010 issue (volume 91) of the *Bulletin of the American Meteorological Society.*

US National Research Council (NRC). 2010. "Verifying Greenhouse Gas Emissions: Methods to Support International Climate Agreements," Prepublication Version. Accessed www.nap.edu on April 8, 2010.

US Office of Management and Budget. 2010a. *The Budget for Fiscal Year 2011, Historical Tables,* Table 3.2, Outlays by Function and Subfunction: 1962–2015, line 272, Energy Conservation. Accessed at www.whitehouse.gov/omb, downloaded on February 6, 2010.

US Office of Management and Budget. 2010b. *Budget of the United States Government, Fiscal Year 2011, Historical Tables,* Table 10.1, Gross Domestic Product and Deflators Used in the Historical Tables. Accessed at www.whitehouse.gov/omb on February 6, 2010.

US Office of Management and Budget. 2011a. *The Budget for Fiscal Year 2012, Historical Tables,* Table 3.2, Outlays by Function and Subfunction: 1962–2015, line 272, Energy Conservation. Accessed at www.whitehouse.gov/omb, downloaded on February 6, 2010.

US Office of Management and Budget. 2011b. *Budget of the United States Government, Fiscal Year 2011, Historical Tables,* Table 10.1, Gross Domestic Product and Deflators Used in the Historical Tables. Accessed at www.whitehouse.gov/omb on February 6, 2010.

US Office of Management and Budget. 2012. *The Budget for Fiscal Year 2013, Historical Tables,* Table 3.2, Outlays by Function and Subfunction: 1962–2015, line 272, Energy Conservation. Accessed at www.whitehouse.gov/omb, downloaded on March 10, 2012.

US Office of Management and Budget. 2013. *Budget of the United States Government, Fiscal Year 2014: Historical Tables.* Washington, DC: GPO. Accessed at www.whitehouse.gov on February 28, 2013.

US Office of the Trade Representative (USTR). 2013. "WTO and Multilateral Affairs." Accessed at www.ustr.gov on January 13, 2013.

US President's Council of Advisors on Science and Technology. 2013. "Letter to the President with Suggestions on Climate Change." Accessed at www.whitehouse.gov on 25 March 2013.

US Securities and Exchange Commission. 2010. "Commission Guidance Regarding Disclosure Related to Climate Change; Final Rule." Accessed at www.sec.gov on June 12, 2011.

US Senate. 1997. Senate Resolution 98. 105th Congress, 1st Session. S. Res. 98. Accessed at www.gpo.gov on January 8, 2011.

US Senate. 2003. 108th Congress, 1st Session, S. 139, Amendment 2028, Vote 420, October 30, 2003. Accessed at www.senate.gov on February 20, 2010.

US Senate. 2005. 109th Congress, 1st Session, H.R. 6, Amendment 826, Vote 148, June 22, 2005. Accessed at www.senate.gov on February 20, 2010.

US Senate. 2008. 110th Congress, 2nd Session, S. 3036, Amendment 4825, Vote 145 (Cloture Vote), June 6, 2008. Accessed at www.senate.gov on February 20, 2010.

US Senate. 2009. S. 1462. American Clean Energy Leadership Act of 2009. Accessed at www.energy.senate.gov on April 26, 2011.

US Senate, Committee on Energy and Natural Resources. 2010. "History." Accessed at http://energy.senate.gov on February 17, 2010.

US Supreme Court. 2007. No. 05–1120, *Massachusetts, et al., Petitioners v. Environmental Protection Agency et al.*, on writ of certiorari to the United States Court of Appeals for the District of Columbia Circuit, April 2. Justice Stevens delivered the opinion of the Court. Accessed at http://www.supremecourtus.gov on November 29, 2009.

US Supreme Court. 2011. Official Transcript – Subject to Final Review. Case 10–174. *American Electric Power Co., et al. v. Connecticut, et al.* Accessed at www.supremecourt.gov on May 11, 2011.

US Trade Representative (USTR). 2013. *2013 Report to Congress on China's WTO Compliance.* Accessed at www.ustr.gov on April 21, 2014.

US White House. 2009. Remarks by the President during Press Availability in Copenhagen. Accessed at www.whitehouse.gov on April 20, 2014.

US White House. 2010. "Boosting Exports of Renewable Energy and Energy Efficiency Technology." Accessed at www.whitehouse.gov on March 24, 2011.

US White House. 2013a. "Remarks by the President on Climate Change." Accessed at www.whitehouse.gov on June 25, 2013.

US White House. 2013b. "United States, China, and Leaders of G-20 Countries Announce Historic Progress Toward a Global Phase Down of HFCs." Accessed at www.whitehouse.gov on September 11, 2013.

USA Today/Gallup. 2009. "Americans Favor US Signature on Copenhagen Treaty." By Lydia Saad. December 15. Accessed at www.gallup.com on April 8, 2013.

Van Asselt, Harro, and Thomas L. Brewer. 2010. "Addressing Competitiveness and Leakage Concerns in Climate Policy: An Analysis of Border Adjustment Measures in the US and the EU," *Energy Policy* 38: 42–51.

Van Asselt, Harro, Thomas L. Brewer, and Michael Mehling. 2009. "Addressing Leakage and Competitiveness in US Climate Policy: Issues Concerning Border Adjustment Measures, Climate Strategies." Available at www.climatestrategies.org.

Venables, Anthony J. 2008. "Economic Geography," in Barry R. Weingast and Donald A. Wittman, editors, *The Oxford Handbook of Political Economy.* Oxford: Oxford University Press.

Verchick, Robert R.M. 2010. "Adaptation, Economics, and Justice," in David M. Driesen, editor, *Economic Thought and US Climate Change Policy.* Cambridge, MA: MIT Press.

Victor, David. 2001. *The Collapse of the Kyoto Protocol and the Struggle to Slow Global Warming.* Princeton, NJ: Princeton University Press.

Victor, David. 2004. *Climate Change: Debating America's Policy Options.* New York, NY: Council on Foreign Relations.

Victor, David G. 2007. "Fragmented Carbon Markets and Reluctant Nations: Implications for the Design of Effective Architectures," in Joseph E. Aldy and Robert N. Stavins, editors, *Architectures for Agreement: Addressing Global Climate Change in the Post-Kyoto World.* Cambridge: Cambridge University Press.

Victor, David. 2011. *Global Warming Gridlock: Creating More Effective Strategies for Protecting the Planet.* Cambridge: Cambridge University Press.

Vig, Norman J., and Michael R. Kraft, editors. 2010. *Environmental Policy: New Directions for the Twenty-First Century.* Washington, DC: CQ Press.

Vogel, David. 1995. *Trading Up – Consumer and Environmental Regulation in a Global Economy.* Cambridge, MA: Harvard University Press.

Vox. 2011. "Thinking Through the Climate Change Challenge, Open Letter." 16 January. Accessed at www.voxeu.org on October 24, 2013.

Vraga, Emily, Connie Roser-Renouf, Anthony Leiserowitz, and Edward Maibach. 2013. "The Political Benefits to Taking a Pro-Climate Stand in 2013," George Mason University, Center for Climate Change Communication. Accessed at www.climatechangecommunication.org on April 8, 2013.

Wagner, Marcus. 2003. "The Porter Hypothesis Revisited: A Literature Review of Theoretical Models and Empirical Tests," Centre for Sustainability Management. Universität Lüneburg. Accessed at www.uni-luenneburg.de on August 23, 2010.

Walker, Gabrielle, and Sir David King. 2008. *The Hot Topic: What We Can Do About Global Warming.* New York, NY: Harcourt.

Washington, Haydn, and John Cook. 2011. *Climate Change Denial.* London and Washington, DC: Earthscan.

Washington Post. 2005. "Beliefs About Climate Change Hold Steady." October 2, p. A16.

Washington Post. 2009. "Obama Unveils Broad Climate Change Plan, to Deliver Major Speech Today." June 25. Accessed at www.projects.washingtonpost.com on June 28, 2009.

Weart, Spencer R. 2008. *The Discovery of Global Warming,* revised and expanded edition. Cambridge, MA: Harvard University Press. For supplementary information, see the associated website at www.aip.org/history/climate/index.htm.

Webster, P.J., G.J. Holland, J.A. Curry, and H.R. Chang. 2005. "Changes in Tropical Cyclone Number, and Intensity in a Warming Environment," *Science,* 309: 1844–1846.

Weingast, Barry R., and Donald A. Wittman, editors. 2008a. *The Oxford Handbook of Political Economy.* Oxford: Oxford University Press.

Weingast, Barry R., and Donald A. Wittman. 2008b. "The Reach of Political Economy," in Barry R. Weingast and Donald A. Wittman, editors, *The Oxford Handbook of Political Economy.* Oxford: Oxford University Press.

Western Climate Initiative. 2010. "Program Design." Accessed at www.westernclimateinitiative.org on December 3, 2010.

Weynant, J.P. 2000. *An Introduction to the Economics of Climate Change.* Arlington, VA: Pew Center on Global Climate Change.

Wigley, Tom M.L. 1999. *The Science of Climate Change.* Arlington, VA: Pew Center on Global Climate Change.

Wigley, Tom M.L. 2011. "Coal to Gas: The Influence of Methane Leakage," *Climatic Change,* 108.

Wildavsky, Aaron, and Naomi Caiden. 2003. *The New Politics of the Budgetary Process,* 5th edition. New York, NY: Longman.

Willer, David, Michael J. Lovaglia, and Barry Markovsky. 1997. "Power and Influence: A Theoretical Bridge," *Social Force,* 76: 571–603.

World Bank. 2009a. "Public Attitudes toward Climate Change: Findings from a Multi-country Poll," prepared for the *World Development Report,* December 3. Accessed at www.worldbank.org on December 6, 2009.

World Bank. 2009b. *World Development Report.* Washington, DC: World Bank.

World Business Council for Sustainable Development (WBCSD). 2010. "Energy and Climate." Accessed at www.wbcsd.org on May 12, 2010.

WorldPublicOpinion.org. 2007a. "Global Warming." August. Accessed at www.americans-world.org on August 6, 2007.

WorldPublicOpinion.org. 2007b. "International Polling on Climate Change." December 6. Accessed at www.worldpublicopinion.org on December 20, 2007.

WorldPublicOpinion.org. 2008. "McCain and Obama Supporters Largely Agree on Approaches to Energy, Climate Change." September 23. Accessed at www.worldpublicopinion.org on September 30, 2008.

WorldPublicOpinion.org. 2012. "Large Majorities in US and Europe Endorse Focus on Renewable Energy." January 18. Accessed at www.worldpublicopinion.org on March 9, 2012.

World Resources Institute (WRI). 2008. "Federalism in the Greenhouse." By Frank Litz and Kathryn Zyla. Accessed at www.wri.org on June 10, 2011.

World Resources Institute (WRI). 2009. "A Closer Look at the American Clean Energy and Security Act." By John Larsen. Accessed at www.wri.org on January 29, 2011.

World Resources Institute (WRI). 2010a. "A Comeback in Cancun: Countries Move Forward with Climate Agreement." Accessed at www.wri.org on May 31, 2010.

World Resources Institute (WRI). 2010b. "Bottom Line on Regional Cap-and-Trade Programs." Compiled by Chris Lau and Nicholas Bianco. Accessed at www.wri.org on January 28, 2011.

World Resources Institute (WRI). 2010c. "Climate Analysis Indicators Tool, version 4.0." Accessed at www.cait.wri.org on February 26, 2010.

World Resources Institute (WRI). 2010d. "Getting to Work: A Review of the Operations of the Clean Technology Fund." By Smita Nakhooda. Accessed at www.wri.org on May 31, 2010.

World Resources Institute (WRI). 2010e. "Reducing Greenhouse Gas Emissions in the United States Using Existing Federal Authorities and State Action." By Nicholas M. Bianco and Franz Litz, Thomas Damassa, and Madeline Gottlieb.

World Resources Institute (WRI). 2010f. "Summary of Climate Finance Pledges Put Forward by Developed Countries." Accessed at www.wri.org on November 3, 2010.

World Resources Institute (WRI). 2013. "Corporate Consultative Group (CCG) Members." Accessed at www.wri.org on January 27, 2013.

World Resources Institute (WRI), CAIS. 2012. "CAIS." Accessed at www.cais.wri.org on January 5, 2013.

World Trade Organization. 2012. "The US Fast-Start Finance Contribution." Accessed at www.wto.org on June 15, 2012.

World Trade Organization. 2013. "Dispute Settlement." Accessible at www.wto.org.

Wren, Anne. 2008. "Comparative Perspectives on the Role of the State in the Economy," in Barry R. Weingast and Donald A. Wittman, editors, *The Oxford Handbook of Political Economy*. Oxford: Oxford University Press.

Wright, John R. 1996. *Interest Groups and Congress: Lobbying, Contributions, and Influence.* Boston, MA: Allyn and Bacon.

Yale Project on Climate Change Communication and George Mason University Center for Climate Change Communication. 2012. "Climate Change in the American Mind." Accessed at www.environment.yale.edu on October 1, 2012.

Yankelovich/Harris Survey. 2001. Report of Results of a Poll Conducted for CNN and *Time* magazine on March 21–22, 2001, *Time* (2001), April 9, p. 32.

Young, Oran R. 1982. "Regime Dynamics: The Rise and Fall of International Regimes," *International Organization*, 36: 277–297.

Young, Oran R. 1986. "Review Article: International Regimes: Toward a New Theory of Institutions," *World Politics*, 39: 104–122.

Young, Oren R. 1989. *International Cooperation: Building Regimes for Natural Resources and the Environment.* Ithaca, NY: Cornell University Press.

Young, Oren R. 2002. *The Institutional Dimensions of Environmental Change: Fit, Interplay and Scale.* Cambridge, MA: MIT Press.

Zedillo, Ernesto, editor. 2008. *Global Warming: Looking Beyond Kyoto.* Washington, DC: Brookings Institution.

Zimmerman, William. 1973. "Issue Area and Foreign-Policy Process: A Research Note in Search of a General Theory," *American Political Science Review* 67: 1204–1212.

Index